A SURVEY OF NUMERICAL METHODS FOR PARTIAL DIFFERENTIAL EQUATIONS

EDITED BY

I. GLADWELL AND R. WAIT

CLARENDON PRESS • OXFORD

1979

Oxford University Press, Walton Street, Oxford OX2 6DP
OXFORD LONDON GLASGOW
NEW YORK TORONTO MELBOURNE WELLINGTON
IBADAN NAIROBI DAR ES SALAAM LUSAKA CAPE TOWN
KUALA LUMPUR SINGAPORE JAKARTA HONG KONG TOKYO
DELHI BOMBAY CALCUTTA MADRAS KARACHI

ISBN 0 19 853351 9

Published in the United States by
Oxford University Press, New York

British Library Cataloguing in Publication Data

A survey of numerical methods for partial
differential equations.
1. Differential equations, Partial-
Numerical solutions
I. Gladwell, I II. Wait, R
515'.353 QA374 79-40602

ISBN 0-19-853351

Printed in Great Britain
by Thomson Litho Ltd
East Kilbride

A SURVEY OF NUMERICAL METHODS FOR PARTIAL DIFFERENTIAL EQUATIONS

PREFACE

The aim of this volume is to provide an introduction to the problems and methods in the numerical solution of partial differential equations together with a guide to some of the more recent developments and their applications.

Part I deals with elliptic problems and it begins with an introduction to the theory of elliptic partial differential equations, followed by a brief introduction to finite element methods. Then follow chapters that cover the recent advances in finite differences, non-conforming finite elements, conformal transformation techniques, regional methods, and methods based on integral equations. In particular, methods for singular problems are discussed in detail. Part II dealing with parabolic problems also starts with an introduction to the theory. It includes a discussion of semi-discrete methods and the numerical solution of diffusion-convection equations, boundary layer problems, and strongly nonlinear equations. In Part III methods for solving the resulting algebraic equations are surveyed. Iterative and direct methods for linear equations are discussed together with methods based on fast Fourier transforms and alternating directions. It also contains a chapter on the solution of nonlinear algebraic equations.

Part IV contains a discussion of methods for free boundary and moving boundary problems. Finally Part V covers hyperbolic equations and includes an introduction to the properties of such equations, a survey of finite difference methods and a discussion of mixed elliptic - hyperbolic problems.

This book is based on material presented at a Joint Summer School in July 1978, organized by the Department of Mathematics, University of Manchester and by the Department of Computational and Statistical Science, University of Liverpool.

I. Gladwell

R. Wait

CONTENTS

PART I: NUMERICAL METHODS FOR ELLIPTIC EQUATIONS

1. Elliptic equations: Analytical background 1
   1.1 Conditions for a well-posed problem 1
   1.2 The maximum principle and monotonicity 4
   1.3 Solution in terms of Green's functions 7
   1.4 Variational formulation of elliptic problems 10

2. Finite element basics 14
   2.1 Introduction 14
   2.2 Variational formulations 16
   2.3 Test functions 17
   2.4 Trial functions 21
   2.5 Structure and solution of algebraic equations 25
   2.6 Finite element analysis 26
   2.7 Isoparametric elements 29
   2.8 Curved elements 35
   2.9 Elements in three dimensions 37
   2.10 Orders of convergence 38

3. Finite differences and singularities in elliptic problems 42
   3.1 Introduction 42
   3.2 Grid refinement 45
   3.3 The deferred approach to the limit 46
   3.4 Global treatment of singularities 51
   3.5 Change of variables 56
   3.6 Local treatment of singularity. Determination of local solutions 58
   3.7 Local treatment of singularity. Method of Motz 62
   3.8 Local treatment of singularity. Method of Furzeland 66

4. Conformal transformation techniques for the numerical solution of Poisson problems 70
   4.1 Introduction 70
   4.2 Conformal transformations 72
   4.3 Conformal mapping techniques for problems containing boundary singularities 74
   4.4 Conclusion 81

5. Non-conforming finite elements 83
   5.1 Introduction 83
   5.2 Analysis of non-conforming elements 84
   5.3 The patch test 87
   5.4 A practical patch test 90
   5.5 Numerical integration 93
   5.6 Finite element treatment of singularities 96
   5.7 Automatic mesh refinement 97
   5.8 Infinite mesh refinement 100
   5.9 Isoparametric elements 103

6. Global and regional methods 106
   6.1 Introduction 106

6.2 The global Galerkin method 106
6.3 Regional methods 111
6.4 Numerical methods 115
6.5 Global methods for problems with singularities 119
6.6 Regional methods for problems with singularities 123
6.7 Discussion 126

7. Two dimensional biharmonic problems 128
7.1 Introduction 128
7.2 Some theoretical aspects 129
7.3 Finite element methods 132
7.4 Finite-difference methods 134
7.5 A biharmonic problem containing a boundary singularity 136

8. Integral equations 140
8.1 Introduction 140
8.2 Integral equations 140
8.3 Fredholm theory 143
8.4 Equations of the first kind 147
8.5 Alternative formulations 150
8.6 Computational effort and complexity 154
8.7 Conditioning 158
8.8 A survey of numerical methods 159
8.9 Further comments 163

PART II: NUMERICAL METHODS FOR PARABOLIC EQUATIONS

9. Parabolic equations and their discretization 167
9.1 Introduction 167
9.2 Properties of the solution 169
9.3 Simple discretizations 170
9.4 Stability 173
9.5 Stability in more complicated situations 178

10. Semi-discretization of parabolic equations and the numerical solution of semi-discretized equations 181
10.1 Semi-discretization techniques 181

11. The diffusion-convection equation 195
11.1 Introduction 195
11.2 Previous work 196
11.3 Particular problems associated with this equation 197
11.4 Some similar computing problems 198
11.5 Finite differences or finite elements? 199
11.6 Spatial semi-discretization 200
11.7 Improved spatial discretization 204
11.8 Time discretization 210
11.9 ODE packages 210
11.10 Conclusions 210

12. Boundary layer and similarity solutions for parabolic equations 212
12.1 Introduction 212
12.2 Boundary-layer equations 213
12.3 Dependence on the Reynolds number 215
12.4 Similar boundary layers 216
12.5 Examples of similar flows 217
12.6 Thermal boundary-layer equation 217

12.7 Nonsimilar boundary layers 219
12.8 Numerical methods 222
12.9 Concluding remarks 223

13. Strongly nonlinear parabolic equations-examples 224
13.1 Introduction 224
13.2 Example 1. Boundary-layer flow 224
13.3 Example 2. Laminar flame propagation 229
13.4 Example 3. Parallel-plate reactor 234

PART III: FINITE-DIFFERENCE AND FINITE ELEMENT EQUATIONS

14. Finite-difference equations and their iterative solution 241
14.1 Finite-difference approximations 241
14.2 The classical iterative methods 243
14.3 The method of conjugate gradients 247
14.4 Stone's strongly implicit method (SIP) 249
14.5 Pre-conditioned conjugate gradients 250
14.6 Multiple grid methods 251

15. Direct solution of finite element and finite-difference equations 252
15.1 Introduction 252
15.2 Band-matrix techniques 253
15.3 Frontal methods 256
15.4 Nested dissection 259
15.5 General sparse techniques and hypermatrices 260
15.6 The frontal approach for more general matrices 261
15.7 Comparison between direct and iterative methods 262

16. Fast direct methods for separable elliptic equations 264
16.1 The fast Fourier transform 264
16.2 Fourier methods for the discrete Poisson equation 266
16.3 Odd/even reduction 269
16.4 Non-rectangular regions 271
16.5 Non-separable equations 274

17. Alternating direction implicit methods 276
17.1 Introduction 276
17.2 Elliptic equations, the model problem 277
17.3 The ADI method 279
17.4 ADI - variable acceleration parameters 283
17.5 Variants of the ADI scheme, three dimensional problems 285
17.6 Practical application of the ADI method for elliptic problems 287
17.7 Parabolic problems 287
17.8 Alternating direction Galerkin methods 291

18. Solution of nonlinear equations 295
18.1 Introduction 295
18.2 Functional iteration 298
18.3 Newton's method 300
18.4 Newton-SOR and SOR-Newton methods 302
18.5 Quasi-Newton methods 305
18.6 Continuation 307

PART IV: FREE AND MOVING BOUNDARY PROBLEMS

19. Free boundary problems in elliptic equations 313
19.1 Introduction 313
19.2 Direct solution of the free boundary problem 318
19.3 Transformation methods 326

20. The Stefan problem: Moving boundaries in parabolic equations 332
20.1 Introduction 332
20.2 Front-tracking methods. (i) Method of lines 334
20.3 Front-tracking methods. (ii) Finite element methods 338
20.4 Coordinate transformation 341
20.5 The isotherm migration method 344
20.6 The enthalpy method 347
20.7 Methods using quadratic programming and variational inequalities 351
20.8 Evaluation of methods 355

PART V: HYPERBOLIC EQUATIONS

21. Hyperbolic equations and characteristics 359
21.1 Nature of the characteristics 359
21.2 Solution along the characteristics 364
21.3 Discontinuities and shock waves 368
21.4 More difficult problems 370

22. Difference methods for hyperbolic equations 372
22.1 Introduction 372
22.2 Simple discretizations 372
22.3 Nonlinear equations in conservation form 376
22.4 Boundary conditions 377
22.5 Conclusions 381

23. Numerical solution of mixed elliptic-hyperbolic equations 382
23.1 Introduction 382
23.2 Symmetric positive differential equations 383
23.3 Finite element method approaches to the solution of mixed equations 385
23.4 Finite-difference approaches to the solution of mixed equations 388
23.5 Nonlinear mixed equations 389

REFERENCES 394

INDEX 425

CONTRIBUTORS

Professor L. Fox (Chapters 3, 20)

Oxford University Computing Laboratory,
19 Parks Road, Oxford OX1 3PL.

Dr. J. A. Hendry (Chapter 23)

The Computer Centre, The University of Birmingham,
P.O. Box 363, Birmingham B15 2TT.

Mr. A. Prothero (Chapters 12, 13)

Shell Research Centre,
Thornton, P.O. Box 1,
Chester, CH1 3SH.

Dr. J. D. Reid (Chapters 14, 15)

Computer Science and Systems Division,
Building 8.9, AERE Harwell,
Oxfordshire OX11 0RA.

Dr. I. M. Smith (Chapter 11)

Department of Civil Engineering, Simon Engineering
Laboratories, The University, Manchester M13 9PL.

Ms. S. Scheffler
Professor J. R. Whiteman (Chapters 4, 7)

Institute for Computational Mathematics,
Brunel University, Uxbridge,
Middlesex UB8 3PH.

Professor L. M. Delves (Chapter 6)
Dr. C. Phillips
Dr. R. Wait (Chapters 2,5)
Mr. J. M. Watt (Chapter 9, 22)

Department of Computational and Statistical Science,
The University,
Liverpool L69 3BX.

Dr. C. T. H. Baker (Chapter 8)
Dr. T. L. Freeman (Chapter 18)
Dr. I. Gladwell (Chapters 10, 19)
Dr. G. Hall (Chapter 16)
Professor J. E. Walsh (Chapters 1, 21)
Dr. J. Williams (Chapter 17)

Department of Mathematics,
The University, Manchester M13 9PL.

PART I

NUMERICAL METHODS FOR ELLIPTIC EQUATIONS

This part contains a description of some necessary introductory theory for second order elliptic equations and a discussion of the treatment of singularities and the use of conformal mapping techniques. Finite element methods are developed from a basic level as are corresponding global and regional techniques. Also included is a discussion of the solution of the biharmonic equation. Finally, integral equation methods are discussed using a number of examples of exterior problems in elliptic equations.

The material of this part is intended to be self-contained. However, the reader may feel some need to consult part III for a discussion of the use of finite differences and the solution of finite-difference equations (the inexperienced reader may also find Smith (1977) or Ames (1969) of help here).

Contents

Chapter	Title	Contributor
1	Elliptic Equations: Analytical Background.	J.E. Walsh
2	Finite Element Basics	R. Wait
3	Finite Differences and Singularities in Elliptic Problems.	L. Fox
4	Conformal Transformation Techniques for the Numerical Solution of Poisson Problems.	J.R. Whiteman
5	Non-conforming Finite Elements.	R. Wait
6	Global and Regional Methods.	L.M. Delves
7	Two Dimensional Biharmonic Problems.	J.R. Whiteman
8	Integral Equations.	C.T.H. Baker

1

ELLIPTIC EQUATIONS: ANALYTICAL BACKGROUND

1.1. CONDITIONS FOR A WELL-POSED PROBLEM

The partial differential equations which arise in practical problems are generally too complex for exact analytical treatment; we have to use various approximate techniques, either numerical or semi-analytical, to obtain the solution, and extensive computations are often needed to evaluate it. The equations of interest cover a very wide range of types, and the boundary conditions and geometrical regions of definition lead to a further range of variants of the problem. The starting point in classification is the analytical theory of well-posed systems, which is discussed in Courant and Hilbert (1962, p. 226). The partial differential equation is considered in a certain region, with boundary conditions specified on all or part of the boundary. The requirements for a well-posed problem are that the equation has a unique solution, and that the solution depends continuously on the data. These conditions enable us to define appropriate forms of the boundary conditions for each of the main types of equation. Problems which are not well-posed in the above sense may still be physically significant, but special difficulties are likely to arise in solving them.

We begin with equations of elliptic type, which usually correspond to the equilibrium state of physical systems. If the system is stable, the solution of the equation should also be stable with respect to the data, so that small perturbations in the given conditions have only a small effect on the solution.

Consider a general linear equation of the second order, in two space variables,

$$Lu \equiv a\frac{\partial^2 u}{\partial x^2} + 2b\frac{\partial^2 u}{\partial x \partial y} + c\frac{\partial^2 u}{\partial y^2} + p\frac{\partial u}{\partial x} + q\frac{\partial u}{\partial y} + su = f\ , \tag{1.1}$$

where the coefficients may be functions of x and y. The equation is said to be elliptic if the following condition is satisfied

$$b^2 - ac < 0 . \qquad (1.2)$$

The best-known example of an elliptic equation is Poisson's equation

$$\nabla^2 u \equiv \frac{\partial^2 u}{\partial x^2} + \frac{\partial^2 u}{\partial y^2} = f . \qquad (1.3)$$

In practical applications, it is very seldom that the general solution of an equation such as (1.1) is required; what is needed is a particular solution satisfying certain boundary conditions. Since the equation is second order, we expect to have function and first derivative values specified, if it is to be solved as an initial value problem. Suppose the values of u, $\partial u/\partial x$, $\partial u/\partial y$ are given on the line $x = 0$, say. Then by the Cauchy-Kowalewski theorem, there is a unique analytic solution of (1.1) in a neighbourhood of this line, provided the coefficients are analytic functions in the neighbourhood, and $a \neq 0$. A similar result holds for initial data given on any line in the (x,y) plane, exceptional cases being excluded by condition (1.2).

However, it does not follow that the solution is stable with respect to the data, and in fact we can show that there is a solution for which the data on the initial line is unbounded, and which tends to zero in its neighbourhood (Courant and Hilbert, 1962, p. 216). This result also depends on condition (1.2), and it follows that an elliptic equation is not well-posed as an initial value problem.

Suppose now that equation (1.1) is specified in a closed region Ω, with a single boundary condition given at each point of the boundary curve Γ. This condition takes one of the three forms

$$\text{(i) } u = g_1 \; ; \quad \text{(ii) } \frac{\partial u}{\partial n} = g_2 \; ; \text{ (iii) } \frac{\partial u}{\partial n} + hu = g_3 . \qquad (1.4)$$

In the case of (i) we have the simple result that the differential equation has a unique solution provided

$$a > 0, \; s \leq 0. \qquad (1.5)$$

If $s > 0$ the homogeneous problem may have an eigensolution, so that the solution of the non-homogeneous problem is not unique. The result also holds for quasi-linear equations, when the coefficients in (1.1) are functions of u, $\partial u/\partial x$, $\partial u/\partial y$, provided conditions (1.2) and (1.5) are satisfied for all u (Courant and Hilbert, 1962, p.322).

For boundary conditions (ii) and (iii) in (1.4), we need additional conditions for a unique solution; in the case of (ii) the solution is only specified to within an additive constant if $s = 0$ and $f = 0$, and in (iii) we need $h > 0$ to guarantee a solution. We note that these results do not establish the stability of the solution, which requires a separate argument. This will be considered in section 1.2 in connexion with the maximum principle.

The problem of solving equation (1.1) with boundary conditions of the form (1.4) is the classical type of elliptic problem, but there are many other forms of the equation and the auxiliary conditions which are of interest in practical applications. We give an example due to Cannon, where the equation contains unknown parameters but additional boundary conditions are specified.

Consider an elliptic equation of the following form

$$\nabla(\underset{\sim}{K}\nabla u) = f, \quad \text{in } \Omega. \tag{1.6}$$

We can interpret this as the steady-state equation for heat conduction, where K is the conductivity and f the source term. Suppose first that K is given, f is unknown, and we have two boundary conditions specified on the closed boundary

$$\left.\begin{aligned} u &= g_1 \\ K\frac{\partial u}{\partial n} &= g_2 \end{aligned}\right\} \quad \text{on } \Gamma\ , \tag{1.7}$$

where g_1 and g_2 are known. If the equation (1.6) is linear, it is easy to see that any function v which vanishes on the boundary together with its first derivatives, may be added to a solution u. This adds a term $\nabla(K \underset{\sim}{\nabla} v)$ to f, so that f cannot be unique. To specify the source term uniquely it is necessary to have additional information about u within the region Ω, and Cannon (1968) gives some cases in which the problem can be solved completely.

A second type of problem arises when f is given, together with the boundary conditions (1.7), and we want to find the conductivity K. If f = 0 the problem may be solved fairly easily (Cannon, 1967), and the solution is unique. Similar elliptic problems arise in fields such as geophysics, where the boundary of the region is accessible but the interior is not, and we want to deduce as much as possible about the region from the data on the boundary. We can try to obtain "best-fitting" solutions by purely numerical methods, but the analytical theory is helpful in indicating when the solutions are uniquely determined by the data.

1.2. THE MAXIMUM PRINCIPLE AND MONOTONICITY

Solutions of Laplace's equation (i.e. equation (1.3) with f = 0) are known as harmonic functions, and they have the property that the maximum value of the function over any closed region occurs on the boundary. Maximum principles of this type can be established for a wide class of elliptic equations, and a full discussion may be found in Protter and Weinberger (1967). We give two examples, for the operator L of equation (1.1). Suppose the coefficients satisfy the conditions (1.2) and (1.5), i.e.

$$b^2 - ac < 0, \; a > 0, \; s \leq 0 \; . \tag{1.8}$$

Then the following results may be proved

(a) If $Lu \geq 0$ in Ω and u is not a constant, then u cannot attain a non-negative maximum in Ω.

(b) If s = 0 and Lu = 0 in Ω, the function u has no maximum or minimum within the region.

Results of this type are useful in proving the uniqueness of the solution of (1.1) as a boundary-value problem, and they can also be used to demonstrate stability. Suppose we have a solution of equation (1.1) with the simple boundary condition

$u = g_1$; if the boundary data is perturbed to $g_1 + \varepsilon$, let the new solution be $u + e$. Then we have

$$Le = 0 \quad \text{in } \Omega \ , \quad e = \varepsilon \quad \text{on } \Gamma \ , \tag{1.9}$$

and it follows from (a) that the magnitude of the perturbation $|e|$ in Ω is bounded by $|\varepsilon|$. Extensions for more general boundary conditions are given by Collatz (1960, p.401); this technique is used in chapter 3.

The maximum principle may be used in obtaining error bounds for certain semi-analytical methods of solving elliptic problems. Suppose we have a set of functions $\{\varphi_i(x,y)\}$ which satisfy the differential equation exactly, but not the boundary conditions. (Such functions are easily found for Laplace's equation, which is satisfied by $u = \text{Re}\{f(z)\}$, for any analytic $f(z)$.) We define an approximate solution of the boundary-value problem to be

$$U_N = \sum_{i=1}^{N} \alpha_i \varphi_i(x,y) \ , \tag{1.10}$$

where the coefficients α_i are determined so that U_N fits the boundary conditions as closely as possible. The fitting can be done by collocation, least squares, or any other suitable method. Thus we obtain an exact solution of the equation, which satisfies perturbed boundary conditions. Consequently the maximum error in the solution can be found from the result (a), as above. This method is particularly useful for problems in which the solution has a singularity on the boundary, provided the appropriate functions $\varphi_i(x,y)$ can be found to represent the form of the singularity. Such problems are discussed by Donnelly (1969), who gives error estimates for various boundary conditions, and applies the results to the computation of eigenvalues for the Laplace operator.

The maximum principle may be extended to certain nonlinear equations, but we shall consider these in the more general context of monotone operators. Let us write equation (1.1) with an associated boundary condition in the form

$$Tu = \begin{cases} -L(u) + f & \text{in } \Omega , \\ u - g_1 & \text{on } \Gamma . \end{cases} \tag{1.11}$$

If the coefficients in L(u) satisfy (1.8), it may be shown that the operator T is monotone, which means that it has the property

$$Tv \le Tw \Rightarrow v \le w \tag{1.12}$$

in $\Omega \cup \Gamma$. A more general problem of this type is

$$Tu = \begin{cases} -L(u) + F(u) & \text{in } \Omega , \\ \dfrac{\partial u}{\partial n} + hu - g_3 & \text{on } \Gamma , \end{cases} \tag{1.13}$$

for which T is monotone if the conditions

$$\frac{\partial F}{\partial u} - s \ge 0, \; h > 0, \tag{1.14}$$

are satisfied, as well as (1.8).

The property may be readily applied to error estimation. We start with an approximate solution $\tilde{U}$ which may or may not satisfy the differential equation and the boundary conditions. Generally $T\tilde{U}$ is not one-signed, so that $\tilde{U}$ does not give an immediate bound for the true solution u. Suppose $T\tilde{U} = d$ say, then we solve the equation

$$Te = |d| \tag{1.15}$$

approximately, and try to find a constant λ such that

$$T(\tilde{U} - \lambda e) \le 0 \le T(\tilde{U} + \lambda e) . \tag{1.16}$$

If (1.16) is satisfied, we have strict bounds on u.

Alternatively we can calculate $\tilde{U}$ so that $T\tilde{U}$ is one-signed, for example by linear programming methods (Barrodale and Young, 1970). For the linear problem (1.11) we take an approximate solution as in (1.10), and choose the coefficients α_i so that

$$TU_N \geq 0, \quad \max(TU_N) \text{ is a minimum}, \tag{1.17}$$

over a finite set of points. This gives an error estimate for U_N, but it is not strict unless the set of points is specially chosen.

1.3. SOLUTION IN TERMS OF GREEN'S FUNCTIONS

An elementary method of solving an elliptic equation is by separation of the variables. The solution is expressed in the form

$$u = \sum_i c_i X_i(x) Y_i(y) , \tag{1.18}$$

and we try to match it to the boundary conditions by suitable choice of the functions X_i, Y_i, and the coefficients c_i. This procedure is only successful for simple equations and simple geometrical regions. For more general elliptic problems, it is known that the solution can be expressed in terms of the data by means of an integral involving the Green's function. This function depends on the equation, the region, and the boundary conditions, and it can be constructed explicitly in a few special cases. Although it cannot be found in the general case, the fact that the solution may be represented in the form of an integral gives some insight into the properties of elliptic equations.

Let us consider first the case of Poisson's equation (1.3), with the boundary condition $u = g_1$. The solution may be written in the form

$$u = \iint_\Omega G(x,y,\xi,\eta) f(\xi,\eta) d\xi d\eta + \int_\Gamma \frac{\partial G}{\partial n} g_1(s) ds , \tag{1.19}$$

where the Green's function G vanishes on the boundary (Courant and Hilbert, 1962, p.262). The Green's function exists for any closed region with a piecewise smooth boundary, but it can only be specified analytically for simple regions such as the square, circle, and annulus. For example, if Ω is a circle of radius

R centre O, then G has the form

$$G(P,Q) = \frac{1}{2\pi}\left\{\log \frac{|PQ|}{|P\tilde{Q}|} - \log \frac{|OQ|}{R}\right\}, \tag{1.20}$$

where P = (x,y), Q = (ξ,η), and $\tilde{Q}$ is the inverse of Q with respect to the circle. We note that G(P,Q) has a logarithmic singularity at P = Q, and that it vanishes on the circle.

For more general elliptic equations and regions, an expression similar to (1.19) still exists, under fairly mild conditions on the boundary, and the singularity in G is of logarithmic type. The form of (1.19) shows that the solution at any point of Ω depends on the boundary values of u at all points of the boundary. Because the integrals are smoothing operators, we see that the solution within the region will generally be smoother than the function on the boundary. In elliptic problems, there are often singularities on the boundary arising from the geometry of the region; for example, the first derivatives of u are generally discontinuous at a re-entrant corner. However, the discontinuities do not extend across the region. Of course, in the neighbourhood of the singular point the solution is badly behaved, and special methods of approximation have to be used (see chapters 3 and 5).

These properties mark some of the differences between elliptic and hyperbolic equations. For equations of the latter type, the solution at any point depends only on a limited range of boundary values, on a segment of the boundary. Further, any discontinuities in the boundary conditions can be propagated across the region.

Returning to the elliptic case, although it is true that the solution at any point depends on all values of f and g_1, the dependence is governed by the spatial variation of the Green's function. If f is modified in a sub-region of Ω, the form of (1.19) shows that the resultant effect on u can be bounded by a quantity which is computable in terms of G. Usually the explicit form of G is not known, but the bound can be estimated either by finding a majorant of G, or by using a discrete approximation to the integral operator in (1.19). In the case of Poisson's equation, it is easy to show that G is negative in Ω; suppose the region has maximum diameter R, then by comparing the Green's function with that for a circumscribing circle, we obtain the following bound from the maximum principle

$$-\,G(P,Q) \le \frac{1}{2\pi}\log\left\{\frac{R}{|PQ|}\right\}. \qquad (1.21)$$

This result is not a sharp bound, and it may not be convenient for practical computation. In practice we can often get a better estimate from the discrete form of the operator. Suppose we set up a finite-difference approximation to Poisson's equation in Ω, with specified boundary values g_1. This gives a matrix equation of the form

$$A\,\underset{\sim}{U} = \underset{\sim}{f} + \underset{\sim}{g}\ , \qquad (1.22)$$

where $\underset{\sim}{U}$ is a vector of approximate function values at mesh-points. The right-hand side involves the inhomogeneous term f at all mesh-points, and the values of g_1 at mesh-points on the boundary, represented by $\underset{\sim}{f}$ and $\underset{\sim}{g}$ respectively. The discrete analogue of (1.19) is obtained by solving (1.22), which gives

$$\underset{\sim}{U} = A^{-1}\underset{\sim}{f} + A^{-1}\underset{\sim}{g}\ . \qquad (1.23)$$

This can be used to estimate the effect of perturbations in f or g_1 on the solution at any point of the region.

This type of local estimation may also be applied to the problem of error control. If we obtain an approximate solution $\underset{\sim}{U}$ from (1.22), we can use it to calculate the leading terms of the local truncation error at all mesh-points, forming a vector $\underset{\sim}{t}$ say. Assuming there is no error in representing the boundary values, an estimate of the error $\underset{\sim}{e}$ in $\underset{\sim}{U}$ is given by $\underset{\sim}{e} = A^{-1}\underset{\sim}{t}$. If we require a more accurate solution at particular points of the region, we can refine the mesh in appropriate sub-regions, in order to reduce $\underset{\sim}{t}$ and thus $\underset{\sim}{e}$. The spatial variation of A^{-1}, which reflects that of the Green's function, indicates where a good local approximation is needed. Although this approach works well for finite-difference methods the position is not so simple when finite elements are used (see chapter 5).

1.4. VARIATIONAL FORMULATION OF ELLIPTIC PROBLEMS

An important class of methods for solving elliptic equations is based on the formulation of the equations as variational problems. In the general theory of the calculus of variations, it is shown that for certain partial differential equations the solution can be expressed as the function which gives a stationary value to a related integral (Courant and Hilbert, 1953, p. 191). This property is used in constructing approximate solutions by the Rayleigh-Ritz method, a particular case of which is the finite element method.

To illustrate the principle, let us take a simple example. Consider the integral

$$I(u) = \iint_{\Omega} \left\{ \left(\frac{\partial u}{\partial x}\right)^2 + \left(\frac{\partial u}{\partial y}\right)^2 + 2fu \right\} dxdy, \tag{1.24}$$

where u is allowed to vary over all functions with appropriate continuity which satisfy the condition $u = g_1$ on Γ. Then it is easy to show that the minimum of I(u) is given by the function which satisfies the differential equation

$$\frac{\partial^2 u}{\partial x^2} + \frac{\partial^2 u}{\partial y^2} = f , \tag{1.25}$$

with the same boundary condition $u = g_1$. Now the solution of (1.25) is a function with continuous second derivatives, so if we minimize I(u) over functions of this class, we shall certainly get the correct result. But the integral I(u) also exists for a wider class of functions, in fact for all functions whose first derivatives are piecewise continuous. In practical work we are concerned with discrete approximations to u, and it is often convenient to use functions with a low degree of continuity (e.g. piecewise linear). By extending the class of functions admitted in this way, we can obtain a sequence of functions which satisfy discrete forms of the minimization problem, and which converge to the solution of equation (1.25) in a suitable norm. A discussion of the function spaces involved in the analysis of problems of both second order and fourth order is given in section 2.6. It is possible to use approximations that are not even piecewise continuous, such non-conforming approximations appear in chapters 5 and 6.

The specification of the boundary conditions associated with (1.25) needs further analysis. In the example above, the condition $u = g_1$ is imposed on the trial functions used in (1.24), and it therefore holds for the minimizing function. Suppose we minimize I(u) without requiring u to satisfy any boundary conditions. By considering a small variation about u, we find

$$I(u+\varepsilon) - I(u) = -2\iint_{\Omega} (\nabla^2 u - f)\varepsilon \, dxdy + 2\int_{\Gamma} \frac{\partial u}{\partial n}\varepsilon \, ds + O(\varepsilon^2). \tag{1.26}$$

If I(u) has a stationary value, the terms in ε must vanish for all choices of ε; so if we take $\varepsilon = 0$ on Γ, we get the Poisson equation (1.25) as before. It is then clear that the second term must vanish independently for all ε, and hence $\partial u/\partial n = 0$ on Γ. This condition is therefore satisfied automatically by the minimizing function, if no other boundary condition is applied, and it is called the "natural" boundary condition. If on the other hand all the trial functions satisfy $u = g_1$ on Γ, then $\varepsilon = 0$ on Γ and the second term still vanishes. This type of boundary condition, which has to be imposed on all trial functions, is called "essential".

The general boundary condition (1.4iii) is introduced by considering

$$I(u) = \iint_{\Omega} \left\{\left(\frac{\partial u}{\partial x}\right)^2 + \left(\frac{\partial u}{\partial y}\right)^2 + 2fu\right\}dxdy + \int_{\Gamma} (hu^2 - 2g_3 u)\,ds . \tag{1.27}$$

By writing down the equation for the first variation as above, we find that the condition for a stationary value gives the natural boundary condition

$$\frac{\partial u}{\partial n} + hu = g_3 \quad \text{on } \Gamma . \tag{1.28}$$

This shows that (1.28) is satisfied if we minimize I(u) with the extra boundary term as in (1.27), without imposing any boundary conditions on u. To guarantee a minimum, we need $h \geq 0$.

The general variational principle is based on the integral

$$J(u) = \iint_{\Omega} F\left(u, \frac{\partial u}{\partial x}, \frac{\partial u}{\partial y}\right) dxdy . \tag{1.29}$$

The condition for a stationary value of J(u) leads to the Euler equation

$$\frac{\partial}{\partial x}\left(\frac{\partial F}{\partial u_x}\right) + \frac{\partial}{\partial y}\left(\frac{\partial F}{\partial u_y}\right) - \frac{\partial F}{\partial u} = 0 , \tag{1.30}$$

which must be satisfied in Ω. Essential boundary conditions can be dealt with by imposing them on the trial functions (which must have appropriate continuity); other types of boundary condition can also be handled as above. A number of examples of (1.29) and (1.30) are given by Mitchell and Wait (1977, p.26).

Let us consider the linear elliptic equation (1.1) and compare it with (1.30); we see that the equation can be put into variational form if the coefficients satisfy

$$p = \frac{\partial a}{\partial x} , \quad q = \frac{\partial c}{\partial y} . \tag{1.31}$$

The integral J(u) then becomes

$$J(u) = \iint_{\Omega} \left\{ a\left(\frac{\partial u}{\partial x}\right)^2 + 2b \frac{\partial u}{\partial x} \frac{\partial u}{\partial y} + c\left(\frac{\partial u}{\partial y}\right)^2 - \right.$$

$$\left. - su^2 + 2fu \right\} dxdy . \tag{1.32}$$

If the stationary value of J(u) is a minimum, the coefficients must also satisfy

$$a > 0, \quad b^2 - ac < 0, \quad s \leq 0 . \tag{1.33}$$

These are the usual conditions for an elliptic problem, as we saw earlier in discussing the uniqueness of the solution, and the maximum principle (cf. (1.8)).

The variational problems associated with the integrals I(u) and J(u) are problems of minimization, and approximate methods of solution will give upper bounds on the integrals. To get some information about the accuracy of the solution, it is useful to have lower bounds as well. These can be derived for some problems by considering a dual variational principle (Courant and Hilbert, 1953, section 4.9), which leads to a corresponding problem of finding the maximum of a certain integral.

We will take a very simple case, the minimization of (1.24) with $f = 0$, and with $u = g_1$ on the boundary. Consider the integral

$$K(v) = -\iint_{\Omega} \left\{\left(\frac{\partial v}{\partial x}\right)^2 + \left(\frac{\partial v}{\partial y}\right)^2\right\}dxdy + \int_{\Gamma} \frac{\partial v}{\partial s}\, g_1 ds , \qquad (1.34)$$

where v is not subject to any boundary conditions. Subtracting (1.34) from (1.24), we can transform the result to get

$$I(u) - K(v) = \iint_{\Omega} \left\{\left(\frac{\partial u}{\partial x} - \frac{\partial v}{\partial y}\right)^2 + \left(\frac{\partial u}{\partial y} + \frac{\partial v}{\partial x}\right)^2\right\}dxdy. \qquad (1.35)$$

The minimum value of this integral is zero, when u satisfies the original problem. Consequently the maximum of K(v) is equal to the minimum of I(u), and it is given by a function v satisfying

$$\frac{\partial u}{\partial x} = \frac{\partial v}{\partial y} , \quad \frac{\partial v}{\partial x} = -\frac{\partial u}{\partial y} . \qquad (1.36)$$

If we solve the dual problems numerically and obtain good agreement between the approximate values of I(u) and K(v), we have close bounds for the exact value. Several examples of dual problems may be found in Mitchell and Wait (1977, p.35), and developments for nonlinear problems are given by Barnsley and Robinson (1977).

2

FINITE ELEMENT BASICS

2.1. INTRODUCTION

The finite element method provides a global approximation based on very simple local representations, thus the domain Ω is partitioned into a large number of small subdomains which are the *elements*. The local representation is then the form of the approximate solution in an individual element. In each element the parameters in the local solution are invariably taken as the value of the solution at various *nodes* in the element. Global continuity of the solution is ensured by careful placement of the nodes, on the sides of the elements and at the corners. Thus for example, assume that the two dimensional region Ω is partitioned into triangles and that in any triangular element the approximate solution is a quadratic, that is

$$U_E(x,y) = a + bx + cy + dx^2 + exy + fy^2 ,$$

where U_E is the restriction of the global solution U to the element E. It is possible to select six points $\underset{\sim}{x}_i \equiv (x_i,y_i)$ $i = 1,\ldots,6$ such the $\underset{\sim}{x}_i \in$ E and then write

$$U_E(x,y) = \sum_{i=1}^{6} U_E^i \, \varphi_E^i(x,y) ,$$

where φ_E^i (known in general as *shape functions*) are quadratic polynomials such that

$$\varphi_E^i(x_j,y_j) = \begin{cases} 1 & i = j \\ 0 & i \neq j \end{cases} \qquad 1 \leq i,j \leq 6.$$

The only restriction on the points $\underset{\sim}{x}_i$ is that the linearly independent functions φ_E^i must exist. If the nodes are selected as the vertices and side mid-points (see Figure 2.1), then it follows that along the interface between elements E_1 and E_2, U_{E_1} is a quadratic in one variable (arc length) defined uniquely by the values at the points A, B and C. Similarly on the same interface U_{E_2} is a quadratic defined in terms of

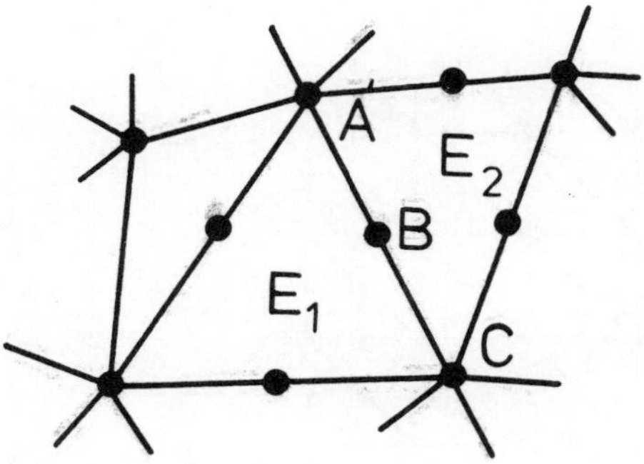

Fig. 2.1. Placement of nodes

the same function values. Thus $U_{E_1} = U_{E_2}$ along the interface and so the global solution is continuous. In addition, shape functions in different elements that are associated with a node on the common boundary reduce to the same form on that boundary. Therefore since shape functions reduce to zero along sides opposite the corresponding nodes, it is possible to combine shape functions associated with the same node to provide a global piecewise polynomial *basis function* (*trial function*) for each node. Thus for any node $\underset{\sim}{x}_i$ the corresponding basis function is defined element-by-element as

$$\varphi_i(\underset{\sim}{x}) = \begin{cases} \varphi_E^i(\underset{\sim}{x}) & \text{if } \underset{\sim}{x}_i \in E \\ 0 & \text{if } \underset{\sim}{x}_i \notin E \end{cases} \qquad \text{when } \underset{\sim}{x} \in E.$$

The union of elements in which $\varphi_i \neq 0$ is termed the *support* and is denoted by $S(\varphi_i)$. It is clear from the construction that such finite element basis functions have *local support* (see Fig. 2.2). Once the piecewise smooth basis

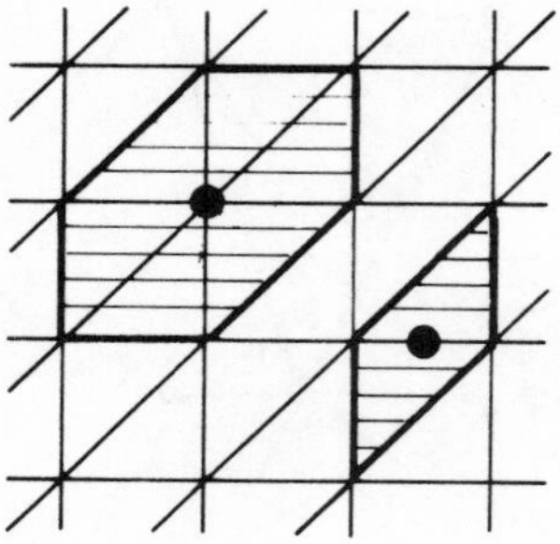

Fig. 2.2. Support of basis functions

functions have been defined it is possible to write the approximate solution as

$$U(x,y) = \sum_i U_i \varphi_i(x,y) .$$

Using this formulation it is possible to describe the equations that define the parameters U_i.

2.2. VARIATIONAL FORMULATIONS

The variational principles enumerated earlier in section 1.4 are just one form that can be utilised to compute finite element approximations. A more general approach which includes such formulations as one of several alternatives is the *method of weighted residuals* otherwise known as the *Petrov-Galerkin method.*

The problem is to solve

$$Lu = f \quad \text{in } \Omega ,$$

subject to certain boundary conditions on Γ.

The approximate solution is to be of the form

$$U(x,y) = \sum_{i=1}^{N} U_i \varphi_i(x,y) ,$$

that is with N unknown parameters. It is necessary to construct N linearly independent *test functions* $\psi_j(x,y)$ and then the approximate solution satisfies

$$(LU - f, \psi_j)_0 = 0 \qquad j = 1,\ldots,N$$

where

$$(u,v)_0 \equiv \iint_\Omega uv \, dxdy.$$

This formulation disguises several technicalities which will be disclosed as and when required. For instance in many cases it is possible to apply Green's theorem (integration by parts) to the integrals. Thus if

$$L \equiv -\frac{\partial^2}{\partial x^2} - \frac{\partial^2}{\partial y^2} ,$$

then

$$-\iint_\Omega \psi \nabla^2 \varphi = \iint_\Omega \left\{ \left(\frac{\partial \varphi}{\partial x}\right)\left(\frac{\partial \psi}{\partial x}\right) + \left(\frac{\partial \varphi}{\partial y}\right)\left(\frac{\partial \psi}{\partial y}\right) \right\} dxdy + \int_\Gamma \psi \frac{\partial \varphi}{\partial n} ds$$

and $(LU,\psi)_0$ can be replaced by a bilinear form $a[U,\psi]$ such that in this example

$$a[U,\psi] = \iint_\Omega \left\{ \left(\frac{\partial U}{\partial x}\right)\left(\frac{\partial \psi}{\partial x}\right) + \left(\frac{\partial U}{\partial y}\right)\left(\frac{\partial \psi}{\partial y}\right) \right\} dxdy + \int_\Gamma \psi \frac{\partial U}{\partial n} ds .$$

Thus the alternative form

$$a[U,\psi_j] = (f,\psi_j)_0 \qquad j = 1,\ldots, N$$

can be used even when the approximate solution does not possess sufficient derivatives to define LU. Since the problem is linear it follows that

and

$$\underset{\sim}{f} \equiv ((f,\psi_1)_0,\ldots,(f,\psi_N)_0)^T \ .$$

The structure of A depends on the form of the test functions ψ_j and the trial functions φ_i. Particular methods of solution will be discussed at a later stage.

2.3. TEST FUNCTIONS

There is clearly an infinite variety of test functions available, but only a few lead to worthwhile numerical methods.

Galerkin Methods

If $\psi_j = \varphi_j$ then the stiffness matrix is symmetric and positive definite when the differential operator is symmetric and positive definite. In addition

$$a[\varphi_i,\varphi_j] \neq 0 \Rightarrow S(\varphi_i) \cap S(\varphi_j) \neq \phi \text{ (the empty set)}$$

and so the number of non-zeros in the i-th row corresponds to the number of nodes in $S(\varphi_i)$. If M_1 is the maximum number of nodes per element and M_2 is the maximum number of elements that meet at a common vertex, the number of non-zeros in any row is bounded by

$$a[U,\psi] = \sum_{i=1}^{N} U_i \ a[\varphi_i,\psi]$$

and thus the approximate problem reduces to the solution of a system of linear algebraic equations

$$A\underset{\sim}{U} = \underset{\sim}{f}$$

where the *stiffness matrix* is defined as

$$A_{ji} = a[\varphi_i,\psi_j]$$

with

$$\underset{\sim}{U} \equiv (U_1,\ldots,U_N)^T$$

$$M_1M_2 - M_2 + 1 .$$

Typically $M_2 \leq 6$ and $M_1 = 6$ or 8 for quadratic elements, while N (the number of unknowns) might be several hundred.

If a variational principle does exist then the Galerkin formulation is equivalent to the so-called Rayleigh-Ritz method.

Collocation Methods

If $\psi_j = \delta(\underset{\sim}{x} - \underset{\sim}{\xi}_j)$ (a Dirac delta function) then the equation

$$(LU - f, \psi_j)_0 = 0$$

is equivalent to

$$LU = f \text{ at } \underset{\sim}{x} = \underset{\sim}{\xi}_j .$$

The significant features of this approach are that the need to compute integrals has been removed and that the collocation points $\underset{\sim}{\xi}_j$ have to be chosen very carefully since the results can be very sensitive to changes in their positions. It is also necessary to define trial functions of sufficient degree so that LU is non-trivial, thus piecewise cubic trial functions are used for second order problems. In general such methods have invariably been applied to one-dimensional problems for which the best positions for the collocation points are known to be the zeros of orthogonal polynomials (Prenter, 1977).

Least-Square Methods

If the trial functions are of sufficiently high degree it is possible to take $\psi_j = L\varphi_j$ and the defining equations then become

$$(LU - f, L\varphi_j)_0 = 0 \qquad j = 1,\ldots,N$$

which is equivalent to

$$\underset{\{U_j\}}{\text{minimise}} \ \|LU - f\|_0^2 \ .$$

As the expression to be minimised is the $\mathcal{L}_2$ norm of the residual, the derivation of the name is obvious. One of the main features of this approach is that the resulting algebraic equations involve a symmetric positive definite matrix irrespective of the properties of the operator L. In addition if the residual to be minimised includes boundary integrals such as

$$\int_\Gamma (U - g)^2 \, ds \ ,$$

it is possible to incorporate 'awkward' boundary conditions in a straightforward manner. The drawbacks of this approach include the large number of non-zeros in the equations resulting from the high degree of trial functions and the possible ill-conditioning of the matrix. Alternative methods of incorporating the boundary conditions into the functional are frequently used in global methods and a description of such modifications is given more appropriately in section 6.2.

Asymmetric Galerkin Methods

The choice $\psi_j = \varphi_j$ leads to a stiffness matrix that corresponds to *central-difference formulae*. It is a rarely acknowledged fact that finite element methods can be viewed as a complicated way of deriving difference equations. There are, however, circumstances in which it is preferable to foresake central differences in favour of backward-difference formulae. The trade off is often between higher order or greater stability. The desired results can be obtained if

$$\psi_j = \varphi_j + \Phi_j \ ,$$

where Φ_j is a suitable function. One simple example of an asymmetric test function in one dimension is shown in Figure 2.3. More examples of this approach will appear later in chapter 11 on diffusion-convection problems.

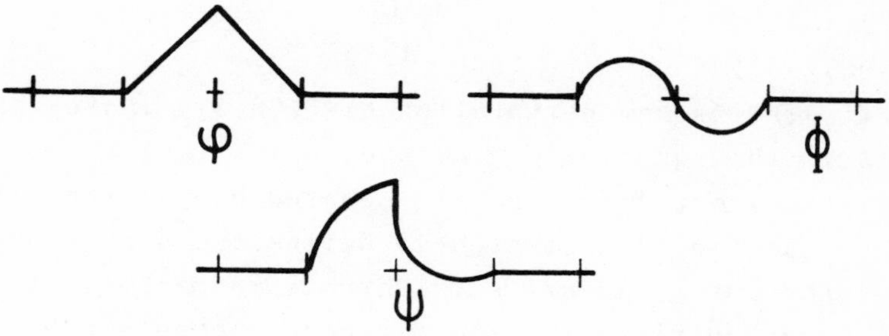

Fig. 2.3.

2.4. TRIAL FUNCTIONS

The shape functions on which nearly all current finite element programs are based are derived from local Lagrange interpolation, the quadratic functions given earlier are just one example. These shape functions can be defined on triangular or rectangular elements.

Triangular Elements

Assume the vertices of the triangle are $\underset{\sim}{x}_i \equiv (x_i, y_i)$ i=1,2,3, let

$$D_{jk} \equiv \det \begin{bmatrix} 1 & x & y \\ 1 & x_j & y_j \\ 1 & x_k & y_k \end{bmatrix} = (x_j y_k - x_k y_j) + (y_j - y_k)x - (x_j - x_k)y$$

and

$$C_{ijk} \equiv \det \begin{bmatrix} 1 & x_i & y_i \\ 1 & x_j & y_j \\ 1 & x_k & y_k \end{bmatrix} \quad , \tag{2.1}$$

then linear shape functions are defined by

$$\varphi_E^i(x,y) = \frac{D_{jk}}{C_{ijk}} \quad ,$$

where (i,j,k) is any permutation of (1,2,3). Note that $D_{jk} = 0$ is the equation of the side opposite $\underset{\sim}{x}_i$. If these normalised linear functions are denoted by p, q and r it follows that they can be used to define the higher degree shape functions in a very straightforward manner. The quadratic shape functions for the six-node element in Figure 2.4a are

$$\begin{aligned} \varphi_E^1 &= p(2p-1) \\ \varphi_E^4 &= 4pq \end{aligned} \tag{2.2}$$

and so on.

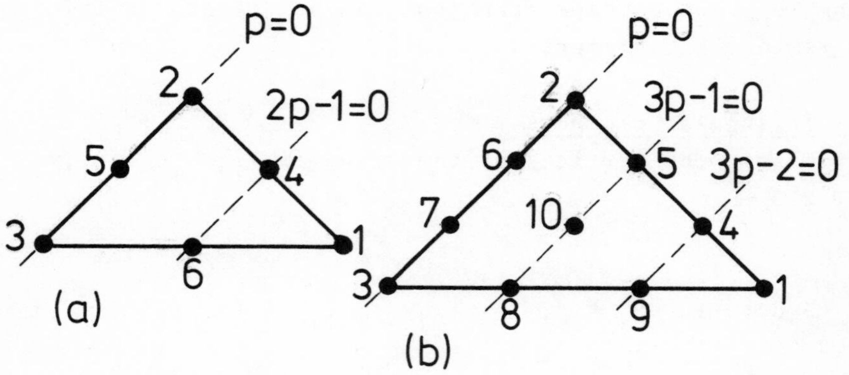

Fig. 2.4.

Cubic shape functions for the ten-node element in Fig. 2.4b follow likewise:

$$\varphi_E^1 = \tfrac{1}{2}p(3p-1)(3p-2)$$

$$\varphi_E^4 = \tfrac{9}{2}pq\ (3p-1)$$

and so on. The shape function corresponding to the internal node

$$\varphi_E^{10} = 27\ pqr$$

is usually eliminated in advance by modifying the functions φ_E^i (i=1,...,9) to ensure an interpolation formula that recovers quadratics. This affects the order of convergence as defined later, but internal nodes are in general avoided since they introduce significant practical difficulties in the curved elements discussed later.

An alternative method of circumverting some of the complications is to use so-called *static condensation*. When the equation corresponding to the internal node is formed it only involves nodes of a single element since the support of that basis function is a single element. Thus, from the stiffness matrix, the equation corresponding to an internal node can be used to specify that internal nodal value in terms of the other nodal values in the triangle so that (in the cubic case) this equation can be written as

$$U_{10} = \sum_{i=1}^{9} \alpha_i\ U_i$$

for some values α_i. Then U_{10} can be eliminated from the equations for the other nodes without altering the structure of the algebraic system. This is an example of a *frontal method* of solution (see chapter 15).

Rectangular Elements

As an alternative to triangles, rectangular elements can be used. Again local Lagrange interpolation is used. In

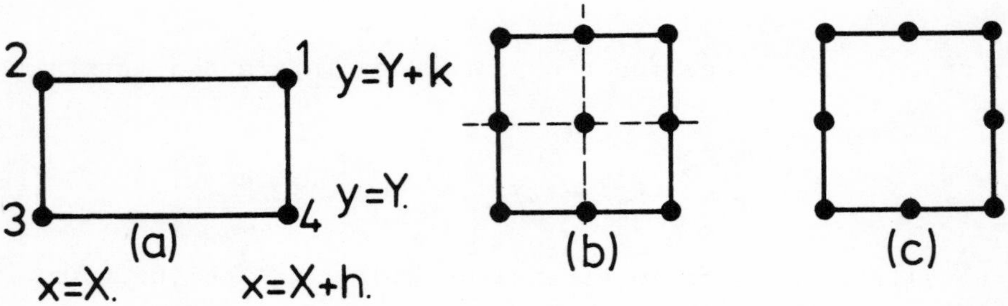

Fig. 2.5.

this case if the sides are aligned with the axes, the simplest form is a bilinear function; thus from Fig. 2.5a

$$\varphi_E^1 = \left(\frac{x-X}{h}\right)\left(\frac{y-Y}{k}\right)$$

and so on. Such a basis function within each element can be written as

$$\varphi_E^i(x,y) \equiv \varphi_E^{i_1}(x)\ \varphi_E^{i_2}(y)$$

this is known as the *tensor product form* and will be used in chapter 17. The nine-node biquadratic is constructed similarly and the eight-node element can be viewed in the same light as the cubic triangular element or as a *bilinear blending* of quadratic boundary data (see section 2.8). The rectangular elements lead to higher degree polynomials but are frequently preferred because numerical integration rules are considered easier to define on rectangles. The lack of freedom in rectangular elements will be removed later when extensions to quadrilaterals are given in section 2.8.

Each of these element types lead to continuous trial functions but normal derivatives are discontinuous. In order to achieve C^1 continuity, it is necessary to construct elements in which some of the degrees of freedom correspond to derivatives or to construct macro-elements such as the Clough and Tocher element. Full details of these and other elements can be found in Mitchell and Wait (1977).

2.5. STRUCTURE AND SOLUTION OF ALGEBRAIC EQUATIONS

It was stated earlier that the number of non-zeros is small compared to the number of unknowns. Thus the matrix is *sparse* and it is usually worthwhile taking account of the sparsity to avoid storing and manipulating too many zeros. In two-dimensional problems a reasonably efficient and simple method is just to order the nodes in a regular manner and so derive a banded matrix. Then it is a simple matter to write a Choleski factorisation routine for a banded matrix. In order to achieve a small (though not necessarily the smallest) band width the *Cuthill-McKee* ordering algorithm should be used. If variable bandwidth solvers are available the *Reverse Cuthill-McKee* ordering provides an additional saving in most cases. If the savings gained in this way are not sufficient more sophisticated procedures may be necessary and these are covered in chapter 15.

It is important to note that the stiffness matrix is constructed element-by-element and not node-by-node as a finite-difference matrix might be. Thus, if, for example, the stiffness matrix A is defined by

$$A_{ij} = \iint_{\Omega} \left\{\left(\frac{\partial\varphi_j}{\partial x}\right)\left(\frac{\partial\psi_i}{\partial x}\right) + \left(\frac{\partial\varphi_j}{\partial y}\right)\left(\frac{\partial\psi_i}{\partial y}\right)\right\} dxdy ,$$

then for each element we construct the *element stiffness matrix* A_E such that

$$(A_E)_{ij} = \iint_{E} \left\{\left(\frac{\partial\varphi_j}{\partial x}\right)\left(\frac{\partial\psi_i}{\partial x}\right) + \left(\frac{\partial\varphi_j}{\partial y}\right)\left(\frac{\partial\psi_i}{\partial y}\right)\right\} dxdy .$$

Thus the element stiffness matrix contains the contributions from a particular element. In all problems it follows that

$$A_{ij} = \sum_{E}(A_E)_{ij} ,$$

where to simplify the notation it is assumed that the $M_1 \times M_1$ element stiffness matrix is embedded in an $N \times N$ zero matrix.

As the components of the stiffness matrix are gradually accumulated in this manner it is possible to use the so-called frontal method of solution in which the matrix is constructed and eliminated simultaneously, further details of this approach are to follow in chapter 15.

2.6. FINITE ELEMENT ANALYSIS

It was shown in section 2.2 that all the common versions of the finite element method have a variational formulation. For this reason it is convenient if the analysis is developed from the weak form of the differential equation. Thus in place of the classical formulation

$$Lu = f \quad \text{in } \Omega , \tag{2.3}$$

subject to boundary conditions on Γ, we have

$$a[u,v] = (f,v)_0 \qquad \text{for all } v \in \mathcal{H} . \tag{2.4}$$

The structure of the bilinear form is governed by the nature of the differential operator and the choice of method. The *space of admissible functions* (or *test space*), denoted above by $\mathcal{H}$, is governed by the order of the differential operator and the form of the boundary conditions. In the remainder of this chapter, attention will be focussed on the Galerkin method and the definitions appropriate for this method may need slight modification to conform to an alternative formulation.

As mentioned in chapter 1, any analysis of the finite element method is frequently done in the general setting of Sobolev spaces (see Adams, 1975, or Kantorovich and Akhilov, 1964). For second order problems these spaces are $\mathcal{H}_1$ and $\mathcal{H}_1^0$ while for fourth order problems they are $\mathcal{H}_2$ and $\mathcal{H}_2^0$. These spaces are most conveniently defined in multi-index notation so that for any $v \equiv v(x,y)$

$$D^{\alpha}v \equiv \frac{\partial^{|\alpha|}v}{\partial x^{\alpha_1}\partial y^{\alpha_2}}$$

where $\alpha \equiv (\alpha_1,\alpha_2)$ and $|\alpha| = \alpha_1+\alpha_2$. Then we derive

$$\mathcal{L}_2 \equiv \{v : (v,v)_0 < \infty\}$$

and

$$\mathcal{H}_k \equiv \{v : v\in\mathcal{L}_2 , \quad D^{\alpha}v \in \mathcal{L}_2 , \quad |\alpha| \leq k\}, \quad k=1,2,$$

with

$$\mathcal{H}_1^0 \equiv \{v : v\in\mathcal{H}_1, \ v = 0 \text{ on } \Gamma\}$$

and

$$\mathcal{H}_2^0 = \{v : v\in\mathcal{H}_2, \ v = \frac{\partial v}{\partial n} = 0 \text{ on } \Gamma\} \quad .$$

The *norms* for these spaces are given by

$$\| v \|_k \equiv \{ \sum_{|\alpha|\leq k} (D^{\alpha}v, D^{\alpha}v)_0\}^{\frac{1}{2}}$$

and the semi-norms on $\mathcal{H}_k$

$$|v|_k \equiv \{ \sum_{|\alpha|=k} (D^{\alpha}v, D^{\alpha}v)_0 \}^{\frac{1}{2}}$$

are norms over the spaces $\mathcal{H}_k^0$.

Since the bilinear form is a continuous mapping onto the real line, it follows that there exists a constant α such that

$$|a[u,v]| \leq \alpha \, \| u \| \, \| v \| \ . \tag{2.5}$$

This inequality holds even when the *trial space* (containing u) and the test space (containing v) are different. However, in Galerkin methods it is assumed that the trial space and the test space coincide. In this case a useful theoretical property that makes the analysis much easier and which is frequently satisfied in practice is that there exists $\gamma > 0$

such that
$$a[u,u] \geq \gamma \| u \|^2 . \tag{2.6}$$

This condition is equivalent to demanding that the operator L be positive definite. If it is satisfied for all $u \in \mathcal{H}$ we say the bilinear form is *$\mathcal{H}$-elliptic*. If (2.5) and (2.6) are both satisfied we have:

Lax-Milgram Lemma

Inequalities (2.5) and (2.6) guarantee that (2.4) is a well-posed problem, i.e. that the solution is unique and depends continuously on the right hand side.$_\Delta$ (The symbol Δ will be used to denote the logical end of a theorem).

Let $\mathcal{H}_h$ denote a space of piecewise polynomial basis functions, since U has N degrees of freedom $\mathcal{H}_h$ will be an N-dimensional space. Assuming that $\mathcal{H}_h \subset \mathcal{H}$, which is a condition that may impose certain boundary conditions on the basis functions, it follows that the Galerkin approximation U satisfies

$$a[U,V] = (f,V)_0 \qquad \text{for all } V \in \mathcal{H}_h . \tag{2.7}$$

This is equivalent to

$$a[U,\varphi i] = (f,\varphi_i)_0 \qquad i=1,\ldots,N,$$

since $V \in \mathcal{H}_h$ can be written as

$$V = \sum_{i=1}^{N} V_j \varphi_j .$$

Cea's Lemma

It follows from (2.4), (2.5), (2.6) and (2.7) that

$$\| u - U \| \leq \frac{\alpha}{\gamma} \inf_{V \in \mathcal{H}_h} \| u - V \| ._\Delta \tag{2.8}$$

The analysis of the convergence of the error in the finite element method has thus been converted to an analysis of the error in approximation by piecewise polynomials. It is not necessary to worry about best approximation, interpolation errors will provide the same estimates of *orders of convergence* it will only be the constant in front that is not necessarily best.

Before proceeding with this analysis it is possible to broaden the scope somewhat by introducing new families of finite element approximations.

2.7. ISOPARAMETRIC ELEMENTS

One method of incorporating complicated element shapes is to provide a local mapping of each element E onto a *reference element* $\mathcal{E}$ and then to define the shape functions in terms of polynomials on the reference element. Since such transformations are local to each element inter-element continuity imposes constraints on the mapping in two ways:

Continuity of the domain, the elements must not overlap or leave gaps as in Figure 2.6.

Continuity of the approximation is then necessary when the inverse images of the local approximations are pieced together.

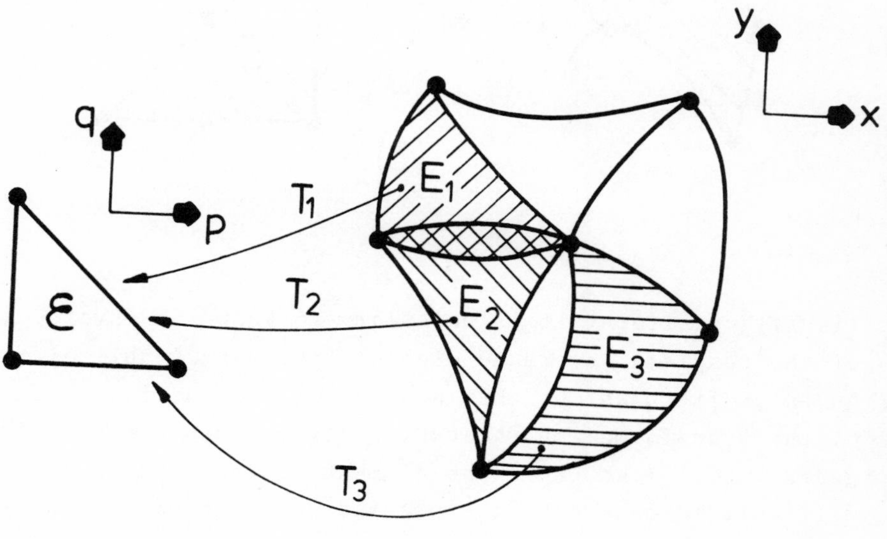

Fig. 2.6.

One method that satisfies both these constraints is an *isoparametric transformation* based on the polynomial Lagrange shape functions of section 2.4. This provides an invertable transformation of a set of nodes in the physical plane onto the regularly spaced nodes on a reference element as illustrated in Figure 2.7. In order to satisfy the domain con-

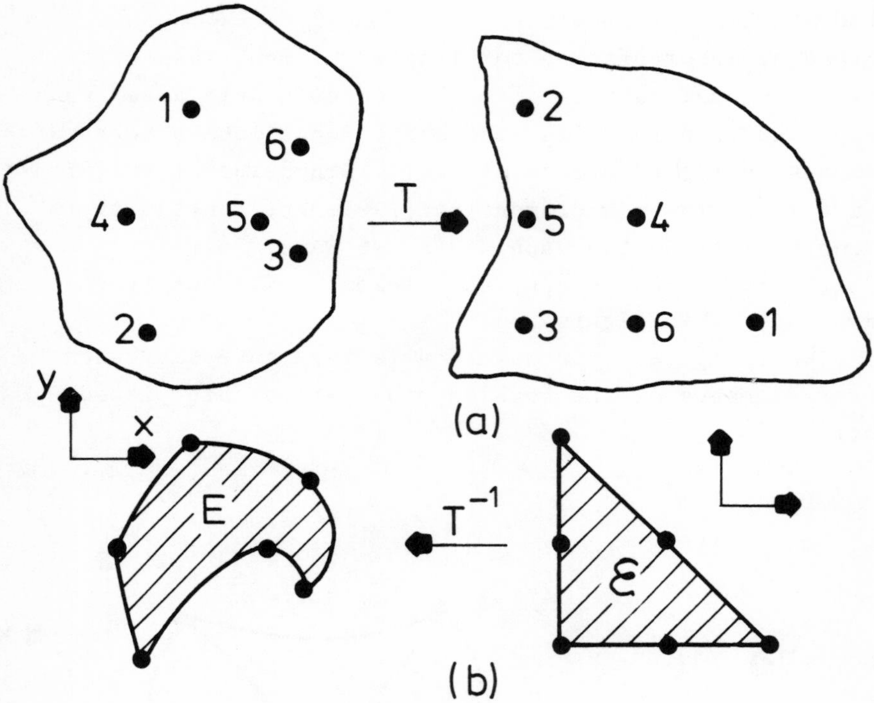

Fig. 2.7.

tinuity constraint it is necessary to know the inverse image of the complete reference element (the convex hull of the nodes). It turns out that because of the particular form of the transformation the continuity between elements is guaranteed, although if the element is too 'fancy' it is difficult to determine the precise position of the inter-element boundary.

If the positions of the physical nodes are $\underset{\sim}{x}_i$ and the positions of their respective images are $\underset{\sim}{p}_i \equiv (p_i, q_i)$ it follows that

$$\underset{\sim}{x} = \sum_{i=1}^{M_1} \underset{\sim}{x}_i \, \varphi_E^i(p,q) \; , \qquad (2.9)$$

where the φ_E^i are Lagrange shape functions of the appropriate degree. These were described in section 2.4. Note that using p and q as the new coordinates leads to a reference element that is a triangle with vertices (1,0), (0,1) and (0,0). Provided that the interior of E is simply connected, the domain continuity constraint is satisfied with this form of local transformation, for the same reason that the Lagrange elements of section 2.4 provide an approximation that is continuous across intra-element boundaries. On any arc of the boundary of an element all non-vanishing shape functions correspond to nodes on that particular arc. Thus for example along the arc $\underset{\sim}{x}_1 \underset{\sim}{x}_4 \underset{\sim}{x}_2$ which corresponds to the side

$$p + q - 1 = 0$$

it follows that the transformation reduces to

$$\underset{\sim}{x} = \underset{\sim}{x}_1 \, \phi_E^1(\underset{\sim}{p}) + \underset{\sim}{x}_4 \, \varphi_E^4(\underset{\sim}{p}) + \underset{\sim}{x}_2 \, \varphi_E^2(\underset{\sim}{p})$$

$$\equiv \underset{\sim}{x}_1 \, p(2p-1) + \underset{\sim}{x}_4 \, 4pq + \underset{\sim}{x}_2 \, q(2q-1) \qquad (2.10)$$

$$= \underset{\sim}{x}_1 \, p(2p-1) + \underset{\sim}{x}_4 \, 4p(1-p) + \underset{\sim}{x}_2 \, (1-p)(1-2p), 0 \le p \le 1,$$

since q = 1 - p. The parametric representation of the boundary arc is independent of the remaining nodes in the element and so it is the same for two elements with a common boundary. The particular quadratic arc given above is a parabola but it shows that in general it is possible to use this approach to approximate a curved boundary by piecewise polynomial arcs.

Although transformation (2.9) is a one-to-one mapping of the point set $\{\underset{\sim}{p}_i\}$ onto the point set $\{\underset{\sim}{x}_i\}$ it will not necessarily be suitable for finite element approximation. For example it is possible to have an element that does not have a connected interior as in Figure 2.8a, this leads to

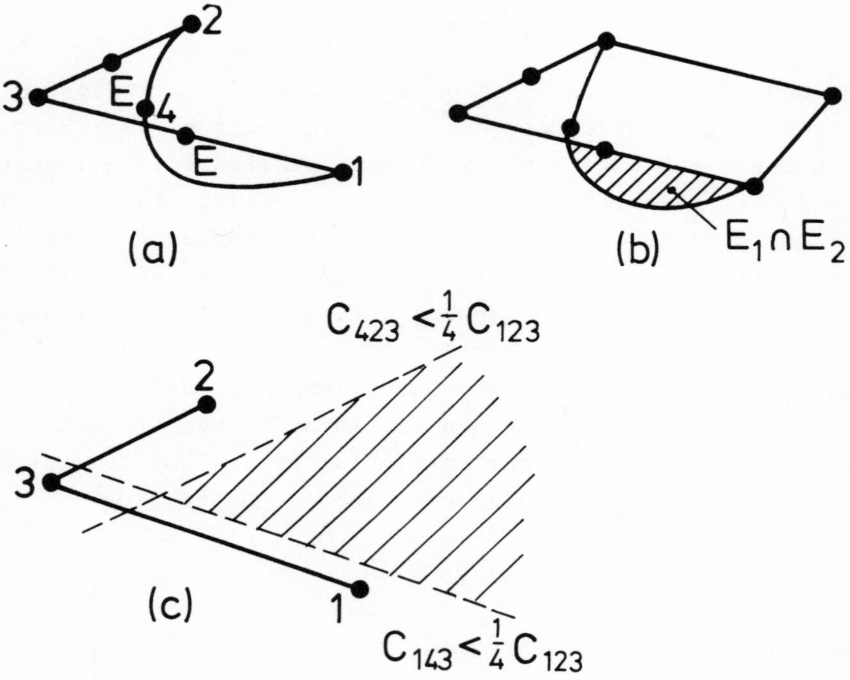

Fig. 2.8.

overlap of adjacent elements as illustrated in Figure 2.8b.

If we consider the special case of an element with only one curved side as in Figure 2.8a it is possible to set down rules that prevent such overlaps, at least in the quadratic case.

Lemma 2.1.

If in the notation of section 2.4

$$0 < C_{143} < \tfrac{1}{4}\, C_{123} \tag{2.11}$$

then the arc $\underset{\sim}{x}_2\ \underset{\sim}{x}_4\ \underset{\sim}{x}_1$ *intersects the side* $\underset{\sim}{x}_3\ \underset{\sim}{x}_1$ *at a point other than* $\underset{\sim}{x}_1$. *In addition there are points in* $\mathcal{E}$ *at which*

$$\frac{\partial(x,y)}{\partial(p,q)} = 0. _\Delta$$

This is equivalent to the *quarter point condition* of Jordan (1970) and is illustrated by Figure 2.8c.

Proof.

Let $\underset{\sim}{x}_s$ be a point on the arc $\underset{\sim}{x}_1\ \underset{\sim}{x}_4\ \underset{\sim}{x}_2$. Thus for some p and q = 1-p

$$\underset{\sim}{x}_s = \underset{\sim}{x}_1\ \varphi_E^1 + \underset{\sim}{x}_2\ \varphi_E^2 + \underset{\sim}{x}_4\ \varphi_E^4 .$$

Thus if $\underset{\sim}{x}_s$ is on the side $\underset{\sim}{x}_3\ \underset{\sim}{x}_1$, in the notation of section 2.4 it follows that

$$C_{1s3} = 0$$

where

$$C_{1s3} = C_{113}\ \varphi_E^1 + C_{123}\ \varphi_E^2 + C_{143}\ \varphi_E^4 .$$

Assuming that

$$C_{143} = (\tfrac{1}{4} - \theta)\ C_{123} \qquad 0 < \theta < \tfrac{1}{4}$$

it follows that

$$0 = C_{1s3} = C_{123}\ \{1 - p(1 + 4\theta)\}$$

which has the non-trivial solution $p = \frac{1}{1+4\theta}$.

The Jacobian of the quadratic transformation is

$$J = C_{123} + 4p\ (C_{143} - \tfrac{1}{2}\ C_{123}) + 4q\ (C_{423} - \tfrac{1}{2}\ C_{123}),$$

and clearly if

$$C_{143} < \tfrac{1}{4}\ C_{123}$$

the equation

$$J = 0$$

has non-trivial solutions with $\underset{\sim}{p} \in \mathcal{E}$.$_\Delta$

Thus, subject to certain constraints on the position of the nodes, it is possible to derive a satisfactory local transformation. The continuity of the global approximation and the justification of the name *iso*parametric approximation follows from the use of a local approximation of the form

$$U_E(\underset{\sim}{x}) = \sum_i U_i \, \varphi_E^i(\underset{\sim}{p})$$

where, to repeat, the transformation is

$$\underset{\sim}{x} = \sum_i \underset{\sim}{x}_i \, \varphi_E^i(\underset{\sim}{p});$$

the connexion is obvious.

Isoparametric elements can be based on the rectangular Lagrange elements; without going through the algebra again it should be clear that:

1) There must be constraints placed on the node positions to ensure continuity.

2) The boundaries of the elements are always polynomial arcs.

3) Elements with or without interior nodes can be accommodated, but the latter are usually preferred.

4) Triangular elements can be (and frequently are) mapped onto rectangular reference elements to simplify numerical integration. This is achieved by assigning all the nodes on a particular side to have the same coordinates.

5) Isoparametric approximations are defined in terms of shape functions that are no longer polynomials

and the shape functions cannot reproduce polynomial solutions exactly.

2.8. CURVED ELEMENTS

The isoparametric elements have been used for several years and appear likely to retain their dominant position in many fields 'by habit and repute' unless alternatives provide dramatic improvements. The significant feature of the most frequently quoted alternatives is the ability to incorporate *exact boundary data* and not merely a piecewise polynomial approximation.

Blending Functions

The use of *transfinite mappings* of arbitrarily curved elements onto a reference element is a very general technique which forms the basis of many sophisticated programs for computer aided design. A serious flaw in the method is that complex arcs lead to complex mappings and for simple arcs the method reduces to isoparametric approximation.

Assume that the element E is to be mapped onto the reference element $\mathcal{E}$ which is the unit square, further assume that E is bounded by four curved arcs as in Figure 2.9.

Since the four arcs form a closed curve it follows that we can define

$$\left.\begin{aligned} \underset{\sim}{x}(0,q) &= \underset{\sim}{X}_4(q) \\ \text{and}\qquad \underset{\sim}{x}(1,q) &= \underset{\sim}{X}_2(q) \end{aligned}\right\}\quad 0 \le q \le 1 ,$$

with

$$\left.\begin{aligned} \underset{\sim}{x}(p,0) &= \underset{\sim}{X}_1(p) \\ \text{and}\qquad \underset{\sim}{x}(p,1) &= \underset{\sim}{X}_3(p) \end{aligned}\right\}\quad 0 \le p \le 1 .$$

Thus it remains to define $\underset{\sim}{x}(p,q)$, $0 < p, q < 1$. This is done using bilinear blending thus

$$\begin{aligned} \underset{\sim}{x}(p,q) = &\{(1-p)\ \underset{\sim}{x}(0,q) + p\ \underset{\sim}{x}(1,q)\} + \\ &+ \{(1-q)\ \underset{\sim}{x}(p,0) + q\ \underset{\sim}{x}(p,1)\} - \\ &- \{(1-p)(1-q)\ \underset{\sim}{x}(0,0) + (1-p)q\ \underset{\sim}{x}(0,1) + \\ &+ (1-q)p\ \underset{\sim}{x}(1,0) + pq\ \underset{\sim}{x}(1,1)\}\ . \end{aligned} \tag{2.12}$$

If the arcs $\underset{\sim}{X}_i$ are quadratics as in (2.10) the element becomes the eight-node isoparametric curved quadrilateral. The approximate solution can then be represented in terms of continuous boundary data or in terms of nodal values. Clearly the most likely combination is continuous data along arcs that correspond to the boundary Γ and nodal values on internal intra-element boundaries. In any situation the solution can be represented in a form compatible with (2.12) as

$$\begin{aligned} U_E(\underset{\sim}{x}) = &\{(1-p)\ U_E(0,q) + p\ U_E(1,q)\} + \\ &+ \{(1-q)\ U_E(p,0) + q\ U_E(p,1)\} - \\ &- \{(1-q)(1-p)\ U_E(0,0) + (1-p)q\ U_E(0,1) + \\ &+ (1-q)p\ U_E(1,0) + pq\ U_E(1,1)\}\ . \end{aligned} \tag{2.13}$$

As isoparametric elements are a special case of blended elements the two types can be combined in the same model. In order for the transformation to be useable in a finite element approximation it is important that the Jacobian is easy to evaluate and have no zeros in $\mathcal{E}$. This is a consideration that does not apply in other applications. Blended elements can be constructed for triangular elements but the algebra becomes very lengthy; details can be found in Barnhill (1976).

Direct Method

It is worth mentioning one attempt to overcome the danger inherent in blended and isoparametric transformations when the Jacobian is not bounded away from zero. So-called direct methods do not involve transformations to a reference element, instead shape functions are constructed explicitly on the curved element. In the simplest case of a triangle with vertices (1,0), (0,1) and (0,0) and curved side $\underset{\sim}{x}_1 \underset{\sim}{x}_2$ that is given by $f(\underset{\sim}{x}) = 0$, where $f(\underset{\sim}{x}_3) = 1$, it follows that φ_E^3 satisfies

$$\alpha(\varphi_E^3)^2 + \{\alpha(x + y - 1) + 1 - \frac{f(x,0)}{1-x} - \frac{f(0,y)}{1-y}\} \varphi_E^3 + f(x,y) = 0.$$

The remaining shape functions are then obtained from consistency conditions such as

$$\sum_i \varphi_E^i = 1.$$

Functions of this type have received a considered amount of theoretical attention (see Mitchell and Wait, 1977) but little or no practical experience has been gathered.

2.9. ELEMENTS IN THREE DIMENSIONS

Excepting the direct methods of the previous section, all the preceding finite element constructions can be generalised to three dimensions. Triangular elements can be generalised to tetrahedra with or without curved sides. Such elements do however have serious shortcomings on a conceptual level - it is an extremely complicated operation to partition an arbitrary three dimensional body into tetrahedra. Regular elements lead to so-called brick elements. As in two dimensions, nodes that are not on edges tend to be viewed as unnecessary complications thus a very popular element is the twenty-node brick illustrated in Figure 2.10.

The most straightforward way to construct such an element is via blending functions. It is a two stage procedure, using bilinear blending to construct each face from the quadratic arcs forming the edges then trilinear blending of faces, by analogy with (2.12), to give

$$\underset{\sim}{x}(p,q,r) = \{(1-p)\, \underset{\sim}{x}(0,q,r) + p\, \underset{\sim}{x}(1,q,r) + \text{similar terms}\} -$$
$$- \{(1-p)(1-q)\, \underset{\sim}{x}(0,0,r) + (1-p)q\, \underset{\sim}{x}(0,1,r) + \text{similar terms}\} +$$
$$+ \{(1-p)(1-q)(1-r)\, \underset{\sim}{x}(0,0,0) + \text{similar terms}\} .$$

An extremely useful third alternative shape for three-dimensional elements is the prism (or wedge). Such an element is most easy to visualise as isoparametric triangles blended together along the axis.

The basic properties of isoparametric elements listed at the end of section 2.7 also generalise to three dimensions:

1) There must be constraints on the node positions to ensure continuity.

2) The boundaries are 'polynomial' surfaces.

3) Elements with interior nodes are not popular.

4) Prisms can be mapped onto reference cubes to facilitate numerical integration.

5) Isoparametric approximations are not in general polynomials in the physical variables.

2.10. ORDERS OF CONVERGENCE

It is possible to provide a reasonably straighforward analysis that is valid for the Lagrange elements of section 2.4 (see Ciarlet and Raviart, 1972). However, the technicalities of such an analysis do not aid the understanding of the proof when nonlinear transformations are involved.

In finite-difference approximation the order of convergence is determined by a local application of Taylor's theorem. One view of convergence might be that *convergence of order* $k+1$ is equivalent to *no errors in approximating solutions that are polynomials of degree* k. There is a suitable analogue of Taylor's theorem in finite element analysis.

The Bramble-Hilbert Lemma

Let F *be a continuous linear form with the property that*

$$F(p) = 0 \quad \textit{for all } p \in P_k$$

Then there exists a constant C *such that*

$$|F(v)| \le C\ \|F\|'_{k+1}\ |v|_{k+1} \quad \text{for all} \quad v \in \mathcal{H}_{k+1}.\Delta$$

Here $\|F\|'_{k+1}$ is the norm in the dual space $\mathcal{H}'_{k+1}$. Further explanation of this notation can be found in Mitchell and Wait (1977). The original proof (Bramble and Hilbert, 1970) uses quotient spaces but a proof that while essentially the same does not mention quotient spaces explicitly can be found in Mitchell and Wait (1977).

With a carefully chosen linear form F(u) such as

$$F(u) \equiv (u - U_I, w)_1$$

where U_I interpolates u and $w \in \mathcal{H}_1$ is arbitrary it is possible to show that

$$\|u - U_I\|_1 \le C|u|_{k+1} \tag{2.14}$$

when the interpolation reproduces polynomials of degree k exactly. Since isoparametric elements and blended elements recover polynomials only on the reference element, (2.14) can only be applied on the reference element. In order to apply this inequality to the physical curved element it is necessary to have a *regularity condition* on each tranformation. This condition can take the form

$$|u|_{k+1,\mathcal{E}} \le Ch^{k+1}\ \{\sup |J|\}^{\frac{1}{2}}\ \|u\|_{k+1,E} \tag{2.15}$$

where

$$J \equiv \frac{\partial(x,y)}{\partial(p,q)} \quad ,$$

together with

$$\|v\|_{1,\mathcal{E}} \geq Ch\{\inf|J|\}^{\frac{1}{2}} \|v\|_{1,E} \tag{2.16}$$

and

$$0 \leq C_1 \leq \frac{\sup|J|}{\inf|J|} \leq \frac{1}{C_1} \quad . \tag{2.17}$$

Using (2.15) on the right hand side of (2.14) and Cea's Lemma followed by (2.16) on the left leads, after summing over all elements, to a result of the form

$$\|u - U\|_1 \leq Ch^k \|u\|_{k+1} \quad . \tag{2.18}$$

That (2.15) and (2.16) are reasonable assumptions is left as an exercise; they certainly are for isoparametric elements. The condition (2.17) clearly implies that $|J| > 0$ throughout $\mathcal{E}$, this is an excessively severe condition which Jamet (1977) has relaxed in the situation when $\mathcal{E}$ is a square and E is a triangle. In this case it is possible to show that

$$\iint_{\mathcal{E}} |J|^{-1} \, dpdq < \infty$$

which is sufficient. As is shown in chapter 7, in fourth order problems the norm on the left in (2.18) becomes $\mathcal{H}_2$ instead of $\mathcal{H}_1$.

In all estimates it has been assumed that the elements are *conforming*; this will be fully explained in chapter 5. It is also assumed that the solution u is sufficiently smooth for $u \in \mathcal{H}_{k+1}$. If this is not true then the value of k on the right hand side of (2.18) has to be replaced by $k^* = \min(k, \ell - 1)$ where $u \in \mathcal{H}_\ell$.

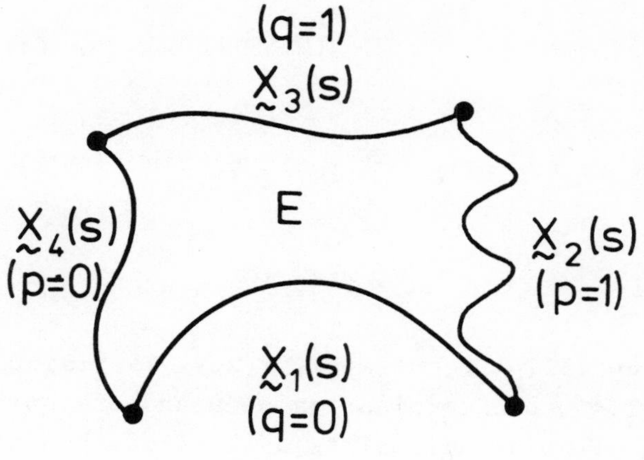

Fig. 2.9.

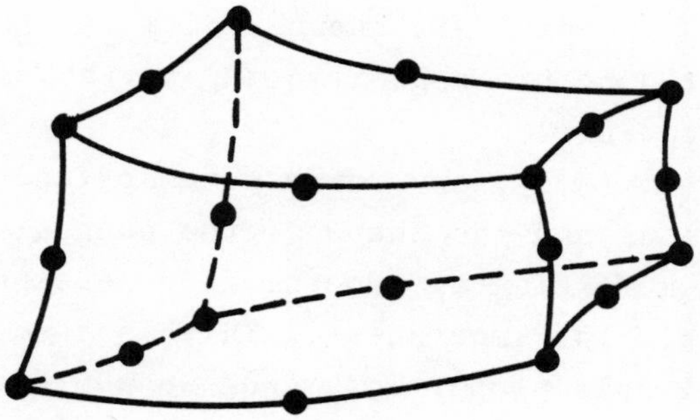

Fig. 2.10. A twenty-node brick.

3

FINITE DIFFERENCES AND SINGULARITIES IN ELLIPTIC PROBLEMS

3.1. INTRODUCTION

Consider the second order elliptic problem in two dimensions given by

$$Lu = f(x,y) \tag{3.1}$$

where L is an elliptic operator, f(x,y) is a specified function, and as in (1.4) some combination of u and its normal derivative is specified at all points of a closed boundary. A finite-difference method, such as the usual five point representation of Lu together with corresponding finite-difference treatment of the boundary conditions, leads to an approximation that satisfies

$$A\underset{\sim}{U} = \underset{\sim}{b} \; . \tag{3.2}$$

The matrix A and vector $\underset{\sim}{b}$ are known, and $\underset{\sim}{U}$ is the vector of values at the grid points approximating to the discrete u(x,y) values at these points.

When f(x,y) in (3.1), any corresponding functions in the boundary conditions and the shape of the boundary of the problem are all sufficiently smooth, then we expect the true u(x,y) to have similar smoothness. In this case, provided the matrix A has suitable properties, then we would expect $\underset{\sim}{U}$ to differ from u by small quantities, typically $O(h^2)$ where h is a typical finite-difference interval, and as h is reduced convergence to the required solution is reasonably rapid. Moreover we can estimate the number of accurate figures obtained in our approximate solutions, by appeal to consistency, without having to use a prohibitively small value of h.

In less smooth situations the difference between $\underset{\sim}{U}$ and u may be much larger as a function of h, and even if we can establish convergence with the use of the simple finite-difference formulae the convergence may be so slow that the labour involved is prohibitive.

Typical singularities occur at one or more boundary points, and arise either through a sudden change in the slope of the boundary or through a sudden change in the nature of the boundary conditions. A problem of the second kind is that investigated originally by Motz (1946), which after taking account of the symmetry requires the solution of Laplace's equation

$$\frac{\partial^2 u}{\partial x^2} + \frac{\partial^2 u}{\partial y^2} = 0 \tag{3.3}$$

in the region of Figure 3.1, in which the boundary conditions are given by

$$\begin{aligned} &u = 0 \quad \text{on} \quad EO, \ \frac{\partial u}{\partial y} = 0 \quad \text{on} \quad OB, \\ &\frac{\partial u}{\partial x} = 0 \quad \text{on} \quad ED, \ \frac{\partial u}{\partial y} = 0 \quad \text{on} \quad DC, \ u = 500 \text{ on } BC, \end{aligned} \tag{3.4}$$

with a sudden change of form at the point O on the boundary EOB.

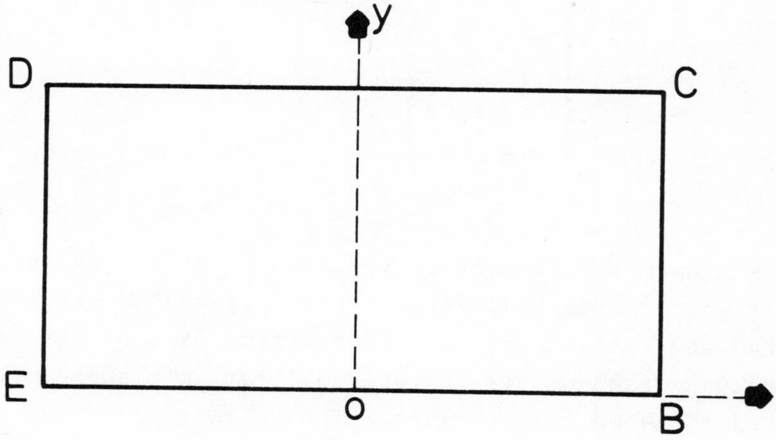

Fig. 3.1.

Convergence of the obvious finite-difference method has been established by Whiteman and Webb (1970), but with the observation that convergence is very slow and without establishing the global error $\underset{\sim}{U} - u$ as a function of h. Another finite-difference solution was provided by Woods (1953). This problem has been solved by many authors and is a useful test problem on which methods in later chapter will be applied, see sections 3.8, 4.3, 5.6 and 6.7.

Another problem which has had much attention is the L-shaped membrane eigenvalue problem, which requires the determination of one or more eigenvalues and possibly eigenfunctions of the equation

$$\frac{\partial^2 u}{\partial x^2} + \frac{\partial^2 u}{\partial y^2} + \lambda u = 0 \tag{3.5}$$

in the region of Figure 3.2, with u = 0 on the boundary.

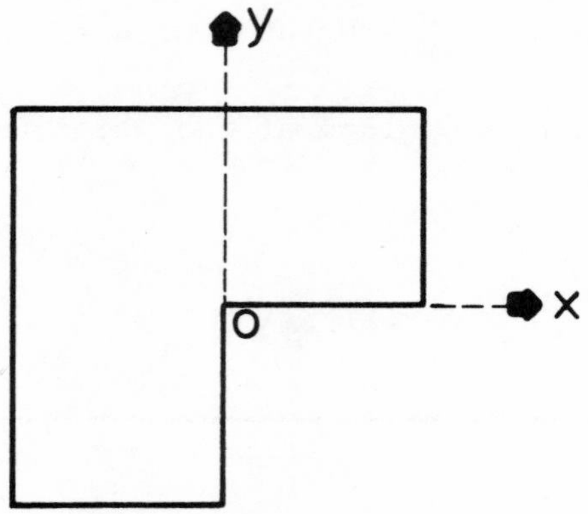

Fig. 3.2.

The simple finite-difference formula for this problem gives rise to an approximating matrix eigenvalue problem, and Forsythe and Wasow (1960) give for estimates μ of the smallest eigenvalue λ the numbers in table 3.1 for an L-shaped region composed of three unit squares. The correct value is $\lambda = 9.6397$, to five figures, and table 3.1 reveals that convergence to this value, as h decreases, is indeed very slow.

We obviously need methods for reducing the labour involved in the production of useful answers, and we discuss below various suggested policies. First, we consider the idea of refining the grid in the neighbourhood of the singular

TABLE 3.1

h	$\frac{1}{3}$	$\frac{1}{4}$	$\frac{1}{6}$	$\frac{1}{8}$	$\frac{1}{10}$	$\frac{1}{12}$
μ	9.5251	9.6414	9.6908	9.6931	9.6883	9.6827

point. Second, we examine the possibilities of accelerating the convergence, as h gets smaller, by adaptations of the method of "the deferred approach to the limit". Third, we look at possibilities of discovering the nature of the singularity, and use this in both "global" and "local" senses, terms which will become obvious in the sequel.

3.2. GRID REFINEMENT

The underlying idea of "grid refinement", used from the earliest "hand-relaxation" days, is based on the principle of St. Venant, well-known to engineers, which for our present purpose can be paraphrased in the rather loose statement that "a disturbance at one point has negligible effect at sufficiently distant points". More mathematically, we know that in most elliptic problems singularities at the boundary do not penetrate into the interior of the region, and for example at all internal points in the two examples of section 3.1 the true solution has continuous derivatives of all orders. Of course the solution may change very rapidly near the singular point and one or more of its higher derivatives consequently gets very large. The local error of finite-difference formulae may involve terms like h^p multiplying some pth derivative, and if the latter get large near the singularity we would like to make h correspondingly smaller in these regions.

A typical graded net in an L-shaped region is shown in Figure 3.3(a), in which the mesh length changes from $h = \frac{1}{4}$ to $h = \frac{1}{16}$ as we approach the singularity. In the special case of the Laplace operator the refinement is simplified by the fact that the five point formula still holds with the coordinate axes rotated through $\pi/4$, so that the grading

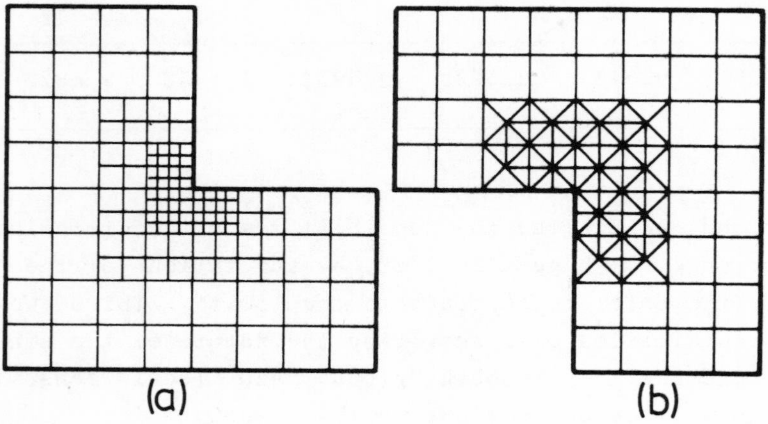

Fig. 3.3.

can be arranged as shown in Figure 3.3(b). The latter is preferable in this favourable case since the value at every mesh point can be related to those of its four obvious near neighbours by a formula which approximately satisfies the differential equation. In Figure 3.3(a), on the other hand, many values can be obtained from vertical or horizontal members only by some form of interpolation.

In both cases, however, there are disadvantages. First, the structure of the finite-difference matrix is less convenient than that obtained with constant h. The bandwidth can be larger and more variable, giving extra complications in the direct solution of the linear equations. With respect to iterative solution, which is likely to be necessary in many problems, we may have corresponding difficulties in finding suitable parameters for the acceleration of convergence of any selected iterative method. A more sophisticated automatic refinement has been attempted for a finite element grid and this is outlined in section 5.7.

3.3. THE DEFERRED APPROACH TO THE LIMIT

It is well known that in the evaluation of a definite integral by a simple formula such as the composite trapezoidal rule, the difference between the true value I and the value T(h) obtained with the trapezoidal rule at interval h is given by

$$I - T(h) = Ah^2 + Bh^4 + Ch^6 + \dots , \qquad (3.6)$$

at least for small enough h and for a sufficiently smooth integrand. Each T(h) differs from I by a term dominated by Ah^2, but by taking two values of h and eliminating the constant A we can obtain a better result with error $O(h^4)$. With estimates at three values of h we can eliminate both A and B and produce a still better result with error $O(h^6)$ and so on.

This device for accelerating convergence and the approach to consistency is even more important when the order of the error is much larger than that of (3.6), and this happens for some kinds of singular integrands. We have a similar situation with our elliptic problems. In sufficiently smooth circumstances the error with our simple finite-difference formulae could be like that of (3.6), where now the A, B and C are constants for a particular mesh point common to successively smaller mesh sizes. In a singular situation the error terms could be much larger, and the "unrefined" convergence therefore much slower.

Fox (1967) showed how to find the error replacements for (3.6) in certain singular situations relevant to quadrature, and similar results would be even more valuable for our singular elliptic problems. We could then use simple formulae throughout, solve with two or three different mesh lengths, and perform the "deferred approach to the limit" with considerable effect. There are, however, very few results of this kind available for the finite-difference treatment of partial differential equations, and the best we can do at this stage is to make some "plausible" suggestions arising from various theories on error bounds.

Hubbard (1966) discusses the problem

$$\nabla^2 u = f(x,y) , \qquad (3.7)$$

with u = g(x,y) on the boundary, solved approximately by

the standard five point difference formulae with local error $O(h^2)$ at internal points, and corresponding formulae near a non-rectangular boundary with local errors of either $O(h)$ or $O(1)$. He shows that if there is a singularity at a point in the interior of the region, at which u is bounded and its derivatives have the behaviour

$$D^{\alpha}u(x,y) = O(|r|^{k-|\alpha|}+1), \ |\alpha| = 1,\ldots,4, \quad k > 0 \qquad (3.8)$$

where r is the distance from the singular point, then the error $e = \underset{\sim}{U} - u$, where $\underset{\sim}{U}$ is the solution of the discrete problem, is given by

$$|e| = O(h^{k-\varepsilon} + h^2), \quad \varepsilon > 0. \qquad (3.9)$$

If the singularity is on the boundary, and (3.8) is still satisfied, then the error is again bounded by (3.9), and in both cases this is true for either of the finite-difference approximations near the boundary. In particular, if the boundary is composed of piecewise analytic arcs, and the singularity occurs at a corner with interior angle π/k, $\frac{1}{2} \le k < \infty$, then if u satisfies Laplace's equation and vanishes on the boundary near this singularity, as for example with a local solution like $r^k \sin k\theta$, so that (3.8) becomes

$$D^{\alpha}u(x,y) = O(|r|^{k-|\alpha|}), \ |\alpha| = 0,1,2,\ldots, \qquad (3.10)$$

then (3.9) again gives the required result with this value of k.

A similar result was given by Bramble and Hubbard (1968) for the eigenvalue problem of type (3.5) in which the boundary is again composed of a finite number of analytic arcs. If π/k is the largest interior angle between adjacent arcs, and finite-difference formulae are used of the type already mentioned, and in which the singular point is not a mesh point, then the difference between the true eigenvalue λ and its finite-difference estimate μ is given by

$$|\mu - \lambda| = O(h^{2k-\varepsilon} + h^2) . \tag{3.11}$$

The difference between suitably normalized true eigenfunctions u and computed eigenfunctions U is also given by

$$|U - u| = O(h^{k-\varepsilon} + h^2) . \tag{3.12}$$

To use these results we must be able to find formulae like (3.8), but this is often quite possible and is discussed in some contexts in section 3.6. Moreover, though the results quoted apply only to the Laplace operator many elliptic problems can be recast in that form by suitable changes of variables, and this is discussed in section 3.5. We know, however, of no results like (3.9) for more general boundary conditions, and we can at present only conjecture that the local behaviour of the solution near the singular point properly reflects an error bound of type (3.9), provided that all the finite-difference equations approximating the differential equation and boundary conditions give rise to a matrix of positive type with suitable bounds on the elements of its inverse.

But even then we cannot conclude that a result like (3.9) automatically produces a formula like

$$u(x,y) - U(x,y,h) = Ah^k + Bh^2 + \ldots , \tag{3.13}$$

where A and B depend on the particular mesh coordinates but are independent of h, such a formula being needed for the success of the deferred approach to the limit. Such a conjecture is plausible, and if for example we assume for the L-membrane problem the formula

$$\mu - \lambda = Ah^{4/3} + Bh^2 + \ldots , \tag{3.14}$$

the terms not given explicitly being unknown, then by applying (3.14) and eliminating the constants A and B from three successive μ values given in table 3.1 we produce the extrapolated sequence

$$9.6304,\ 9.6360,\ 9.6382,\ 9.6390, \tag{3.15}$$

which appear to be converging quite nicely and rapidly to the known accurate value of 9.6397 to this precision.

There are some dangers, however, in *relying* on conjectures which lack rigorous justification, since it is quite possible to achieve what Fox and Mayers (1968) call "inaccurate consistency". Even in the unrefined values table 3.1, for example, the non-monotonic behaviour of $\mu(h)$ indicates that for some h between 1/8 and 1/10 there is a $\mu(h)$ identical with that for $h = 1/6$, whereas the true solution is considerably smaller.

It may also be true that there is no simple and explicit formula of type (3.13) for the error, and much more research in this area would be very useful. In this connexion the most valuable result is probably that of Hofmann (1967) who considered the Dirichlet problem (given boundary values) and the Neumann problem (given normal derivatives) for Laplace's equation in a rectangular region. Without loss of generality he assumes a square region, with the function (or its normal derivative) being zero on three sides and specified on the fourth (with its integral along this side being zero for the Neumann problem).

For the Dirichlet problem he considers the unit square with one corner at the origin; and takes $u = g_1(x)$ on $y = 0$, $0 \le x \le 1$ and zero on the other boundaries. He uses the simplest five point finite-difference formula with $h = 1/N$, and assumes that $g_1(x)$ is bounded and piecewise continuous for $0 \le x \le 1$ and satisfies in $0 < x < 1$ the conditions

$$g_1(0) = \lim_{t\to 0+} g_1(t),\ g_1(1) = \lim_{t\to 1-} g_1(t),$$

and

$$g_1(x) = \{\lim_{t\to x+} g_1(t) + \lim_{t\to x-} g_1(t)\}. \tag{3.16}$$

He then postulates the formula

$$\int_0^1 g_1(x)\sin n\pi x\,dx - h\sum_{i=1}^{N-1} g_1\left(\frac{i}{N}\right)\sin\frac{n\pi i}{N} = \sum_{i=1}^{t} d_i(n)n^{p_i}h^{\tau_i} + O(h^{\tau_i+1}), \tag{3.17}$$

where p_i and τ_i are positive real numbers independent of n, t is an integer greater than 1 and independent of n, and the functions $d_i(n)$ are bounded for $n \geq 1$. He then shows that the required formula for the error U(x,y) - u(x,y) is given by

$$U(x,y) - u(x,y) = \sum_{i=1}^{r} s_i(x,y)h^{\sigma_i} + O(h^{\sigma_{r+1}}) , \qquad (3.18)$$

where the s_i are bounded, and the σ_i are members of the set

$$\{2j + \tau_k : j \geq 0,\ k \geq 0,\ j+k \neq 0,\ 2j + \tau_k \leq \tau_{k+1}\} \qquad (3.19)$$

ordered such that $\sigma_{i+1} > \sigma_i$ with $\tau_0 < \tau_1 < \tau_2 < \ldots$.

For the Neumann problem he finds similar results, the exponents in the expression corresponding to (3.18) being determined uniquely from the numbers corresponding to the τ_i in a formula like (3.17) relating to the quadrature error in the evaluation of $\int_0^1 g_2(x) \cos nx \; dx$, where $g_2(x)$ is the specified normal derivative.

These results, of course, are very useful indeed, and could possibly be extended to cases in which the right-hand side of (3.17) contains terms other than powers of h. Even more important would be extensions of this analysis to cover boundary conditions of Dirichlet type in some regions, Neumann type in other regions, and mixed type on the rest of the boundary.

3.4. GLOBAL TREATMENT OF SINGULARITIES

Subtracting out the singularity

From the foregoing it is clear that we should prefer, when this is possible, to "subtract out" the singularity and leave ourselves with the task of solving a nonsingular problem for which numerical methods of finite-difference type have a much sounder basis. For this purpose we need to know the true behaviour of the singularity in closed form, though we can achieve this only in rare cases.

For the equation

$$\nabla^2 u = f(x,y) , \qquad (3.20)$$

with specified boundary conditions, where $f = 0$ except at some point inside the boundary, where its value is unity, if we take the origin at this point we can be sure that the function

$$v = \frac{1}{2\pi} \log r, \quad r^2 = x^2 + y^2 , \tag{3.21}$$

satisfies (3.20) and has all the ingredients of the singularity. Then the function $w = u-v$ satisfies Laplace's equation, its boundary conditions are calculable from those of u and v, and we are left with a nonsingular problem for w, soluble by finite-difference methods.

More commonly the singularity occurs on the boundary. Suppose, for example, that the x-axis is part of the boundary and $y > 0$ in the region of interest in which we are solving Laplace's equation. Suppose further that the boundary condition near the origin specifies

$$\begin{aligned} u &= g_1(x) \quad \text{for} \quad x < 0 , \\ u &= g_2(x) \quad \text{for} \quad x > 0 . \end{aligned} \tag{3.22}$$

Then it is not difficult to show that

$$v = I_m\phi(z), \quad \phi(z) = \frac{1}{\pi} \sum_{s=0}^{N} \frac{a_s}{s!} z^s \log z , \tag{3.23}$$

where

$$a_s = g_1^{(s)}(0-) - g_2^{(s)}(0+), \quad s = 0,1,\ldots,N , \tag{3.24}$$

satisfies Laplace's equation and has the required behaviour at the origin if the a_s are finite. In fact

$$v(x,0) = \begin{cases} \sum_{s=0}^{N} \frac{a_s x^s}{s!} , & x > 0, \\ 0, & x < 0 , \end{cases} \tag{3.25}$$

since

$$\lim_{y\to 0+} \tan^{-1} y/x = \pi, \ x < 0 \quad . \tag{3.26}$$

Then w = u-v satisfies Laplace's equation, has continuous derivatives up to order N at the origin, and its boundary conditions are computable.

Volkov (1963) obtained corresponding results for Laplace's equation in more general cases, notably when the boundary is curved but sufficiently smooth, and when the function g(s) involved in boundary conditions like

$$u = g(s), \quad \frac{\partial u}{\partial n} = g(s) \quad \text{or} \quad \frac{\partial u}{\partial n} - \phi(s)u = g(s) \ , \tag{3.27}$$

is m times continuously differentiable on the whole boundary except at one point. He takes this point to be the origin, the x-axis along the tangent to the boundary at this point, and s the length of the boundary measured from this point. Near enough to the origin s = s(x) is smooth and single-valued, and

$$g(s) = g(s(x)) = g^*(x) \ . \tag{3.28}$$

He then allows $g^*(x)$ to take the form

$$g^*(x) = p_m(x)\log x^2 + q_m(x)\sigma(x) + r_m(x), \ \sigma(x) = \begin{cases} 1, & x \le 0, \\ 0, & x > 0, \end{cases} \tag{3.29}$$

where $p_m(x)$, $q_m(x)$ and $r_m(x)$ are polynomials of degree m, and the equation for the boundary near the origin is

$$y = y(x) = ax^2 + bx^3 + cx^4 + dx^5 + ex^6 + \text{higher order terms.} \tag{3.30}$$

He then constructs the function v(x,y) which satisfies Laplace's equation and has at the origin the same singularity as f(s) for any of the boundary conditions (3.27). These functions are all linear combinations of

$$F_{1k}(x,y) = \mathrm{Re}(z^k \log z)$$
$$F_{2k}(x,y) = I_m(z^k \log z) \qquad (3.31)$$

where $z = x + iy$.

An infinite series representation

More commonly the singularity occurs where the shape of the boundary changes abruptly, or where the nature of the boundary condition changes abruptly, or both these phenomena occur simultaneously, and in such cases we are generally unable to find the multiples of any singular solutions without considering the boundary conditions at all boundary points. We show in section 3.6 how to produce these solutions in some particular cases, and the result is that the required solution of our problem, which satisfies the differential equation and the boundary conditions near a singular point, is given by

$$u = \varphi_0(x,y) + \sum_{j=1}^{\infty} c_j \varphi_j(x,y) . \qquad (3.32)$$

The c_j are not known in advance, and must be computed as part of our numerical method.

One possible method, of a "global" nature, replaces infinity by some reasonably large number N, and computes the coefficients $c_1, c_2, \ldots, c_N$ by collocation, in which (3.32) is made to satisfy the boundary condition at N points at which it is not already satisfied. Alternatively one could attempt to satisfy the conditions at many boundary points, obtaining the coefficients for example in a least-squares sense.

For various reasons this method is not used very widely. Numerically, it usually happens that the linear equations for the c_j, obtained either by collocation or least-squares, are rather badly ill-conditioned. Analytically, we do not know *a priori* what value of N will give satisfactory results, and again we usually have to rely on consistency, taking successively larger values of N and comparing the results obtained. Collocation and least-squares methods using finite elements are outlined in section 2.3.

For some eigenvalue problems, however, the method can be very convenient numerically and can be associated with some very practical error analysis. For equation (3.5) and the L-membrane of Figure 3.2, for example, the function

$$u(r,\theta) = \sum_{j=1}^{\infty} c_j \, J_{\frac{2}{3}j}(r\surd\lambda) \sin(\tfrac{2}{3}j\theta) , \tag{3.33}$$

where r,θ are polar coordinates with origin at the singular point, satisfies the differential equation and vanishes on the boundary lines which intersect at the origin. Fox *et al.* (1967) used this fact, replaced infinity by N, made (3.33) vanish at N other boundary points, and hence reduced the problem to the approximating algebraic form

$$A(\surd\lambda)\underset{\sim}{c} = \underset{\sim}{0} , \tag{3.34}$$

and obtained a λ which satisfies

$$\det(A(\surd\lambda)) = 0 . \tag{3.35}$$

The arithmetic is here rather nicely minimized by certain symmetries, which make it necessary to include only the sequence

$$j = 1, 5, 7, 11, 13, \ldots, \tag{3.36}$$

all even numbers and all multiples of three being omitted. But in more general situations the computation is by no means prohibitive.

There is also a very satisfactory error analysis. Fox *et al.*(1967) show that if $\bar{\lambda}$ is an approximate eigenvalue and U is an approximate suitably normalized eigenfunction which satisfy the differential equation but not necessarily the boundary condition at all points, and if

$$\varepsilon = \max|U| \quad \text{on the boundary} , \tag{3.37}$$

then there is an eigenvalue λ which satisfies the inequality

$$\frac{|\lambda-\bar{\lambda}|}{\bar{\lambda}} \leq \frac{\sqrt{2}\varepsilon+\varepsilon^2}{1-\varepsilon^2} . \qquad (3.38)$$

All we need is to make ε sufficiently small, and this is effectively accomplished by taking enough terms in (3.33). The theory, incidentally, is valid for any self-adjoint operator in any number of dimensions. Nichol (1967) removed the restriction that $\bar{\lambda}$ and U satisfy the differential equation exactly and produced a bound similar to (3.38) in which ε is replaced by another easily computable number. Donnelly (1969) also extended the analysis to the treatment of the corresponding eigenvalue problem for an H-shaped region, in which symmetry gives rise to boundary conditions in which the eigenfunction is zero on some boundaries and its normal derivative is zero on other boundaries.

In many problems, however, we find it more convenient to use the singular solution like (3.33) only locally, which normally permits the inclusion of far fewer terms in the infinite series. This is most convenient, like many other things in elliptic problems, when the differential operator is of the Laplace type, and in the next section we discuss transformations which produce this form from a more general elliptic equation. In section 3.6 we show how to obtain local solutions in certain cases, and numerical treatment of this local approximation is then treated in sections 3.7 and 3.8.

3.5. CHANGE OF VARIABLES

The general two-dimensional linear elliptic problem

$$a\frac{\partial^2 u}{\partial x^2} + 2b\frac{\partial^2 u}{\partial x \partial y} + c\frac{\partial^2 u}{\partial y^2} + p\frac{\partial u}{\partial x} + q\frac{\partial u}{\partial y} + su = f , \qquad (3.39)$$

in which the coefficients are functions of x and y, can be transformed into a simpler form, in which the second derivative terms are those of Laplace's equation, with the change of independent variable to (ξ,η), where

$$\frac{\partial\xi}{\partial x} = (b\frac{\partial\eta}{\partial x} + c\frac{\partial\eta}{\partial y})/\surd(ac - b^2) ,$$
$$-\frac{\partial\xi}{\partial y} = (a\frac{\partial\eta}{\partial x} + b\frac{\partial\eta}{\partial y})/\surd(ac - b^2) . \tag{3.40}$$

For example, the equation

$$\frac{\partial^2 u}{\partial x^2} + x\frac{\partial^2 u}{\partial y^2} = 0 , \quad x > 0 , \tag{3.41}$$

is transformed with the change of variables

$$\xi = -x^{3/2} , \quad \eta = \frac{3}{2}y , \tag{3.42}$$

to the form

$$\frac{\partial^2 u}{\partial\xi^2} + \frac{\partial^2 u}{\partial\eta^2} + \frac{1}{3\xi}\frac{\partial u}{\partial\xi} = 0 . \tag{3.43}$$

The terms involving first derivatives can in certain cases be eliminated with a further change of dependent variable. The equation

$$\nabla^2 u + p\frac{\partial u}{\partial x} + q\frac{\partial u}{\partial y} + su = f , \tag{3.44}$$

whenever p is a function of x only and q a function of y only, is transformed to the still simpler form

$$\nabla^2\phi + s_1 = f_1, \tag{3.45}$$

with

$$u = \phi\exp\{\tfrac{1}{2}\int pdx + \tfrac{1}{2}\int qdy\} . \tag{3.46}$$

For example, (3.43) is reduced to

$$\nabla^2\phi + \frac{5}{36}\xi^{-2}\phi = 0 \tag{3.47}$$

with the substitution

$$u = \xi^{-1/6}\phi . \tag{3.48}$$

Though we shall here be concerned only with problems in two independent variables we remark that some of these results are also true for equations in more dimensions, and the ideas are discussed in Courant and Hilbert (1962) and Sankar (1967). The former also treat the solution of equations like (3.40).

It is also worthy of note that the transformations are often quite involved even in two dimensions, especially if they are associated with the useful technique of translating a particular point in the (x,y)-plane to the origin in the (ξ,η)-plane. For example, Sankar (1967) treated the equation

$$x^2 \frac{\partial^2 u}{\partial x^2} + x \frac{\partial u}{\partial x} + (1 + x^2) \frac{\partial^2 u}{\partial y^2} = 0 , \tag{3.49}$$

and with the transformations

$$\begin{aligned} \xi &= (1+x^2)^{\frac{1}{2}} - (1+a^2)^{\frac{1}{2}} - \tfrac{1}{2}\log \frac{(1+x^2)^{\frac{1}{2}}+1}{(1+a^2)^{\frac{1}{2}}-1} + \tfrac{1}{2}\log \frac{(1+x^2)^{\frac{1}{2}}-1}{(1+a^2)^{\frac{1}{2}}-1}, \\ \eta &= y, \quad \phi = (1 + x^2)^{\frac{1}{4}} u \end{aligned} \tag{3.50}$$

produced the required simpler form

$$\nabla^2 \phi = \frac{x^2(4-x^2)}{4(1+x^2)^3}\, \phi \tag{3.51}$$

with the point $y = 0$, $x = a$ translated to the origin $\eta = 0$, $\xi = 0$.

3.6. LOCAL TREATMENT OF SINGULARITY. DETERMINATION OF LOCAL SOLUTIONS

When we cannot find the nature in closed form of the singularity at a boundary point, we are left with the possibility of constructing a solution which has satisfactory "local" behaviour, and matching this with the usual finite-difference approximation computed at points at a reasonable distance from the singularity. If we can find a local solution like (3.32), for example, the aim is to replace infinity by N in this equation, use the resulting (3.32) in the neighbourhood of the singular point, and compute the coefficients c_j, $j = 1,2,\ldots,N$, together with values obtained by finite-difference methods at more remote points. We discuss the numerical technique in sections 3.7 and 3.8, but here we show, in favourable circumstances, how solutions of type (3.32) can be constructed when the operator is essentially of Laplace type.

Fox and Sankar (1969) consider an equation of the form

$$\nabla^2 u + s(r,\theta)u = 0, \quad \nabla^2 \equiv \frac{\partial^2}{\partial r^2} + \frac{1}{r}\frac{\partial}{\partial r} + \frac{1}{r^2}\frac{\partial^2}{\partial \theta^2}, \tag{3.52}$$

in polar coordinates with origin at a boundary point which is the intersection of two boundary lines $\theta = 0$ and $\theta = w$. On these straight lines the boundary conditions may have one of the various combinations given by

$$u = g_1(r) \quad \text{on} \quad \theta = 0, \quad u = g_2(r) \quad \text{on} \quad \theta = w,$$

$$\frac{1}{r}\frac{\partial u}{\partial \theta} = g_1(r) \quad \text{on} \quad \theta = 0, \quad \frac{1}{r}\frac{\partial u}{\partial \theta} = g_2(r) \quad \text{on} \quad \theta = w, \tag{3.53}$$

$$u = g_1(r) \quad \text{on} \quad \theta = 0, \quad \frac{1}{r}\frac{\partial u}{\partial \theta} = g_2(r) \quad \text{on} \quad \theta = w.$$

The various functions in (3.52) and (3.53) are also assumed to have convergent expansions of the form

$$s(r,\theta) = \sum_{n=0}^{\infty} s_n(\theta) r^n, \quad g_1(r) = \sum_{n=0}^{\infty} e_n r^{n+\beta},$$

$$g_2(r) = \sum_{n=0}^{\infty} h_r \, r^{n+\gamma}, \quad \beta, \gamma \geq 0. \tag{3.54}$$

They then look for solutions in one or other of the forms

$$u = \sum_{j=0}^{\infty} A_{\alpha,j}(\theta) r^{\alpha+j},$$

$$u = \sum_{j=0}^{\infty} r^{\alpha+j}\{\log r \, A_{\alpha,j}(\theta) + B_{\alpha,j}(\theta)\}, \tag{3.55}$$

which are known (Wigley, 1964) to cover all possibilities.

Substituting the first of (3.55) into (3.52) and equating to zero successive powers of r gives the equations

$$A''_{\alpha,0} + \alpha^2 A_{\alpha,0} = 0$$

$$A''_{\alpha,1} + (\alpha+1)^2 A_{\alpha,1} = 0 \qquad (3.56)$$

$$A''_{\alpha,m+2} + (\alpha+m+2)^2 A_{\alpha,m+2} = -\sum_{j=0}^{m} s_{m-j} A_{\alpha,j}, \quad m = 0,1,2, \ldots .$$

These differential equations, with θ as independent variable, must satisfy the relevant boundary conditions at $\theta = 0$ and $\theta = w$.

The solutions are obtained as linear combinations of partial solutions. With the first of (3.53) as boundary conditions, for example, we solve in succession for

$$u = e_n r^{n+\beta} \text{ on } \theta = 0, \quad u = 0 \text{ on } \theta = w,$$

$$u = 0 \text{ on } \theta = 0, \quad u = h_n r^{n+\gamma} \text{ on } \theta = w, \qquad (3.57)$$

$$u = 0 \text{ on } \theta = 0, \quad u = 0 \text{ on } \theta = w.$$

For the first of (3.57) we can take

$$\alpha = n + \beta,$$

$$A_{\alpha,0}(0) = e_n, \quad A_{\alpha,0}(w) = 0, \qquad (3.58)$$

$$A_{\alpha,j}(0) = 0, \quad A_{\alpha,j}(w) = 0, \quad j = 1,2,\ldots ,$$

and there follows

$$A_{\alpha,0}(\theta) = -e_n \frac{\sin(n+\beta)(\theta-w)}{\sin(n+\beta)w} . \qquad (3.59)$$

The $A_{\alpha,1}$, $A_{\alpha,2}$, ..., are then obtained in succession from (3.56) and the last of (3.58). The second of (3.57) is treated similarly, and for the third of (3.57) we merely take

$$\alpha = \frac{m\pi}{w}, \quad A_{\alpha,0} = \sin\frac{m\pi\theta}{w}, \quad m = 1,2, \ldots \tag{3.60}$$

and again find the $A_{\alpha,1}$, $A_{\alpha,2}$, ..., in succession.

Fox and Sankar (1969) denote by

$$u_n^{(1)}(r,\theta) = \frac{-e_n}{\sin(n+\beta)w} D[r^{n+\beta}\sin(n+\beta)(\theta-w)] \tag{3.61}$$

the complete solution of (3.56) with the first of (3.57), where "D" refers to Dirichlet conditions, and the complete solution for the first of (3.53) is then given by

$$u(r,\theta) = \sum_{n=0}^{\infty} \{u_n^{(1)}(r,\theta) + u_n^{(2)}(r,\theta)\} + \sum_{m=0}^{\infty} c_m D[r^{m\pi/w}\sin\frac{m\pi\theta}{w}] \tag{3.62}$$

with obvious meaning for the symbols.

The second of (3.55), for the solution $u(r,\theta)$ containing the $\log r$ term, is necessary when any denominator vanishes, such as $\sin(n+\beta)w$ in (3.59). The equations (3.56) follow as before for the $A_{\alpha,j}(\theta)$, and those for $B_{\alpha,j}(\theta)$ are

$$\begin{aligned} B''_{\alpha,0} + \alpha^2 B_{\alpha,0} &= -2\alpha A_{\alpha,0}, \\ B''_{\alpha,1} + (\alpha+1)^2 B_{\alpha,1} &= -2(\alpha+1)A_{\alpha,1}, \\ B''_{\alpha,m+2} + (\alpha+m+2)^2 B_{\alpha,m+2} &= -\sum_{j=0}^{m} s_{m-j}B_{\alpha,j} - 2(\alpha+m+2)A_{\alpha,m+2}, \\ m &= 0,1, \ldots . \end{aligned} \tag{3.63}$$

The solution corresponding to (3.54) is then given by

$$A_{\alpha,0}(\theta) = \frac{-e_n\sin(n+\beta)(\theta-w)}{w\cos(n+\beta)w}, \quad B_{\alpha,0}(\theta) = \frac{-e_n(\theta-w)\cos(n+\beta)(\theta-w)}{w\cos(n+\beta)w} \tag{3.64}$$

Similar results are easily obtainable, and are given by Fox and Sankar, for both the other combination of boundary conditions in (3.53). In particular the third (mixed) type of boundary conditions will cause (3.60) to be replaced by

$$\alpha = (m + \tfrac{1}{2})\,\frac{\pi}{w}\,, \quad A_{\alpha,0} = \sin(m + \tfrac{1}{2})\,\frac{\pi\theta}{w}\ . \qquad (3.65)$$

We note in passing that a term independent of u on the right of (3.52) introduces no significant complications, merely adding known terms to the right of equations of type (3.56).

3.7. LOCAL TREATMENT OF SINGULARITY. METHOD OF MOTZ

Having obtained an expansion for the solution in the neighbourhood of the singular point we must now consider how it can be incorporated in the method of solution. In the method of Motz (1946), also used by Sankar (1967) and Fox and Sankar (1969), the complete region is divided into two subregions. In subregion I, an area containing the singular point, we assume that the solution has the form of (3.32), with infinity replaced by a reasonably small number N, and in which the forms of the functions $\varphi_j(x,y)$ have been obtained by the methods of the last section. The constants c_j are as yet unknown. In subregion II, the rest of the region remote from the singular point, we use standard finite-difference formulae.

Consider, for example, the solution of (3.52) with boundary conditions represented by the second of (3.53), in a region in which the singularity is at a point O on the boundary which is a straight line near O, so that $\theta = 0$ on OA and $\theta = \pi$ on OB. The full region is as shown in Figure 3.4 in the (x,y)-plane.

Suppose that in (3.32) we take the three leading terms in the summation, so that in subregion I we assume the solution in the form

$$u = \varphi_0(r,\theta) + \sum_{j=1}^{3} c_j\,\varphi_j(r,\theta)\ . \qquad (3.66)$$

Suppose this also holds at any three points, Q_1, Q_2 and Q_3 just in subregion II. It follows that we can express the coefficients c_1, c_2 and c_3 as linear combinations of the u values at Q_1, Q_2 and Q_3. We can therefore likewise express

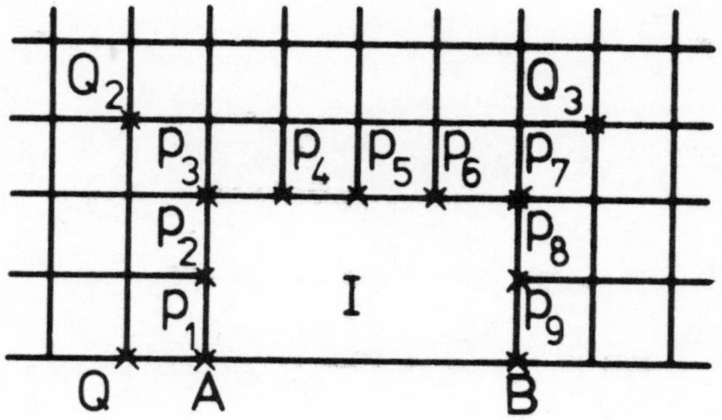

Fig. 3.4.

the values at P_1, $P_2,\ldots,P_9$, on the boundary of subregion I, as linear combinations of these values. Application of the standard finite-difference equations at all internal points in subregion II, including of course the points Q_1, Q_2 and Q_3, then gives linear equations for the computation of all the P_r values and all the mesh values in subregion II. The c_j coefficients are then computable, and hence also the values at any required points in subregion I.

Just as in the grid refinement technique of section 3.2, here the full set of linear equations has a matrix whose structure is somewhat less satisfactory than the standard form, and in particular the bandwidth can be quite a bit larger than that of the standard form in rows relating to the equations for the P_r and the finite-difference equations for points of which the P_r are neighbours. This can be mitigated to some extent. For example we could manipulate the equations so that the value of P_1, say, is expressible in terms of a few neighbouring P_r, and the value at P_3 in terms of say Q_2 and a few neighbouring P_r. However we know of no computer program which achieves this automatically, and so far each case has been tested empirically. As far as we know, moreover, iterative methods have been applied only by Walsh (1960), and again we know of no work of S.O.R. type for accelerating convergence.

With these limitations, however, the method works very well on quite complicated problems. For example, Fox and Sankar (1969) solved (3.49) in the region of Figure 3.5, with the conditions

$$\begin{aligned} &\frac{\partial u}{\partial y} = -\frac{x^2}{1+x^2} \quad \text{on OA}, \quad u = 0 \quad \text{on AB, DC, OD}, \\ &\qquad \frac{\partial u}{\partial x} \to 0 \quad \text{as} \quad x \to \infty \quad \text{on BC}. \end{aligned} \tag{3.67}$$

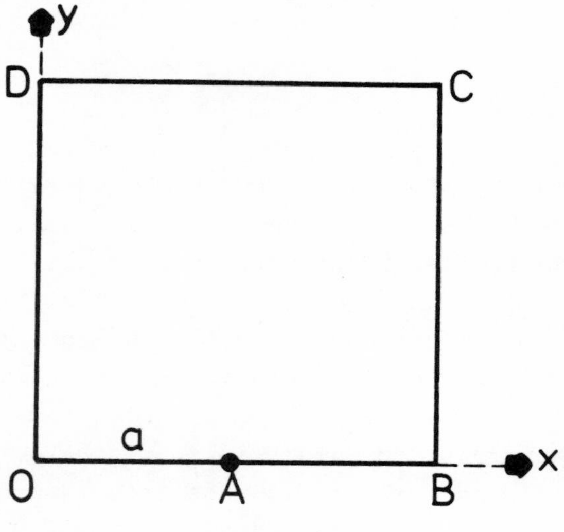

Fig. 3.5.

They made the transformation which produces (3.51) and found the "special" singular solution φ_s in terms of polar co-ordinates, but effectively performed the rest of the computation in the original plane, using the special solution

$$u_s = (1+x^2)^{-\frac{1}{4}} \varphi_s(r,\theta) . \tag{3.68}$$

Their special solution neglected terms of $O(r^{7/2})$, and the special solution in subregion I would appear to be locally more accurate than the finite-difference solution in subregion II. They found, however, that they could improve the latter by incorporating third and fourth differences in a "difference correction" type of improvement, with very good results.

Fox (1971) considered a problem of somewhat similar type which involved two boundary singularities, and also claimed good results without a prohibitive amount of analysis and computation. There are obvious complications, of course, unless the singular points are reasonably well separated.

Walsh (1960) solved by similar methods the eigenvalue problem for the H-shaped region, one quarter of which is shown in Figure **3.6**. The function vanishes on the boundary and has zero normal derivative on the lines of symmetry. The singularity at the point P produces a local solution of type (3.33), and Walsh used the first 10 terms of this in a special region near to P. With a finite-difference interval of 1/12 she found a value of 7.8065 for the smallest eigenvalue without allowing for the singularity and corrected this to 7.7332 with allowance for the singularity and application of the difference correction in the region remote from the singularity. The correct result to this number of figures is 7.7331 (Donnelly, 1969), so that the method again worked very well.

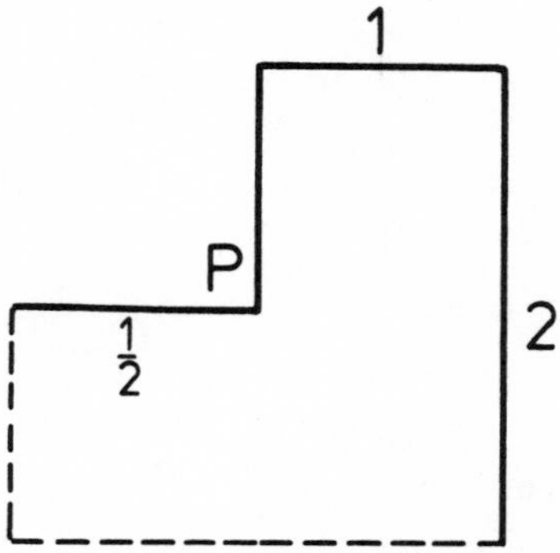

Fig. 3.6.

We know of no rigorous error analysis for these methods, so that again we are left with the appeal to consistency. One can usually take the subregion II far enough from the singularity to make the finite-difference equations quite accurate, especially with the use of the difference correction, and repeat the process with the retention of more terms in the singular solution. Higher terms, of course, can become quite complicated in form!

3.8. LOCAL TREATMENT OF SINGULARITY. METHOD OF FURZELAND

More recently (Furzeland, 1977a, and Crank and Furzeland, 1976) a different method has been suggested for dealing with the singularity in the type of problem treated by Sankar. Furzeland considers the equation

$$\nabla^2 u - s(x,y)u = f(x,y) , \qquad (3.69)$$

and following Sankar finds a truncated singular solution of type

$$u_I = \varphi_0(r,\theta) + \sum_{i=1}^{N} c_i \varphi_i(r,\theta) \qquad (3.70)$$

with reference to polar coordinates with origin at the singular boundary point. Sankar, as we have seen, aimed to compute the c_i for use in his subregion I, with no explicit intention of using anything like a finite-difference grid in this region. Furzeland, however, deliberately uses a rectangular grid at all points, with standard five point finite-difference equations at points corresponding to Sankar's subregion II, but with special five point formulae in what corresponds to Sankar's subregion I which deliberately allow for the local solution of type (3.70). Effectively he computes a separate set of coefficients c_i, not only for each mesh value in subregion I but also for each second derivative belonging to that mesh point. Furzeland, however, computes explicitly just the mesh values at all points of the region, and the coefficients c_i are somewhere implied in the calculation.

Consider, for example, the determination of such a five point formula at mesh point 1 within the shaded subregion I in the neighbourhood of a singular point O as shown in Figure 3.7.

Given the special solution (3.70) we can write

$$\frac{\partial^2 u_I}{\partial x^2} = w_0(r,\theta) + \sum_{i=1}^{N} c_i w_i(r,\theta) \qquad (3.71a)$$

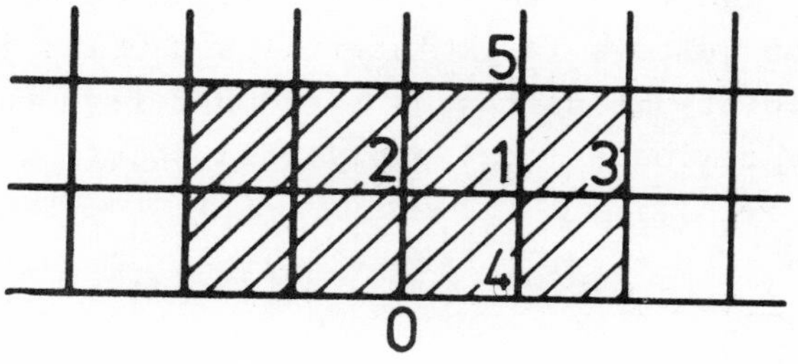

Fig. 3.7.

and

$$\frac{\partial^2 U_I}{\partial y^2} = \bar{w}_0(r,\theta) + \sum_{i=1}^{N} c_i \bar{w}_i(r,\theta) \qquad (3.71b)$$

where the w_i and $\bar{w}_i$ are obtained by differentiating (3.70). Now with finite-difference formulae we would approximate to $\partial^2 u_I/\partial x^2$ at the point 1 in Figure 3.7 by a linear combination of the values $u_{I,2}$, $u_{I,1}$ and $u_{I,3}$ in obvious notation (and for ease of writing we now delete the suffix I in these symbols). Trying to do a similar thing, but taking into account the singular nature of the solution in this region, we compute some estimates of c_1, c_2 and c_3, taking N = 3 in (3.70), by satisfying (3.70) at points 2, 1 and 3 and finding equations of the form

$$c_i = A_i + B_i u_1 + C_i u_2 + D_i u_3 , \quad i = 1, 2, 3. \qquad (3.72)$$

Substitution of these into the first of (3.71) then gives an approximation to $\partial^2 u/\partial x^2$ at point 1 in terms of u_1, u_2 and u_3. Similar treatment in the y-direction gives an approximation to $\partial^2 u/\partial y^2$ at point 1 in terms of u_1, u_4 and u_5. Finally substituting into the differential equation (3.69), and using either (3.72) or its partner, or perhaps an average thereof, together with (3.70) for the expression for u_1, we finally arrive at a "special" five point approximation to the differential equation at point 1.

Furzeland solved by this method the problem of Motz defined in (3.2) and (3.3) and Figure 3.1, and obtained very good results at quite a large interval and using just the four special points at distances h and $h\sqrt{2}$ from the singular point. The approximate local analytical solution is

$$u = \sum_{i=1}^{3} \alpha_i \, r^{(i-\frac{1}{2})} \cos(i-\tfrac{1}{2})\theta \quad , \tag{3.73}$$

and in (3.71) we find

$$w_1 = -\frac{1}{4} r^{-3/2} \cos\frac{3}{2}\theta, \; w_2 = \frac{3}{4} r^{-\frac{1}{2}} \cos\frac{1}{2}\theta, \; w_3 = \frac{15}{4} r^{\frac{1}{2}} \cos\frac{1}{2}\theta \; , \tag{3.74}$$

with $\bar{w}_i = -w_i$. The truncation error in the special five point formulae is therefore $0(r^{3/2})$, and Furzeland suggests that the special points should be taken such that $r^{3/2}$ is nowhere an order of magnitude greater than h^2, the order of the error in the standard five point formula.

It is not always possible to find estimates for the c_i from the "best" three points, and in the neighbourhood of the singular point O in this example Furzeland used the five point formulae indicated in Figure 3.8.

This method has one great advantage over that of Sankar, in that the bandwidth of the linear equations is fairly constant and not much larger than usual even with the "one-sided" formulae at three of the special points in Figure 3.8. It is not quite so easy, on the other hand, to match something like the difference correction improvement in the finite-difference region with a corresponding higher accuracy formula in the special region, and there is room for further research on this method. As with Sankar's method there is no rigorous error analysis, and an appeal to consistency is again the main guide to the accuracy of computed results.

Nevertheless the method is obviously of considerable value, and Furzeland applied it successfully to quite a few non-trivial problems.

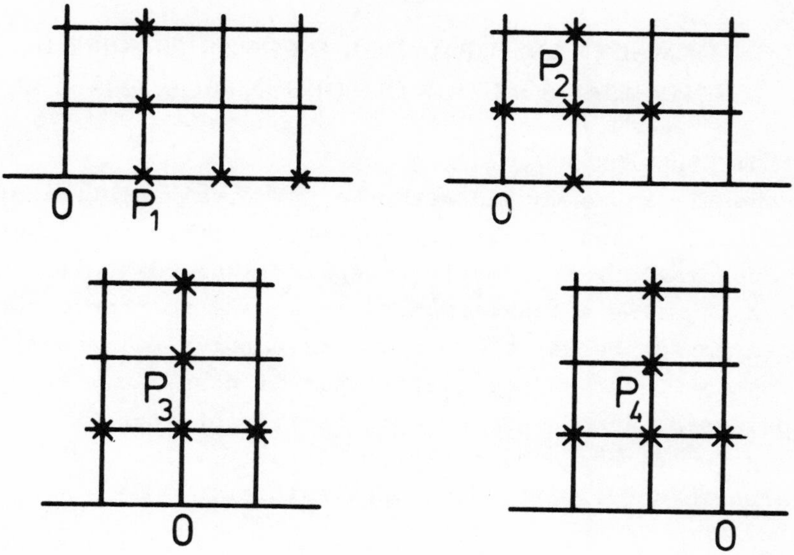

Fig. 3.8.

Alternative numerical methods for elliptic problems with singularities are discussed in chapters 4, 5 and 6.

4

CONFORMAL TRANSFORMATION TECHNIQUES FOR THE NUMERICAL SOLUTION OF POISSON PROBLEMS

4.1. INTRODUCTION

This chapter is concerned with the numerical solution of two dimensional Poisson problems, and it is within this context that conformal transformation techniques are discussed. Our purpose in using a tranformation is always to change the problem in such a way that an accurate numerical approximation to the exact solution, together if possible with a theoretical error bound, may be more easily obtained.

Any two dimensional elliptic boundary value problem contains three distinct defining features:

the differential equation,
the region in which the problem is defined,
the boundary conditions.

A conformal transformation can be applied to the region Ω, in which the problem is defined, to map it onto a second region Ω'. Under this mapping not only is the region Ω affected, but so also are the differential equation and the boundary conditions. In choosing the transformation one has to balance these three separate effects. The mapped region may be further transformed using another mapping or succession of mappings. The intent is always to produce finally a mapped problem which can be readily solved using a numerical technique; for example either a semi-analytic method, or a finite-difference or finite element method.

Two classes of Poisson problems where conformal transformations can be effective are those of problems defined in infinite regions, for example flow of an ideal fluid past an obstacle, and those where the boundary has a shape such that the solution contains a singularity, for example potential, diffusion and torsion problems in regions containing re-entrant corners. In the exterior problem the mapping can either be used to transform the obstacle into a more convenient shape or to change the problem into an interior problem; see the standard texts of Churchill (1960), Henrici (1974) or Southwell (1956).

We shall confine ourselves here to treatment of interior problems defined in awkwardly shaped regions. It is assumed that for any problem the differential equation, the boundary and the boundary conditions are such that they ensure the existence and uniqueness of the solution. However, whilst this may be so, the *regularity* of the solution is dependent on the shape of the boundary and the smoothness of the boundary conditions. Thus, if a boundary contains a re-entrant corner or if the boundary conditions change from Dirichlet to Neumann at a point where the boundary is smooth, then the solution frequently contains a singularity. This usually has the form that certain derivatives of the solution are unbounded at the singular point and the presence of the singularity causes standard numerical techniques for solving the problem to produce inaccurate approximations, especially in the neighbourhood of the singular point. The conformal transformations are thus used to change the problem so that the singularity is at best removed, or at least changed into a form less harmful to a numerical procedure.

The Poisson problems considered are defined in the (x,y)-plane and it is required to find u(x,y) such that

$$\nabla^2 u(x,y) = f(x,y) , \quad (x,y) \in \Omega , \qquad (4.1)$$

$$u(x,y) = g_1(x,y) , \quad (x,y) \in \Gamma_1 , \qquad (4.2)$$

and

$$\frac{\partial u(x,y)}{\partial n} = g_2(x,y) , \quad (x,y) \in \Gamma_2 , \qquad (4.3)$$

where $\Omega \subset \mathbb{R}^2$ is a simply connected open bounded domain with boundary $\Gamma_1 \cup \Gamma_2$, $\bar{\Omega} \equiv \Omega \cup \Gamma$, and f, g_1 and g_2 satisfy suitable smoothness conditions. The parts Γ_1 and Γ_2 may each consist of a number of disjoint subarcs. The normal derivative at a point of Γ_2 is denoted by $\partial/\partial n$.

Some appropriate properties of conformal transformations are given briefly in section 4.2. Two mapping techniques are discussed in section 4.3 and are applied to two mixed boundary value problems containing singularities. The merits of these methods in the context of numerical methods for solving singular problems of the form (4.1) - (4.3) are discussed in section 4.4.

4.2. CONFORMAL TRANSFORMATIONS

Some properties of conformal transformations are now given. For more detail the reader is referred to classical texts on complex analysis such as Ahlfors (1953), Copson (1955) or Henrici (1974). Let $F : w \to w' = F(w)$ be a function of w which maps $\Omega \subset w = x + iy$ plane onto $\Omega' \subset w' = \xi + i\eta$ plane. Under this mapping $(x,y) \to (\xi(x,y), \eta(x,y))$. The transformation is one-one if J does not vanish on Ω, where

$$J \equiv \frac{\partial(\xi,\eta)}{\partial(x,y)} = \det \begin{bmatrix} \frac{\partial \xi}{\partial x} & \frac{\partial \xi}{\partial y} \\ \frac{\partial \eta}{\partial x} & \frac{\partial \eta}{\partial y} \end{bmatrix} . \tag{4.4}$$

A mapping with the property that the size and sense of angles are preserved is conformal. If $F(w)$ is analytic and $dF(w)/dw \neq 0$ at a point of Ω, then the mapping $w' = F(w)$ is conformal at this point. The real and imaginary parts of the analytic function $F(w) = \xi + i\eta$ satisfy the Cauchy-Riemann equations

$$\frac{\partial \xi}{\partial x} = \frac{\partial \eta}{\partial y} , \quad \frac{\partial \xi}{\partial y} = - \frac{\partial \eta}{\partial x} . \tag{4.5}$$

<u>Riemann Mapping Theorem</u>

Let Γ be a simple closed curve in the w-plane forming the boundary of Ω. There exists a function $w' = F(w)$, analytic in Ω which maps each point of Ω conformally into a corresponding point of the unit disc in the w'-plane and each point of Γ into a point of $|w'| = 1$. The correspondence is one to one. Δ

This is an existence theorem but does not give any information as to the form of the mapping. A useful transformation which maps the real axis and upper half of the w-plane onto the boundary and interior of a simply connected polygon in the w'-plane is the Schwarz-Christoffel transformation, see Henrici (1974). If the polygon has more than four corners, however, the Schwarz-Christoffel formula is of theoretical value and has in general to be implemented by numerical means.

Consider the Poisson problem (4.1) - (4.3) and let the transformation $w' = F(w)$ map the region $\overline{\Omega} \subset$ w-plane of the problem onto a region $\overline{\Omega}' \subset$ w'-plane ($w' = \xi + i\eta$). Then the solution $u(x,y)$ of the original problem is related to the solution $v(\xi,\eta)$ of the transformed problem by

$$v(w') = u(F^{-1}(w')) = u(x(\xi,\eta),y(\xi,\eta)) ,$$

and

$$u(x,y) = v(\xi(x,y),\eta(x,y)) .$$

Using the Cauchy-Riemann equations (4.5) we find that

$$\begin{aligned} \nabla^2_{xy}\, u \equiv \frac{\partial^2 u}{\partial x^2} + \frac{\partial^2 u}{\partial y^2} &= \left(\frac{\partial^2 v}{\partial \xi^2} + \frac{\partial^2 v}{\partial \eta^2}\right) \left|\frac{dF(w)}{dw}\right|^2 \\ &= \nabla^2_{\xi\eta}\, v \left|\frac{dw'}{dw}\right|^2 \end{aligned} \qquad (4.6)$$

and that

$$\left|\underset{\sim}{\nabla}_{xy}\, u\right| = \left|\underset{\sim}{\nabla}_{\xi\eta}\, v\right| \left|\frac{dw'}{dw}\right| . \qquad (4.7)$$

For the problem (4.1) - (4.3), if under the conformal transformation $w' = F(w)$, $f(x,y) \to f'(\xi,\eta)$, $g_i(x,y) \to g_i'(\xi,\eta)$, $i = 1, 2$, $\Gamma_1 \to \Gamma_1'$ and $\Gamma_2 \to \Gamma_2'$, then the transformed problem is

$$\nabla^2_{\xi\eta}\, v = \left|\frac{dw}{dw'}\right|^2 f'(\xi,\eta) , \quad (\xi,\eta) \in \Omega'$$

$$v(\xi,\eta) = g_1'(\xi,\eta) , \quad (\xi,\eta) \in \Gamma_1'$$

$$\frac{\partial v(\xi,\eta)}{\partial n'} = g_2'(\xi,\eta) \left|\frac{dw}{dw'}\right| , \quad (\xi,\eta) \in \Gamma_2' ,$$

where $\partial/\partial n'$ is the derivative in the outward normal direction to Γ_2'.

Clearly the invariance of the form of the Laplacian operator under conformal transformation is important. Higher order operators are in general not invariant; for example with the biharmonic operator

$$\begin{aligned}\nabla^4 u(x,y) &= \nabla^2_{xy}\ (\nabla^2_{xy}\ u)\\ &= \left|\frac{dw'}{dw}\right|^2 \nabla^2_{\xi\eta}\ \left(\left|\frac{dw'}{dw}\right|^2 \nabla^2_{\xi\eta}\ v(\xi,\eta)\right).\end{aligned}$$

The biharmonic character is preserved only in the case of dw'/dw = constant, i.e. $w' = \alpha w + \beta$; see Kantorovich and Krylov (1964).

4.3. CONFORMAL MAPPING TECHNIQUES FOR PROBLEMS CONTAINING BOUNDARY SINGULARITIES

Some of the reasons for using conformal transformation techniques for Poisson problems have been outlined in section 4.1, whilst the invariance of the form of Laplacian operator has been shown in section 4.2. Thus conformal mappings are now applied to some Laplacian problems defined in regions containing re-entrant corners and for which the solutions possess singularities. In each case the aim is to transform the region of the problem so that the solution of the mapped problem does not contain a singularity. Two approaches are presented for achieving this, which may be summarised as

(i) simple transformation, complicated final region,
(ii) complicated transformation, simple final region.

In (i) a simple transformation is chosen specifically to remove the re-entrant corner. This has at the same time the effect of changing the *whole* boundary of the problem and one has to accept the mapped region which results. Under (ii) a suitable mapped region, which will produce a problem not containing a singularity, is chosen and a mapping, or succession of mappings, has to be found which will produce this from the original region.

Simple transformation, complicated final region

In the w-plane let us consider a boundary which has a sharp corner with straight sides and interior angle $\alpha\pi$ with $1 < \alpha \leq 2$ at the point w_0. Under the mapping

$$w' = (w - w_0)^{1/\alpha} \tag{4.9}$$

this angle is transformed to become in the w'-plane an angle of interior size π. This mapping is conformal throughout the w-plane except at w_0. If a problem of the type (4.1) - (4.3) is defined in a region Ω containing a corner of the above type with, say, homogeneous Neumann boundary conditions on both arms of the corner, it can be shown, see Lehman (1959) or Kondrat'ev (1968), that in terms of polar coordinates, centred at the corner and having zero angle along one of the arms of the corner, the solution locally has the form

$$u(r,\theta) = \sum_{i=0}^{\infty} b_i r^{\frac{1}{\alpha}} \cos \frac{i\theta}{\alpha} , \tag{4.10}$$

where the b_i are unknown coefficients to be determined from the remaining boundary conditions. The function $u(r,\theta)$ of (4.10) has unbounded derivatives at the corner. Similarly, since in the mapped region the size of the interior angle at the mapped corner is π, it is seen from (4.10) that the image function has bounded derivatives at this point. The singularity has thus been removed. However, a price has to be paid for this in that the remaining parts of the boundary become of irregular shape. Two regions with re-entrant corners and their mapped images are shown in Figures 4.1 and 4.2 to illustrate this situation.

The well known Motz problem (see section 3.1) is of the form (4.1) - (4.3), and is defined on the region of Figure 4.1(a). The solution can be written as

$$u(r,\theta) = \sum_{n=1}^{\infty} \alpha_n r^{n-\frac{1}{2}} \cos(n-\tfrac{1}{2})\theta \quad . \tag{4.11}$$

This problem has been solved by many numerical methods not involving conformal transformation; finite differences in chapter 3; dual series, Whiteman (1968); finite elements

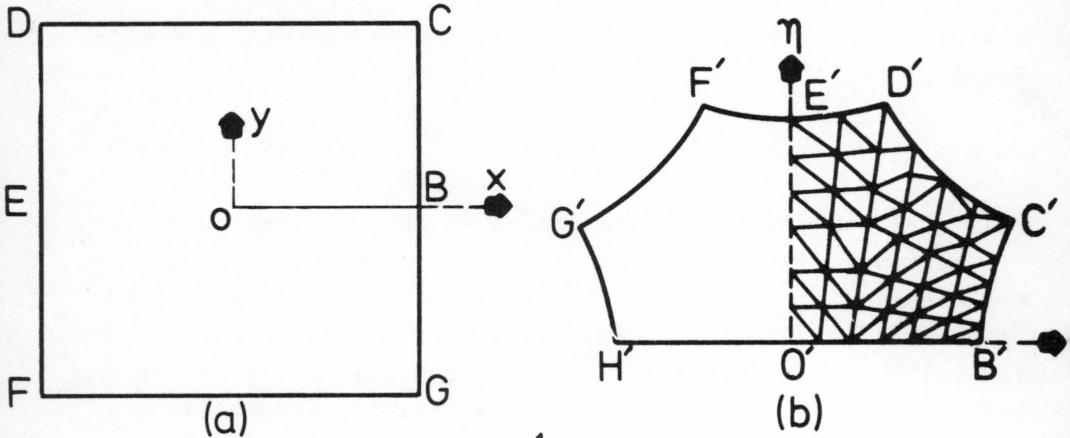

Fig. 4.1. $w' = w^{\frac{1}{2}}$

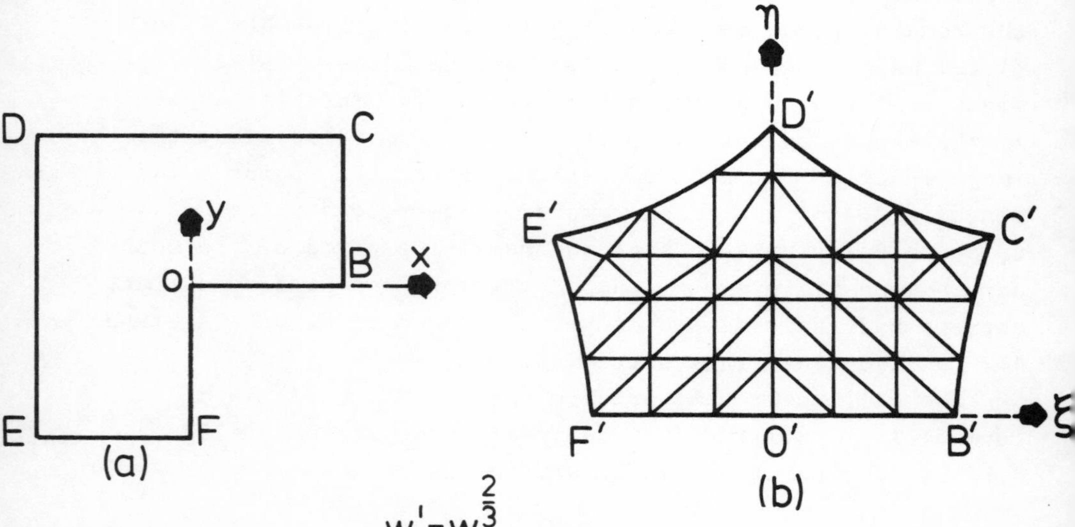

Fig. 4.2. $w' = w^{\frac{2}{3}}$

in chapter 5; global elements in chapter 6. The Motz problem has also been solved by Whiteman (1970) using a method involving conformal transformation, dual series and finite differences.

Reid and Walsh (1965) have used the mapping $w' = w^{2/3}$ to map an eigenvalue problem defined on the L-shaped region of Figure 4.2(a) onto a problem defined on the region of Figure 4.2(b), which they solve using finite differences. The region in Figure 4.1(a) has been mapped onto the region O'B'C'D'E'O' of Figure 4.1(b) with the transformation $w' = w^{1/2}$ by Whiteman (1967) who applies a finite-difference method in the (ξ,η)-plane. Accuracy of the numerical solution is now good in the neighbourhood of O', but less so near the curved boundaries. However, a complication arises in that, although the finite-difference mesh is for the most part square, appropriate special difference stars have to be used at points on and near the curved boundaries. The method of producing these stars is simple but awkward. Accuracy near the boundary can be improved by local mesh refinement. A finite element technique, similar to that of Cheng, has been applied by Scheffler and Whiteman to the mapped form using the mesh of Figure 4.1(b) based on six noded isoparametric triangular elements (see section 2.7). In this method the curved boundaries B'C', C'D' and D'E' are approximated by curves which are piecewise parabolae, and each element is then mapped onto the standard six noded triangle. Accuracy is again much better in the neighbourhood of O' than near to the curved boundaries. A similar problem is defined in the L-shaped region of Figure 4.2(a), where

$$\nabla^2 u = 0 \quad \text{in} \quad OBCDEFO$$

$$u = 1 \quad \text{on} \quad CD, \quad u = 0 \quad \text{on} \quad EF,$$

$$\frac{\partial u}{\partial n} = 0 \quad \text{on} \quad OB \cup BC \cup DE \cup FO.$$

The region is mapped onto that of Figure 4.2(b) by the transformation $w' = w^{2/3}$ and the finite element method as above is applied to the transformed problem on the mesh of Figure 4.2(b) with the same effect as before near the curved boundaries. However, since expressions for these boundary curves are known explicitly a strong case can be made for using, adjacent to the boundary, curved elements of the blending type (see section 2.8).

When either finite-difference or finite element methods are applied to a mapped problem in the (ξ,η)-plane, a way has to be found of transferring the results to the original (x,y)-plane. Since in any finite-difference method the values are only computed at discrete points, Whiteman (1967) maps the points of Ω at which the solution is required onto image points in Ω'. These points are not in general mesh points so that a high accuracy interpolation formula is used to calculate the relevant values. The situation is different with finite elements in that the numerical solution is defined over the whole of Ω'. Thus the required points in are mapped onto the Ω'-plane, where in any triangle the solution is defined in terms of basis functions. In practice a further local mapping is performed in each case onto the standard triangle so that *standard* basis functions can be used.

Complicated transformation, simple final region

In this section a technique of choosing the form of the final mapped region and problem, and of achieving these through a sequence of transformations, is now described. This technique, known as the *conformal transformation method*, was first proposed by Papamichael and Whiteman (1972), (1973), has since been steadily refined, and its application has been extended to a wider class of problems, by Papamichael together with Levin, Rosser, Sideridis and Symm; see papers in (1973), (1977) and (1978). It consists of three conformal transformations which are used in succession to map the region $\bar{\Omega}$ in the w-plane onto a rectangle $\bar{\Omega}'$ in the w'-plane. Let w_i, i = 1, 2, 3, 4 be four distinct points of Γ. These will in practice often coincide with ends of subarcs of Γ_1 and Γ_2 in problems of the type (4.1) - (4.3). The three transformations of the method are (see Whiteman and Papamichael (1972) and Papamichael and Sideridis (1978)):

(i) The tranformation

$$z = T(w) \tag{4.12}$$

where $T(w)$ is a function of w, analytic in Ω, so that (4.12) is a conformal mapping of Ω onto the upper half z-plane.

(ii) The bilinear transformation

$$t = M(z) = k\left(\frac{z-T(w_1)}{z-T(w_2)}\right) \tag{4.13}$$

where k is a constant.

(iii) The Schwarz-Christoffel transformation

$$w' = S(t) = \frac{1}{K(m)}\, \mathrm{sn}^{-1}(t^{\frac{1}{2}}, m) , \tag{4.14}$$

where $0 < m < 1$, sn denotes the Jacobian elliptic sine and $K(m)$ is the complete elliptic integral of the first kind with modulus m. The combined effect of (4.12) - (4.14) is to map $\bar{\Omega}$ in the w-plane onto the rectangle $\bar{\Omega}'$ in the w'-plane, where

$$\bar{\Omega}' = \{(\xi,\eta) : 0 \le \xi \le 1,\ 0 \le \eta \le \frac{K\{(1-m^2)^{\frac{1}{2}}\}}{K(m)} \equiv H\}$$

in such a way that w_1 becomes $(0,0)$ and w_2 becomes $(0,H)$. The points w_1 and w_2 divide Γ into two distinct arcs; the major and minor arcs between w_1 and w_2. If the two points w_3 and w_4 both belong to one of these arcs, they may be mapped respectively into $(1,H)$ and $(1,0)$ by suitable choice of k and m in (4.13) and (4.14). The transformed problem produced in $\bar{\Omega}'$ can be solved either by inspection or more generally by a standard numerical technique.

Finding the initial transformation $z = T(w)$, (4.12), which maps $\bar{\Omega}$ onto the real axis and upper half of the z-plane is crucial to the success of the method. Once $T(w)$ is constructed and the values of k and m are prescribed, the image $P' \equiv (\xi,\eta) \in \bar{\Omega}'$ of $P \equiv (x,y) \in \bar{\Omega}$ is determined accurately from (4.13) and (4.14) by means of a standard procedure. Whilst it follows from the Riemann mapping theorem that transformation (4.12) exists, its form clearly depends on the geometry of Ω. In certain cases the transformation can be obtained in closed form. This is however not usually possible so that (4.12) must be performed numerically, as for example by Papamichael and Whiteman (1973) using the method of Symm (1966). This has been used for harmonic problems with only limited success. More recently, and for a wider class of problems, better results have been obtained by Papamichael and Sideridis (1978). They use the approximation

$$\tilde{T}(w) = i\left(\frac{1+\tilde{\phi}(w)}{1-\tilde{\phi}(w)}\right)$$

to $T(w)$ in (4.12), where $\tilde{\phi}(w)$ is an approximation to the mapping function $\phi(w)$ which maps Ω conformally onto the unit disc. The approximation $\tilde{\phi}(w)$ is then computed using the Bergman kernel function defined over Ω; see Levin *et al.* (1978). The method involves the use of a finite series of orthogonalised functions, the convergence of which is slow on account of the Bergman kernel function possessing singularities outside Ω. This slow convergence is overcome by building into the basis of the approximating space functions which reflect the dominant part of the singular behaviour. It is important to note that in some applications the method itself introduces singularities into the solution of the transformed problem. However, for Poisson problems the damaging effects of these singularities can be removed; see Levin *et al.* (1977).

In order to illustrate the method we give details of its application to the Motz problem as performed by Whiteman, Papamichael and Martin (1971) and Whiteman and Papamichael (1972). Here the first transformation (4.12) can be performed exactly using the Jacobi elliptic sine

$$T(w) = \mathrm{sn}(Kw, 1/\sqrt{2}) ,$$

where $K = K(1/\sqrt{2})$ is the complete elliptic integral of the first kind with modulus $1/\sqrt{2}$. The Jacobi elliptic sine can be expressed in terms of infinite series, see Copson (1955, p.412) the convergence of which is so rapid that numerical evaluation with truncation after only four terms gives ten figure accuracy. In (4.13) and (4.14) respectively

$$t = \frac{2z}{1+z} \quad , \qquad \text{and} \qquad m = \left(\frac{1}{2} + \frac{1}{2\sqrt{2}}\right)^{\frac{1}{2}} \quad .$$

The method produces the exact solution to the Motz problem, because T(w) is known exactly, the boundary and boundary conditions can be subdivided into four segments and the transformed problem has a simple solution. The series expansion involving unknown coefficients α_n of the solution u of the Motz problem is given in (4.11). Rosser and Papamichael (1973), using the transformations outlined above have evaluated the first few α_n to very high accuracy. Since these values, and in particular the value of the singularity intensity factor α_1, are of interest, seven figure approximations to the first five coefficients are given below

$$\begin{aligned} \alpha_1 &= 401.1625 \\ \alpha_2 &= 87.65592 \\ \alpha_3 &= 17.23792 \\ \alpha_4 &= -8.071222 \\ \alpha_5 &= 1.440273 \quad . \end{aligned}$$

4.4. CONCLUSION

The purpose of this chapter has been to outline methods based on conformal transformations for the numerical solution of singular Poisson problems. The methods have been defined using basic properties of this class of problems and thus rely heavily on the form of the differential equation and on analytic properties of the solution. In the context of Poisson problems these methods are effective and the conformal transformation method in particular is a very competitive method producing extremely accurate numerical solutions. However, although it has been generalised to deal with problems involving equations of the type $\nabla^2 u = au + b$, both approaches using conformal mapping are not as versatile as the modified finite-difference and finite element methods listed in section 4.3, and with which they are in competition.

The simple transformations of section 4.3, when used together with finite element techniques, produce results which overcome the effect of the singularity in as much as they are accurate near the singular point. The possibility of exploiting this effect in embedded subregions of a problem is currently being investigated.

5

NON-CONFORMING FINITE ELEMENTS

5.1. INTRODUCTION

In chapter 2 the analysis of the finite element method assumed that the solution u of

$$a[u,v] = (f,v)_0 \qquad \text{for all } v \in \mathcal{H} \tag{5.1}$$

was approximated by U such that

$$a[U,V] = (f,V)_0 \qquad \text{for all } V \in \mathcal{K}_h \tag{5.2}$$

where

$$\mathcal{H}_h \subset \mathcal{H} \;. \tag{5.3}$$

In the solution of second order equations, for example, the inclusion (5.3) implies that $\mathcal{H}_h \subset \mathcal{H}_1(\Omega)$. The domain is emphasised in this notation since (5.3) is a global condition on the approximation. In many situations it is convenient to violate this inclusion, in others it is necessary to modify the functionals in (5.2); numerical integration is one modification, approximating the boundary Γ by polynomial arcs is another. These modifications are all examples of so-called *variational crimes*. Elements that violate inclusion (5.3) are referred to as *non-conforming elements* but the name has been also used to describe any solution involving variational crimes.

<u>Why should non-conforming elements be used?</u>

1) To provide better answers. In linear elasticity a solution based on linear shape functions with the nodes placed at the vertices as in Figure 5.1(a). leads to a solution that is too rigid. A solution based on linear shape functions with nodes placed in the middle of the sides as in Figure 5.1(b) is

discontinuous and too flexible but is often as accurate as the continuous solution. Together the two approximations often provide a bound on the true solution.

2) Conforming elements are unnecessarily complicated. In fourth order problems, such as the biharmonic equation

$$\nabla^4 u = 0 ,$$

conforming elements must have continuous derivatives across inter-element boundaries. The simplest polynomial elements in such problems are an 18 degrees of freedom triangle with quintic shape functions defined by the function values together with first and second derivatives given at the vertices, a 16 degrees of freedom bicubic rectangle or a macro-element such as the Clough and Tocher element. Further details of all these and other elements can be found in Mitchell and Wait (1977).

5.2. ANALYSIS OF NON-CONFORMING ELEMENTS

The non-conforming elements illustrated in Figure 5.1(b) are not continuous across inter-element boundaries, the shape

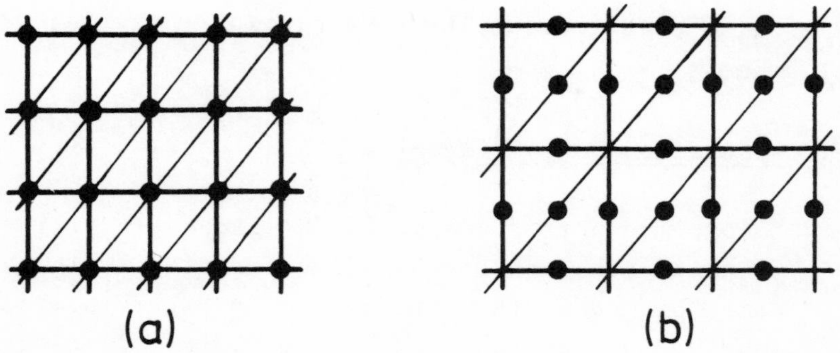

Fig. 5.1.

functions in adjacent elements only match up at the mid-side nodes. The derivatives of such piecewise continuous functions involve Dirac delta functions corresponding to the discontinuities. In chapter 2 the global stiffness matrix was taken as the sum of element stiffness matrices, for example

$$A_{ij} = \iint_\Omega \left\{\left(\frac{\partial \varphi_j}{\partial x}\right)\left(\frac{\partial \psi_i}{\partial x}\right) + \left(\frac{\partial \varphi_j}{\partial y}\right)\left(\frac{\partial \psi_i}{\partial y}\right)\right\} dxdy.$$

Then

$$A = \sum_E A_E$$

for conforming elements because the functions φ_j and ψ_i are at least continuous. If the elements are non-conforming the definition of A involves delta functions but A_E does not since no integration across boundaries is involved. Therefore it is possible to define a matrix

$$A_h \equiv \sum_E A_E$$

that corresponds to a bilinear form

$$a_h[u,v] \equiv \sum_E \iint_E \left\{\left(\frac{\partial U}{\partial x}\right)\left(\frac{\partial V}{\partial x}\right) + \left(\frac{\partial U}{\partial y}\right)\left(\frac{\partial V}{\partial y}\right)\right\} dxdy .$$

It is then necessary to estimate the error

$$\| u - U_h \|$$

or alternatively the perturbation

$$\| U - U_h \|$$

where u satisfies (5.1), U satisfies (5.2) and U_h satisfies

$$a_h[U_h,V_h] = (f,V_h)_0 \qquad \text{for all } V_h \in \mathcal{H}_h^* \tag{5.4}$$

where $\mathcal{H}_h^* \not\subset \mathcal{H}$. It is possible to define norms for spaces of non-conforming functions by replacing $\iint_\Omega$ in the Sobolev spaces by $\Sigma_E \iint_E$. If such norms are denoted by a subscript h, it is possible to write an analogue of the Lax-Milgram Lemma for the existence of the solution U_h, provided

$$|a_h[U_h,V_h]| \leq \alpha \|U_h\|_h \, \|V_h\|_h, \qquad \text{for all } U_h,V_h \in \mathcal{H} \cup \mathcal{H}_h^* \tag{5.5}$$

and

$$a_h[U_h,U_h] \geq \gamma \, \|U_h\|_h^2 , \qquad \text{for all } U_h \in \mathcal{H}_h^* . \tag{5.6}$$

Then we define for any w_1 and w_2

$$E_h[w_1,w_2] \equiv a_h[w_1,w_2] - (Aw_1,w_2)_0 .$$

It is more convenient to write the last term without applying integration by parts to avoid boundary integral terms. Corresponding to Cea's Lemma we have:

Lemma 5.1.

It follows from (5.4), (5.5) *and* (5.6) *that*

$$\|u - U_h\|_h \leq \left(1 + \frac{\alpha}{\gamma}\right) \inf_{V_h \in \mathcal{H}_h^*} \|u - V_h\|_h + \sup_{W_h \in \mathcal{H}_h^*} \frac{1}{\gamma} \left\{\frac{|E_h[u,W_h]|}{\|W_h\|_h}\right\} . \Delta \tag{5.7}$$

Proof

For any $V_h \in \mathcal{H}_h^*$

$$\|u - U_h\|_h \leq \|u - V_h\|_h + \|U_h - V_h\|_h \tag{5.8}$$

and

$$\begin{aligned}\gamma \|U_h - V_h\|_h^2 \leq \; & a_h[U_h - V_h, U_h - V_h] \\ = \; & a_h[u - V_h + U_h - u, U_h - V_h] \\ = \; & a_h[u - V_h, U_h - V_h] + a_h[U_h, U_h - V_h] - \\ & - a_h[u, U_h - V_h].\end{aligned}$$

For any W_h

$$a_h[U_h, W_h] = (f, W_h)_0 = (Au, W_h)_0 \ .$$

Thus

$$\gamma \| U_h - W_h \|_h^2 \leq \alpha \| u - V_h \|_h \, \| U_h - V_h \|_h + |E_h[u, U_h - V_h]| \quad (5.9)$$

and so combining (5.8) and (5.9) leads to the desired result.$_\Delta$

In order to transform (5.7) into an estimate of the order of convergence, we proceed as in chapter 2 and investigate the approximation of polynomial solutions.

5.3. THE PATCH TEST

The bound given by (5.7) is clearly in two parts. Apart from a different value for the constant, the first term corresponds to the term in Cea's Lemma and it can be estimated using the Bramble-Hilbert Lemma. Although the results given in chapter 2 are for second order problems only they can be generalised. In this more general form for equations of order 2r they can be interpreted as follows:

Criterion 1

When the bilinear form a[u,v] *contains derivatives of order at most* r *the finite element approximation converges in the sense that*

$$\lim_{h \to 0} \| u - U \|_r = 0 \ ,$$

provided that interpolation *using the finite element trial functions recovers all polynomials of degree* r.$_\Delta$

For an estimate of the *order* of convergence the degree of interpolation has to be defined more precisely:

Criterion 2

If in addition finite element interpolation recovers all polynomials of degree at most $k (\geq r)$ *then*

$$\| u - U \|_r = O(h^{k+1-r}) \ ._\Delta$$

The interpolation conditions can be written as

$$\inf_{V_h \in \mathcal{H}_h} \| u - V_h \| = 0$$

for all $u \in P_r$ and $u \in P_k$ respectively. Analogous results can be constructed for the second (perturbation) term in 5.7. These were first suggested by Irons (see Irons and Razzaque, 1972) on the basis of experimental evidence and were later used as the basis for a mathematical proof of convergence (Ciarlet, 1977).

The Patch Test

When the bilinear form $a_h[u,v]$ *contains derivatives of order at most* r *and the non-conforming trial space contains all polynomials of degree at most* r, *the test is passed provided that* the finite element method *recovers all solutions that are polynomials of degree at most* r. Δ

Lemma 5.2.

If the patch test is passed then

$$\lim_{h \to 0} \| u - U_h \|_h = 0 \; . \Delta$$

It is also possible to modify the criterion to yield an estimate of the order of convergence.

The Super Patch Test

This test is passed if for some $k > r$ *the non-conforming finite element method recovers all solutions that are polynomials of degree at most* k. Δ

The mathematical statement of the patch test was first investigated by Strang (see Strang and Fix, 1973) and it can be written as:

The Revised Patch Test

If the non-conforming trial space $\mathcal{H}_h^*$ *satisfies the inclusion* $\mathcal{H}_h^* \supset P_k$ *and if for all* $W_h \in \mathcal{H}_h^*$

$$E_h[u, W_h] = 0$$

for any solution $u \in P_k$ *(k = r in the original patch test), then the patch test is satisfied.* $_\Delta$

It is then possible to provide an estimate of the order of convergence.

Theorem 5.1.

If a finite element method passes the revised patch test then

$$\| u - U_h \|_h = O(h^{k+1-r}) \; ._\Delta$$

The proof uses the Bramble-Hilbert Lemma applied to the approximation error term and to the perturbation term assuming the revised patch test. Separate proofs have been constructed for second order problems r = 1 and for fourth order problems r = 2; the technicalities involved will not be repeated here but they can be found in full in Ciarlet (1977). It only remains to verify that the two forms of the patch test are consistent.

Lemma 5.3

If the revised patch test is satisfied, then the original patch test is satisfied. $_\Delta$

The proof follows directly from Lemma 5.1.

Lemma 5.4

If the original (super) patch test is satisfied then the revised patch test is satisfied. $_\Delta$

Proof

The result follows directly from the inequality

$$\| u - U_h \|_h \geq \frac{1}{\alpha} \sup_{W_h \in \mathcal{H}_h^*} \left\{ \frac{|E_h[u, W_h]|}{\| W_h \|_h} \right\} \; . \qquad (5.10)$$

In order to verify (5.10) we observe that

$$\alpha\| u - U_h\|_h \|W_h\|_h \geq |a_h[u - U_h, W_h]|$$

$$= |a_h[u,W_h] - a_h[U_h,W_h]|$$

$$= |E_h[u,W_h]|$$

since

$$a_h[U_h,W_h] = (f,W_h)_0 = (Au,W_h)_0$$

and the proof is complete.$_\Delta$

5.4. A PRACTICAL PATCH TEST

The statements of the patch test in the previous section are short; unfortunately neither the original test nor the revised form lead directly to a simple method of applying the test. Both definitions refer to *all* polynomial solutions which over an arbitrary domain provides a considerable number of alternatives.

Once the forms of the equation Lu = f and of the bilinear form $a_h[u,v]$ have been specified it is possible to manipulate the revised version into a readily applicable test.

A test for second order problems

Let

$$L = -\frac{\partial^2}{\partial x^2} - \frac{\partial^2}{\partial y^2}$$

and

$$a_h[u,v] = \sum_E \iint_E \left\{\left(\frac{\partial u}{\partial x}\right)\left(\frac{\partial v}{\partial x}\right) + \left(\frac{\partial u}{\partial y}\right)\left(\frac{\partial v}{\partial y}\right)\right\} dxdy ,$$

then applying Green's theorem element by element leads to

$$a_h[u,v] = \sum_E \left\{ \iint_E (Au)v \; dxdy + \int_{\partial E} \left(\frac{\partial u}{\partial n}\right) v \; ds\right\} .$$

Then if u is sufficiently smooth

$$E_h [u,W_h] = \sum_E \int_{\partial E} \left(\frac{\partial u}{\partial n}\right) W_h \; ds .$$

The patch test therefore leads to the condition

$$\sum_E \int_{\partial E} \left(\frac{\partial u}{\partial n}\right) W_h \, ds = 0 \qquad \text{for all } u \in P_1 . \tag{5.11}$$

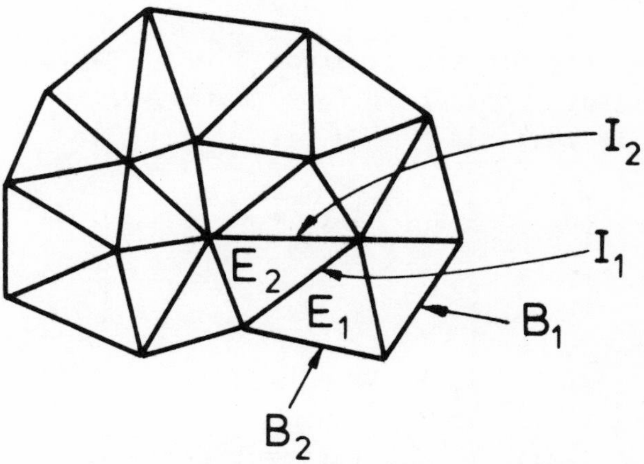

Fig. 5.2.

The boundary ∂E of an element E consists of arcs that are either internal boundaries between E and another element or part of the boundary Γ. Thus if a typical boundary arc is denoted by B and a typical interface by I (see Figure 5.2) it follows that the summation in (5.11) can be reordered as

$$\sum_I \int_I \left[\left\{\left(\frac{\partial u}{\partial n}\right) W_h \right\}^{(1)} + \left\{\left(\frac{\partial u}{\partial n}\right) W_h\right\}^{(2)}\right] ds + \\ + \sum_B \int_B \left(\frac{\partial u}{\partial n}\right) W_h \, ds = 0 \quad \text{for all } u \in P_1 . \tag{5.12}$$

The superscripts in the first summation indicate limits taken in elements on opposite sides of the interface. If the arcs are *all straight lines* then on any arc $u \in P_1$ and thus $\partial u/\partial n$

is a constant. In addition on the interfaces it follows that

$$\left(\frac{\partial u}{\partial n}\right)^{(1)} + \left(\frac{\partial u}{\partial n}\right)^{(2)} = 0$$

and so (5.12) reduces to

$$\sum_I C_I \int_I \left\{W_h^{(1)} - W_h^{(2)}\right\} ds + \sum_B C_B \int_B W_h \, ds = 0 , \qquad (5.13)$$

where on any arc the constant C is the value of the normal derivative. It then follows that a *sufficient* condition for (5.13) is that

$$\int_I \left\{W_h^{(1)} - W_h^{(2)}\right\} ds = 0, \qquad \text{for all I} ,$$

$$(5.14)$$

$$\int_B W_h \, ds = 0, \qquad \text{for all B.}$$

This is clearly a much easier test to apply.

Examples and Extensions

1) Piecewise linear triangular elements with mid-side nodes (Figure 5.1). On each interface $W_h^{(1)} - W_h^{(2)}$ is a linear function that is zero at the mid-point of the domain hence the integral is zero. The second condition in (5.14) is satisfied if the nodal values on the boundary are zero.

2) Higher degree non-conforming elements satisfy (5.14) if the nodes are placed at the zeros of Legendre polynomials (Brown, 1976).

3) Elements that are to pass a super patch test should satisfy

$$\int_I v(s) \left\{W_h^{(1)} - W_h^{(2)}\right\} ds = 0 \qquad \text{for all I} ,$$

for any $v \in P_{k-1}$ and a similar condition on each B. Curved elements also involve additional complications.

4) A similar analysis can be followed for the biharmonic equation using the weak form $a_3[u,v] = (f,v)$ given in section 6.2. The result is a set of conditions involving integrals along sides of elements such as

$$\int_I \left\{ \left(\frac{\partial W_h}{\partial n}\right)^{(1)} - \left(\frac{\partial W_h}{\partial n}\right)^{(2)} \right\} ds = 0, \qquad \text{for all I}$$

and

$$\int_I \left\{ \left(\frac{\partial W_h}{\partial s}\right)^{(1)} - \left(\frac{\partial W_h}{\partial s}\right)^{(2)} \right\} ds = 0 \qquad \text{for all I},$$

together with similar boundary terms. Further details can be found in Brown (1976) or Ciarlet (1976).

5.5. NUMERICAL INTEGRATION

In this section an alternative (or additional) variational crime is analysed. It is assumed that all integrals on the reference element are evaluated using a numerical quadrature scheme of the form

$$I(u) = \sum_i w_i u(\underset{\sim}{p}_i) \qquad (\approx \iint_{\mathcal{E}} u d\underset{\sim}{p})$$

where $\underset{\sim}{p}_j \in \mathcal{E}$. It follows that integrals on physical elements are first transformed and then approximated so that

$$\iint_E u d\underset{\sim}{x} = \iint_{\mathcal{E}} u\, J\, d\underset{\sim}{p} \approx I(uJ) \ ,$$

where J is the Jacobian of the transformation.

Thus, for example, the bilinear form $a[u,v]$ used previously in this chapter would be replaced by

$$a_h[u,v] = \sum_E \sum_i w_i \left[\left\{ \left(\frac{\partial u}{\partial x}\right)\left(\frac{\partial v}{\partial x}\right) + \left(\frac{\partial u}{\partial y}\right)\left(\frac{\partial v}{\partial y}\right) \right\} J \right]_{\underset{\sim}{p} = \underset{\sim}{p}_i}$$

Assuming that the right hand side is also approximated

using numerical quadrature it follows that the approximate (conforming) solution satisfies

$$a_h[U,V] = (f,V)_h \qquad \text{for all } V \in \mathcal{H}_h \ . \tag{5.15}$$

For any W_1 and W_2 we define

$$E_h^1[W_1] \equiv (f,W_1)_h - (f,W_1)_0$$

and

$$E_h^2[W_1,W_2] \equiv a_h[W_1,W_2] - a[W_1,W_2] \ .$$

If it were possible to provide analogues of both the inequalities (5.5) and (5.6) then it would be possible to construct the direct analogue of Lemma 5.1 in which there would be an additional perturbation term. Unfortunately, since numerical integration involves sampling on a point set, (5.5) is no longer valid in general. But as the trial functions are based on piecewise polynomials it is possible to construct quadrature schemes that satisfy

$$a_h[U_h,U_h] \geq \gamma \, \|U_h\|^2 \qquad \text{for all } U_h \in \mathcal{H}_h \ . \tag{5.16}$$

It is then possible to derive a bound using (5.16) together with the continuity condition on the analytic bilinear form, that is with

$$|a[u,v]| \leq \alpha \, \|u\| \, \|v\| \qquad \text{for all } u, \ v \in \mathcal{H} \ . \tag{5.17}$$

Lemma 5.5

It follows from (5.15), (5.16) *and* (5.17) *that*

$$\|u - U_h\| \leq \inf_{V_h \in \mathcal{H}_h} \left\{ \left(1 + \frac{\alpha}{\gamma}\right) \|u - V_h\| + \frac{\alpha}{\gamma} \sup_{W_h \in \mathcal{H}_h} \left\{ \frac{|E_h^2[V_h,W_h]|}{\|W_h\|} \right\} \right\} +$$

$$+ \frac{1}{\gamma} \sup_{W_h \in \mathcal{H}_h} \left\{ \frac{|E_h^1[W_h]|}{\|W_h\|} \right\} \ . \tag{5.18}$$

Δ

The proof is similar to the proof of Lemma 5.1 and it can be found in Ciarlet (1977). As with the perturbation arising from non-conforming elements, it is necessary to convert the bound into an estimate of the order of convergence by considering polynomial approximation. The numerical quadrature scheme is said to be *consistent* if

$$\lim_{h\to 0} |E_h^2[V_h, W_h]| = 0$$

and

$$\lim_{h\to 0} |E_h^1[W_h]| = 0$$

for all $W_h \in \mathcal{H}_h$, when V_h interpolates a sufficiently smooth function u. The quadrature is *optimal* if

$$E_h^2\ [V_h, W_h] = O(h^{k+1-r}),$$

$$E_h^1\ [W_h] = O(h^{k+1-r}) \ , \tag{5.19}$$

where k+1-r is the same index as in the approximation term.

It is only possible here to indicate very briefly how estimates such as (5.19) are derived. Assume a typical term in the bilinear form is

$$\iint_\Omega t(\underset{\sim}{x}) \left(\frac{\partial V}{\partial x}\right)\left(\frac{\partial W}{\partial x}\right) d\underset{\sim}{x} \ ,$$

then the error $E_h^1[V,W]$ consists of the sum over all elements of terms such as

$$E_{\mathcal{E}}^2 \equiv \iint_{\mathcal{E}} t \left(\frac{\partial V}{\partial x}\right)\left(\frac{\partial W}{\partial x}\right) J \ d\underset{\sim}{p} - \sum_i w_i \left[t\left(\frac{\partial V}{\partial x}\right)\left(\frac{\partial W}{\partial x}\right) J\right]_{\underset{\sim}{p} = \underset{\sim}{p}_i} .$$

It is possible to show that if the transformation onto the reference element is a polynomial of degree s and $W \in P_s$ on $\mathcal{E}$ then $\{(\partial W/\partial x)\ J\} \in P_{2s-2}$.

This is of vital importance in all proofs provided up

to now. It follows from the Bramble-Hilbert Lemma that if the numerical integration is exact for polynomials of degree 2s-2+k then

$$E_h^2[V_h, W_h] = O(h^{k+1}) .$$

Thus to be consistent on linear elements (s=1) the integration rule need be only first order accurate even when the coefficients $t(\underset{\sim}{x})$ are variable. This is the so-called *constant strain condition* which is equivalent to a numerical quadrature patch test. Further details of this and the analysis of the second term $E_h^1[W_h]$ can be found in Mitchell and Wait (1977).

5.6. FINITE ELEMENT TREATMENT OF SINGULARITIES

In chapters 3 and 6 it is shown how to make use of a knowledge of the form of the solution in order to achieve high accuracy results. It is possible to devise finite element methods that take the form of the singularity into account; this is particularly true in *fracture mechanics* (see Whiteman and Akin, 1978, for an extensive survey).

It is known that the displacements in an elastic solid behave as $r^{\frac{1}{2}}$ in the neighbourhood of the crack tip and it is necessary in any study of brittle fracture to have an accurate determination of the coefficient of the dominant term $r^{\frac{1}{2}}\sin\theta/2$. This coefficient is then used to determine the *stress intensity factor*. In three dimensional problems, there are three such factors to evaluate but, again, they all assume that the coefficients of the leading terms can be computed accurately.

It is possible to modify the shape functions of a finite element approximation either locally (see Strang and Fix 1973; Barnhill and Whiteman, 1973; or Fix, Gulati and Wakoff, 1973) or globally (Morley, 1973). The addition of such singular functions may lead to an illconditioned stiffness matrix. One effect of such illconditioning is the poor accuracy with which the stress intensity factor is computed (see Table 6.2). In the following sections three alternatives are given that rely on forms of mesh refinement. The key to any local refinement is the positioning of the nodes and as mentioned in chapter 3, unless there is some analysis available there are many risks inherent in this approach. On the other hand, some form of mesh refinement is vital if effective use is to be made of the available computer resources. It is a simple matter to show that a uniform grid is a very uneconomic method of approximation (Szabo and Mehta, 1978).

5.7. AUTOMATIC MESH REFINEMENT

Unlike finite differences, finite elements have a mathematical basis that permits a form of mesh refinement for which it is possible to provide a basis in theory.

Consider the problem

$$a[u,v] = (f,v)_0, \qquad \text{for all } v \in \mathcal{H}, \tag{5.20}$$

then if the finite element solution satisfies

$$a[U,V] = (f,V)_0, \qquad \text{for all } V \in \mathcal{H}_h, \tag{5.21}$$

and if the bilinear form satisfies (2.5) and (2.6), the following bound on the error $e \equiv u - U$ has been derived by Babuška and Rheinboldt (1977a).

<u>Theorem 5.2</u>

For a given type of mesh, there exist constants D_1 *and* D_2*, independent of* u *and* U *such that*

$$D_1\eta \le \| e \| \le D_2\eta$$

where

$$\eta^2 \equiv \sum_i \eta_i^2$$

with

$$\eta_i^2 \equiv \sup_{v\in\mathcal{H}} \left\{\frac{|a[e,\varphi_i v]|}{\|\varphi_i v\|}\right\} \tag{5.22}$$

where φ_i *are the trial functions used in (5.21).*$_\Delta$

It is not in general possible to compute D_1 and D_2 but it is assumed that they are approximately equal so that $\|e\|$ behaves like η. In addition it is not possible, in general to compute η_i^2 exactly and so it is necessary to provide an estimate.

Assume that (5.20) is the weak form of the differential equation

$$Lu = f \qquad \text{in } \Omega ,$$

with u given on the boundary Γ. It follows that if $\Omega_i \equiv S(\varphi_i)$ with boundary Γ_i then the function $\hat{u}_i$ such that

$$L\hat{u}_i = f \qquad \text{in } \Omega_i$$

with

$$\hat{u}_i = U \qquad \text{on } \Gamma_i$$

also satisfies

$$a[\hat{u}_i - U, \varphi_i v] = a[e, \varphi_i v] \qquad \text{for all } v \in \mathcal{H}.$$

Thus the analytic solution in (5.22) can be replaced by the solution of a local problem. It is now possible to determine the solution of this local problem exactly. Fortunately this is not necessary and an approximation $\hat{U}_i$ is sufficient, provided that in Ω_i

$$|\hat{u}_i - \hat{U}_i| \ll |u - U| .$$

The procedure is therefore to subdivide Ω_i into smaller elements $E_{i,j}$ $j = 1, \cdots, k(i)$ and then compute a local finite element approximation that agrees with U on Γ_i. Once all the local approximations $\hat{U}_i$ have been computed it is possible to provide an algorithm for the mesh refinement.

The refinement uses the estimates

$$\eta_i^2 \approx \hat{\eta}_i^2 \equiv \| \hat{U}_i - U \|^2_{\mathcal{H}(\Omega_i)}$$

$$= \sum_{\substack{j \\ E_j \subset \Omega_i}} \| \hat{U}_i - U \|^2_{\mathcal{H}(E_j)}$$

and

$$\| e \| \approx \eta \approx \hat{\eta},$$

where

$$\hat{\eta}^2 = \sum_i \hat{\eta}_i^2 \quad .$$

In order to obtain a useful criterion for mesh refinement, observe that

$$\hat{\eta}^2 = \sum_i \sum_{\substack{j \\ E_j \subset \Omega_i}} \| \hat{U}_i - U \|^2_{\mathcal{H}(E_j)}$$

$$= \sum_j \sum_{\substack{i \\ \Omega_i \supset E_j}} \| \hat{U}_i - U \|^2_{\mathcal{H}(E_j)}$$

$$\equiv \sum_j \bar{\eta}_j^2 \quad .$$

It is the values of $\bar{\eta}_j$ that are used to decide whether or not to subdivide element E_j. If it is assumed that locally

$$\bar{\eta} \approx \bar{\eta}(h) = ch^{\lambda}$$

then

$$\bar{\eta}\left(\frac{h}{2}\right) = \frac{\bar{\eta}(h)^2}{\bar{\eta}(2h)} \tag{5.23}$$

and this provides an estimate of the probable value of $\bar{\eta}$ after a further subdivision. The next step is then to subdivide all elements in which the current value of $\bar{\eta}$ is greater than the largest predicted value after subdivision.

In spite of all the assumptions it has been found (Babuška

and Rheinboldt, 1977b, 1978) that this method of automatic mesh refinement is very effective, further details can be found in the references.

5.8. INFINITE MESH REFINEMENT

An alternative form of grid refinement that can be applied in many problems, is the form of infinite refinement suggested by Thatcher (1975). Assume the region near the singularity can be divided up by means of rays emanating from the singularity. It is then possible to construct bands of elements using as vertices the intersections of the rays with circular arcs of radius r_0, ar_0, a^2r_0,... where $0 < a < 1$ and r_0 is chosen such that the region $r > r_0$ is sufficiently far from the singularity, that is, as far as possible. The construction of a typical ring of elements as illustrated in Figure 5.3. It follows from this symmetric construction that if the

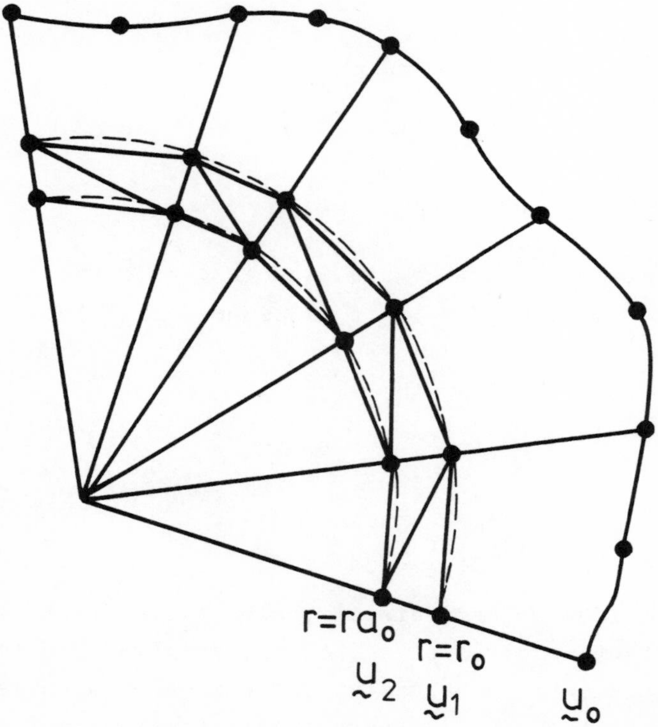

Fig. 5.3.

approximation involves the arcs $r = a^i r_0$ ($i = 0,\ldots,p$), the nodal values on $r = a^{i-1} r_0, i = 1,\ldots,p-1$ are denoted by $\underset{\sim}{U}_i$, the nodal values for $r > r_0$ by $\underset{\sim}{U}_0$ and for $r \leq a^p r_0$ by $\underset{\sim}{U}^*_p$ then the nodal values satisfy

$$\begin{bmatrix} A_0 & B_0 & & & \\ B_0^T & A & B & & \\ & B & A & B & \\ & \cdot & \cdot & \cdot & \\ & & \cdot & \cdot & \cdot \\ & & & \cdot & \cdot & \cdot \\ & & & & \cdot & \cdot & \cdot \\ & & & & \cdot & \cdot & \cdot \\ & & & & B^T & A & B^*_p \\ & & & & & B^{*T}_p & A^*_p \end{bmatrix} \begin{bmatrix} \underset{\sim}{U}_0 \\ \underset{\sim}{U}_1 \\ \cdot \\ \cdot \\ \cdot \\ \cdot \\ \cdot \\ \underset{\sim}{U}_{p-1} \\ \underset{\sim}{U}^*_p \end{bmatrix} = \begin{bmatrix} \underset{\sim}{b}_0 \\ \underset{\sim}{b}_1 \\ \cdot \\ \cdot \\ \cdot \\ \cdot \\ \cdot \\ \underset{\sim}{b}_{p-1} \\ \underset{\sim}{b}^*_p \end{bmatrix} . \tag{5.24}$$

The size of the blocks A and B are governed by the number of rays in the mesh. It can be shown that as $p\to\infty$ the solution of (5.2) converges to a solution of

$$\begin{bmatrix} A_0 & B_0 & & & \\ B_0^T & A & B & & \\ & B^T & A & B & \\ & \cdot & \cdot & \cdot & \\ & & \cdot & \cdot & \cdot \\ & & & \cdot & \cdot & \cdot \\ & & & & \cdot & \cdot & \cdot \end{bmatrix} \begin{bmatrix} \underset{\sim}{U}_0 \\ \underset{\sim}{U}_1 \\ \underset{\sim}{U}_2 \\ \cdot \\ \cdot \\ \cdot \\ \cdot \end{bmatrix} = \begin{bmatrix} \underset{\sim}{b}_0 \\ \underset{\sim}{b}_1 \\ \underset{\sim}{b}_2 \\ \cdot \\ \cdot \\ \cdot \\ \cdot \end{bmatrix} . \tag{5.25}$$

The solution of (5.25) is found from a particular solution of the recurrence relation

$$B^T \underset{\sim}{U}_{i-1} + A\underset{\sim}{U}_i + B\underset{\sim}{U}_{i+1} = \underset{\sim}{b}_i \quad i = 2,3\ldots \tag{5.26}$$

together with the general solution of

$$B^T \underset{\sim}{W}_{i-1} + A\underset{\sim}{W}_i + B\underset{\sim}{W}_{i+1} = \underset{\sim}{0} \qquad i = 2,3,\ldots. \tag{5.27}$$

It is then possible to select the particular solution that satisfies the first two sets of equations in (5.25) and leads to an admissible approximation. Observe that if

$$Z \equiv - \begin{bmatrix} B & O \\ A & B \end{bmatrix}^{-1} \begin{bmatrix} B^T & A \\ O & B^T \end{bmatrix}$$

then

$$\begin{bmatrix} \underset{\sim}{W}_{2i+1} \\ \underset{\sim}{W}_{2i+2} \end{bmatrix} = Z \begin{bmatrix} \underset{\sim}{W}_{2i-1} \\ \underset{\sim}{W}_{2i} \end{bmatrix}$$

and so the solution of (5.27) is derived from the eigensystem of Z. This has been studied in Thatcher (1976).

If it is assumed that in the neighbourhood of the singularity the solution is of the form

$$u = \alpha_1 r^{\gamma_1} f_1(\theta) + \ldots.$$

then it is possible to use this method to provide estimates for the coefficients α_1 and γ_1. If the eigenvalues of Z that are less than one in modulus, are denoted by $\lambda_j (j=1,\ldots N)$ (see Thatcher, 1975) and the corresponding eigenvectors by $\underset{\sim}{V}_j$ then the solution can be written as

$$\underset{\sim}{U}_i = \sum_j c_j \lambda_j^{i-2} \underset{\sim}{V}_j + \text{boundary terms} \qquad i=2,\ldots$$

Then ignoring boundary terms and all the sub-dominant eigenvalues and considering only the nodal values of $r = a^{i-1} r_0$ it is possible to compare the terms

$$c_1 \lambda_1^{i-2} \underset{\sim}{V}_1 \quad \text{and} \quad \alpha_1 (a^{i-1} r_0)^{\gamma_1} f_1(\theta).$$

The vector $\underset{\sim}{V}_1$ is a discrete approximation to the function

$f_1(\theta)$, in addition

$$\lambda_1^i \approx a^{\gamma_1 i}$$

and

$$c_1 \approx \alpha_1 (ar_0)^{\gamma_1}$$

Thus

$$\gamma_1 \approx \frac{\log \lambda_1}{\log a}$$

and

$$\alpha_1 \approx \frac{c_1}{(ar_0)^{\gamma_1}} \quad .$$

The accuracy of these approximations is governed by the accuracy of the approximation to $f_1(\theta)$, which in turn is governed by the number of rays through the singularity.

5.9. ISOPARAMETRIC ELEMENTS

In chapter 2 the analysis of isoparametric elements was based on the Jacobian being bounded away from zero because the inequality

$$0 \le C_1 \le \frac{\sup|J|}{\inf|J|} \le \frac{1}{C_1}$$

was used; although this condition was slightly relaxed later. In this section two types of element will be considered that violate the conditions imposed by Lemma 2.1, they are the quarter point triangles illustrated in Figure 5.4.

The transformations are respectively

$$\underset{\sim}{x} - \underset{\sim}{x}_3 = \{(\underset{\sim}{x}_1 - \underset{\sim}{x}_3)p + (\underset{\sim}{x}_2 - \underset{\sim}{x}_3)q\}(p + q) \qquad (5.28)$$

and

$$\underset{\sim}{x} - \underset{\sim}{x}_3 = \{(\underset{\sim}{x}_1 - \underset{\sim}{x}_3)(1-q) + (\underset{\sim}{x}_2 - \underset{\sim}{x}_3)q\}p^2 \; . \qquad (5.29)$$

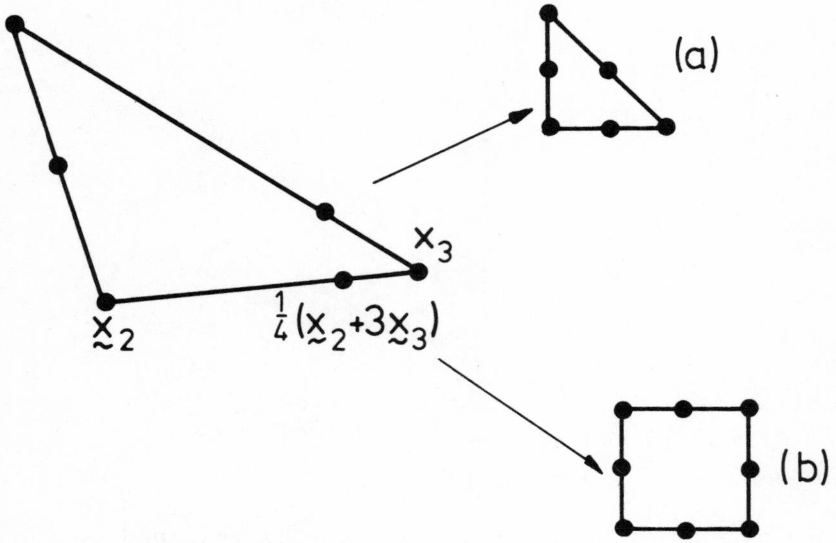

Fig. 5.4.

Both these element types have the property that along any line through $\underset{\sim}{x}_3$, a linear function in p and q behaves as $r^{\frac{1}{2}}$. Thus we have an element that can successfully approximate singularities of the form

$$u = \alpha_1 \; r^{\frac{1}{2}} f_1(\theta) + \alpha_2 \; r^{3/2} f_2(\theta) + \ldots,$$

since higher order polynomials in p and q provide approximations to the higher powers of $r^{\frac{1}{2}}$. As for the infinite mesh approximation, the accuracy with which the coefficient α_1 can be approximated depends on how accurately the function $f_1(\theta)$ can be approximated (see for example, Wait, 1978c).

In order to approximate solutions of the form

$$u = \alpha_1 \; r^{1/m} \; f_1(\theta) + \ldots$$

it is necessary to use transformations of the form

$$\underset{\sim}{x}-\underset{\sim}{x}_3 = \{(\underset{\sim}{x}_1-\underset{\sim}{x}_3)p + (\underset{\sim}{x}_2-\underset{\sim}{x}_3)q\}(p+q)^{m-1} \tag{5.30}$$

and

$$\underset{\sim}{x}-\underset{\sim}{x}_3 = \{(\underset{\sim}{x}_1-\underset{\sim}{x}_3)(1-q) + (\underset{\sim}{x}_2-\underset{\sim}{x}_3)q\}\ p^m \tag{5.31}$$

in place of (5.28) and (5.29) respectively where $m \geq 1$ is not necessarily an integer. Approximations to the coefficient α_1 can be found in each case as linear combinations of the nodal values (see Wait, 1978a). The technical problems of how to accommodate curved boundaries (since formulae (5.30) and (5.31) only produce the described behaviour for specific straight sided elements) cannot be dealt with here and details can be found in Morris and Wait (1978). Another problem that this method shares with the 'special shape function' approach is in providing a realistic estimate of the size of the singular element. Once this size has been decided, the grid in the neighbourhood of the singularity can be refined using the macro-element approach outlined in Wait (1978c). This reference also describes how the technique of singular isoparametric elements can be extended to three dimensional problems.

6

GLOBAL AND REGIONAL METHODS

6.1. INTRODUCTION

As in the preceding chapters we are concerned with a linear elliptic differential equation

$$Lu = f \tag{6.1}$$

over a region Ω, subject to certain conditions on the boundary Γ. For definiteness we consider a two dimensional region and assume L to be a second order differential operator. In this chapter we consider approximations of the form

$$U_N = \sum_{i=1}^{N} \alpha_i \varphi_i \tag{6.2}$$

but it is no longer assumed that the trial functions φ_i are piecewise polynomial finite element basis functions.

A *global* method chooses by contrast a set of functions φ_i which have support at the whole region Ω; a typical choice for a two dimensional region would be a product of orthogonal polynomials in x and y. The resulting algorithm is similar in outline to, but very different in detail from, a finite element calculation, and the two methods each have their strong and weak points. In succeeding sections we outline the structure of a global calculation, illustrate its strengths and show how some of the weaknesses can be overcome by considering *regional* approximations.

6.2. THE GLOBAL GALERKIN METHOD

To simplify discussion, assume that homogeneous boundary conditions are imposed on u and that the functions $\{\varphi_i\}$ each satisfy these conditions exactly. Then so does U_N for any choice of $\{\alpha_i\}$ and as in section 2.3 the α_i may be determined from the Galerkin equations

$$a[U_N, \varphi_i] = (f, \varphi_i) \qquad i = 1,2,\ldots,N \tag{6.3}$$

in global calculations it is often useful to modify the inner product so that

$$(f,g) \equiv \iint_\Omega fg\omega \, dxdy$$

where $\omega(x,y) > 0$ is a suitable weight function. We illustrate the main attraction of a global scheme with an example from Gawley and Delves (1973) where

$$\nabla^2 u = -2 \quad \text{in } \Omega$$

and

$$u = 0 \quad \text{on } \Gamma \tag{6.4}$$

with Ω the ellipse with boundary Γ defined by

$$r = R(\theta) \equiv (1-e^2)a/(1+e\cos\theta) \; . \tag{6.5}$$

The exact solution is

$$U(r,\theta) = \frac{a^2b^2}{a^2+b^2}\left[1 - \left(\frac{r\cos\theta}{a} + e\right)^2 - \frac{r^2\sin^2\theta}{b^2}\right]$$

where $e = (a^2-b^2)^{\frac{1}{2}}/a$ is the eccentricity of the ellipse. We choose the trial functions

$$\varphi_i(\underset{\sim}{x}) = p_j(\rho)q_k(\theta) \qquad j,k=1\ldots n;\; N = n^2 \tag{6.6}$$

and weight function

$$\omega(\underset{\sim}{x}) = r$$

where

$$\rho = r/R(\theta),$$

$$p_j(\rho) = (1-\rho)^j \tag{6.7a}$$

and

$$q_k(\theta) = \begin{cases} \cos\left[\frac{k}{2}\right]\theta & k \text{ odd} \\ \sin\left[\frac{k}{2}\right]\theta & k \text{ even} . \end{cases} \qquad (6.7b)$$

The set (6.6) is zero as required on the boundary of the ellipse. The use of polar coordinates and (ρ,θ) variables effectively maps the region onto a circle and the integrals can be evaluated numerically in a straightforward manner using a product quadrature rule in ρ and θ. We assume that enough quadrature points are taken to make the quadrature errors negligible, see Gawley and Delves (1973) or Yates (1975) for numerical examples of the effects of quadrature error. Figure 6.1 shows the results obtained for various values of eccentricity e, for the (relative) mean square error σ_N^2, defined as

$$\sigma_N^2 = \frac{\Sigma(u-U_N)^2}{\Sigma u^2} \qquad (6.8)$$

where the sums are taken over a rectangular grid in (ρ,θ) coordinates.

It is clear from the graph that in each case, the points are well fitted by a straight line; on the logarithmic scale used. This implies that σ_N^2 has the functional dependence

$$\sigma_N^2 \sim c\ \lambda^N \qquad (6.9)$$

that is, an exponentially fast convergence rate. The values of λ shown in Figure 6.1 tend to unity as $e \to 1$ and the ellipse degenerates to a straight line. For high eccentricity, the results are very poor despite the nominal "fast" convergence and we should not use this expansion in this case.

Inhomogeneous boundary conditions can be handled in one of two ways. In one we constrain the approximate solution U_N to satisfy the boundary conditions. In the framework of a global calculation, we are required to find a function U_B, a defined everywhere in Ω, and satisfying the boundary conditions. We may then solve the equation

$$L(u-U_B) = f - LU_B \quad \text{in } \Omega \qquad (6.10)$$

for the function $u - U_B$. Constructing a suitable smooth U_B is straightforward if only Dirichlet boundary conditions are required, but is very difficult in general.

A more flexible approach is to introduce additional terms into the equations which ensure the boundary conditions are approximately satisfied, the degree of approximation improving with increasing N. This is achieved most satisfactorily in methods which start from a variational functional F(v) whose stationary point is at u = v, the solution of (6.1). The functional is

$$F(v) = a[v,v] - 2(f,v) \tag{6.11}$$

provided that v is constrained to satisfy the boundary conditions exactly. Modified functionals which do not require this constraint have been given by, Arthurs (1970), Yates (1975), Davies and Hendry (1976). Yates' modified Galerkin equations take the form

$$A\underset{\sim}{\alpha} = \underset{\sim}{b}$$

with

$$A_{ij} = \iint_\Omega \omega\varphi_i \, L\varphi_j \, dxdy + \gamma\int_\Gamma \varphi_i\left(h_1\varphi_j + h_2 \frac{\partial\varphi_j}{\partial n}\right) ds \tag{6.12}$$

and

$$b_i = \iint_\Omega \omega\varphi_i f \, dxdy + \gamma\int_\Gamma \mu\varphi_i g \, ds$$

where ω,μ are weight functions, γ is an arbitrary constant and it is assumed that the boundary conditions are of the form

$$h_1 u + h_2 \frac{\partial u}{\partial n} = g \qquad \text{on } \Gamma.$$

We illustrate this approach by means of another example in which f_1, f_2, f_3 satisfy the equation

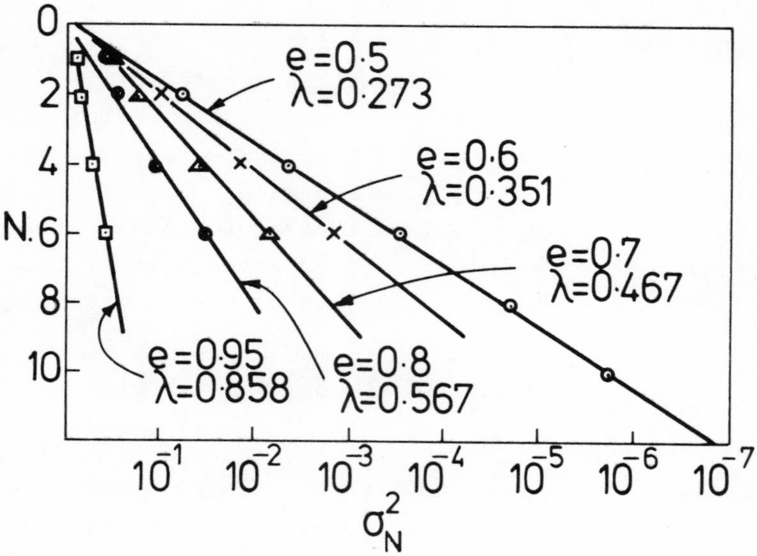
e=0·5
λ=0·273
e=0·6
λ=0·351
e=0·7
λ=0·467
e=0·95
λ=0·858
e=0·8
λ=0·567
N.
0
2
4
6
8
10
10⁻¹ 10⁻² 10⁻³ 10⁻⁴ 10⁻⁵ 10⁻⁶ 10⁻⁷
σ²_N

Fig. 6.1.

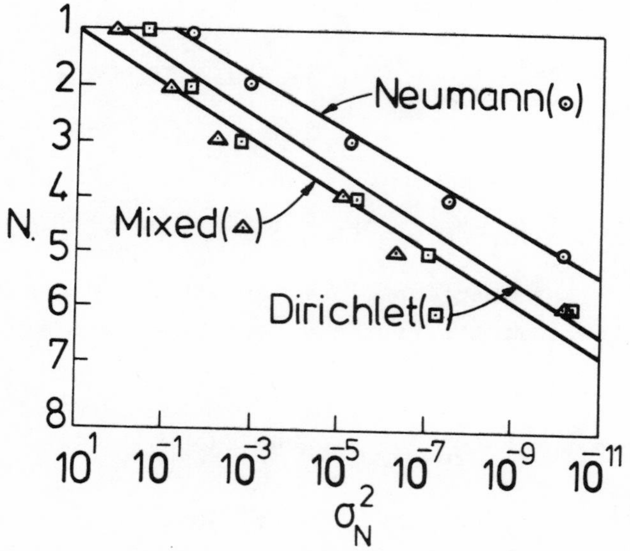
Neumann(⊙)
Mixed(△)
Dirichlet(⊡)
N.
1
2
3
4
5
6
7
8
10¹ 10⁻¹ 10⁻³ 10⁻⁵ 10⁻⁷ 10⁻⁹ 10⁻¹¹
σ²_N

Fig. 6.2.

$$\nabla^2 f_i(r,\theta) = \frac{5r+4(\cos\theta-1)}{4r}\, e^{-r/2(2+\cos\theta)} \tag{6.13}$$

inside the ellipse $r = 1/(1+\frac{1}{2}\cos\theta)$ and subject to boundary conditions on Γ in the form

$$f_1(r,\theta) = e^{-1} \tag{6.14}$$

$$\left(\frac{5}{4} + \cos\theta\right)^{\frac{1}{2}} \frac{\partial f_2}{\partial n} + \frac{5}{4} f_2 = -e^{-1}\cos\theta \tag{6.15}$$

and

$$\frac{\partial f_3}{\partial n} = -\left\{\frac{5}{4} + \cos\theta\right\}^{\frac{1}{2}} e^{-1}. \tag{6.16}$$

Each of these problems has the same solution:

$$f_i(r,\theta) = e^{-r(\cos\theta+2)/2} \qquad i = 1,2,3 \tag{6.17}$$

although f_3 is unique only within an additive constant. We set $\omega = \mu = \gamma = 1$ in (6.12) and with the trial functions

$$\varphi_i(r,\theta) = \begin{cases} r^j \sin k\theta \\ r^j \cos k\theta \end{cases} \tag{6.18}$$

we obtain the mean square errors σ_N^2 shown in Figure 6.2. It appears once again that exponentially fast convergence is achieved, in this case with $\lambda \approx 0.01$. Approximating the boundary conditions causes no problems and the convergence rate achieved appears insensitive to the type of condition imposed. However, the Neumann conditions which are "natural" for this problem are easier to approximate, leading to a smaller value of C in (6.9).

6.3. REGIONAL METHODS

The two model problems in the previous section are defined on simple, smooth regions and have smooth solutions. A standard finite element calculation has an error of $O(h^s)$

for some small s, where h is a typical element dimension, and, at least if high accuracy is required, the global method appears to be superior. However, on more complicated regions, the position changes radically. It may prove difficult to construct a sensible global basis; or the convergence, although exponential, may nonetheless be slow as for the highly eccentric ellipses.

One possible compromise is to use regional methods, that is methods which split the given region into a few, more manageable, subregions, and then use a global expansion over each subregion, with suitable arrangements for ensuring continuity across the subregion boundaries. Different methods are characterised by the nature of these arrangements, as well as by the choice of subregions and of trial functions within each subregion. Finite element methods may be viewed as a regional methods, choosing a large number of small regions; within each, a low order expansion is normally chosen. With conforming elements interface and boundary conditions are imposed explicitly, and this may make the construction of high order approximations very difficult.

Regional methods have been quite widely used in the field of reactor engineering, where they were introduced by Wachspress; see, for example, (Wachspress and Becker, 1965; Yasinsky and Kaplan, 1976; Henry and Yeng, 1976). A number of variants of the general method have been tried with success, but all involve the use of a first approximation φ_0 within each region with either multiplicative or additive correction terms:

$$U_N = \varphi_0 + \sum_{i=1}^{N} \alpha_i \varphi_i \qquad \text{(additive)} \qquad (6.19)$$

$$U_N = \varphi_0 \left(1 + \sum_{i=1}^{N} \alpha_i \varphi_i\right) \qquad \text{(multiplicative).} \qquad (6.20)$$

Here, the trial functions φ_i are taken to be of finite element type on a fine mesh, the function φ_0 may be constructed globally within each region, or using a coarse mesh finite element basis and a separate initial calculation. A major advantage of the technique used in these papers is in the way that the interface matching conditions are satisfied

The calculations proceed by minimising an appropriate functional (Yasinsky and Kaplan, 1967) over the parameter space of the approximate solutions (6.19) or (6.20). This functional is designed to include the effect of the interface conditions, so that these need not be imposed explicitly.

Regional methods have also been used in the field of electrical engineering (see, e.g., Bates, 1969, and references cited there). These authors construct a global expansion within each subregion, the expansion being chosen to satisfy the differential equation involved. The parameters are then chosen to satisfy the boundary and interface conditions either in a least squares sense or by collocation on a point set. The method works well for the problems involved, but it is restricted essentially to Laplace or similar equations for which complete solution sets are available.

The global element method (Delves and Hall, 1976 ; Delves and Phillips, 1977) attempts to give a generally applicable method capable of yielding fast convergence for a variety of geometries and problems. The method uses a systematic construction for the trial space in each element. The subregion is mapped onto a standard element (a square) and a product of modified orthogonal polynomials in the mapped coordinates used as basis functions. Both boundary and interface conditions are treated implicitly, via the choice of variational functional. Finally, the basis functions chosen also allow efficient construction and solution of the algebraic equations for the expansion coefficients (Delves and Phillips, 1977).

As an example of the use of the global element method we consider the problem (Delves and Phillips, 1978)

$$\frac{\partial^2 U}{\partial r^2} + \frac{1}{r}\frac{\partial^2 U}{\partial \theta^2} = f(r,\theta) \qquad \text{in } \Omega \tag{6.21}$$

where

$$f(r,\theta) \equiv \frac{2}{3}\left\{1 + 3\cos\theta + \frac{15\cos\theta}{r} + e^{r-1}[5+2r-3\cos\theta(r+1)+\tfrac{1}{2}r^2]\right\}$$

and Ω is the region of Figure 6.3(a). We impose the boundary conditions

$$u = \begin{cases} (11-10\cos\theta)e^4 & \text{on } \Gamma_1 \\ 11-10\cos\theta & \text{on } \Gamma_2, \Gamma_3 \end{cases} . \tag{6.22}$$

The exact solution to this problem is

$$u(r,\theta) = \tfrac{1}{3}\{(r^2-6r\cos\theta+8)(e^{r-1}-1) + 3(11-10\cos\theta)\} . \tag{6.23}$$

It is most convenient to solve the equations in (r,θ) variables; symmetry about the line $y = 0$ allows a solution on the upper half of Ω only, and in (r,θ) coordinates this has the form shown in Figure 6.3(b).

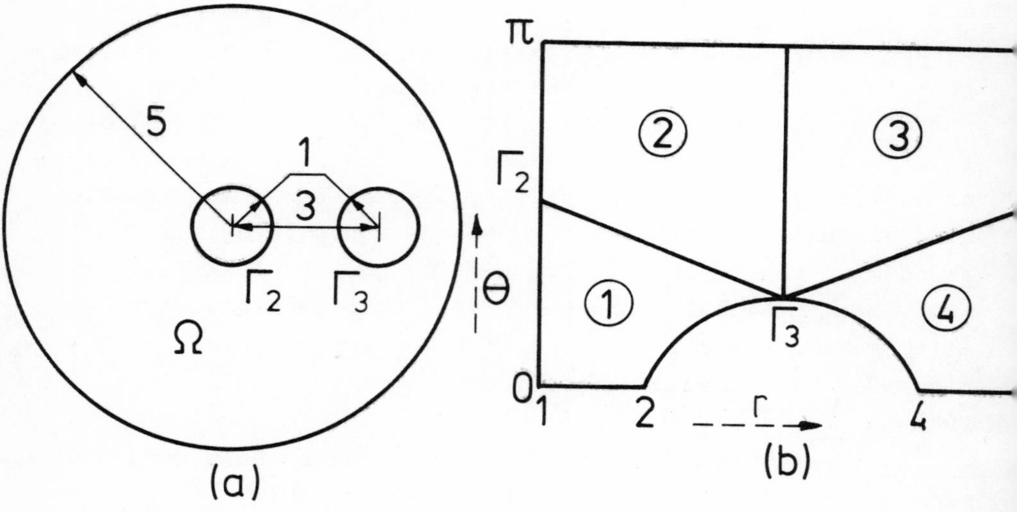

Fig. 6.3.

We split the region into four global elements as shown. Then the standard technique described in the next section maps each element onto the square $[-1,1] \times [-1,1]$ and in the mapped (s,t)-space uses the trial functions

$$\varphi_k(s,t) = p_i(s)p_j(t) \qquad i,j = 1\ldots,N$$

$$p_1(s) = 1; \quad p_2(s) = s; \quad p_k(s) = (1-s^2)T_{k-3}(s), k > 2 \tag{6.24}$$

where $T_\ell(s)$ is a Chebyshev polynomial.

There are therefore $4N^2$ parameters in the calculation for a given N. For any finite N the approximate solution is not continuous across the element boundaries; but the discontinuities tend rapidly to zero with N. Table 6.1 shows maximum relative error $\| U_N(x,y)-u(x,y)\|_\infty / \| u(x,y)\|_\infty$ in each element. As N increases, convergence is exponentially fast as with the Global method, and this exponentially fast convergence depends crucially on the implicit handling of the interface continuity conditions. Also shown along with the calculated errors are the error estimates returned by the program, which are based on the observed behaviour of the coefficients α_i and depend on the orthogonality properties of the set (6.24). The ability to yield such estimates is a free bonus with the global element method.

TABLE 6.1

N	Element 1		Element 2		Element 3		Element 4	
	calc.	est.	calc.	est.	calc.	est.	calc.	est.
3	0.15	0.12	0.35	0.33	0.78	0.32	1.5	1.3
4	1.5E-2	5.3E-3	2.1E-2	8.5E-3	4.0E-2	7.2E-2	6.1E-2	4.4E-2
5	6.9E-4	3.0E-4	6.2E-3	1.1E-3	3.5E-3	1.0E-2	6.6E-3	3.3E-3
6	2.4E-4	5.0E-5	5.3E-5	1.0E-4	1.8E-4	1.4E-3	2.0E-3	4.6E-4

calc.= computed error est. = estimated error returned by program

6.4. NUMERICAL METHODS

For a given global or regional method, there are two major and related numerical problems; the choice of trial functions and the calculation of the double integral in the stiffness matrix. Global calculations in the past have typically been ad-hoc; the choice of the set $\{\varphi_i\}$ being tailored to the problem at hand and the methods of integration being tailored to the region involved. It is possible to get a highly efficient method in this way, at the cost of making each problem special. It is possible to make these choices systematically and we describe briefly how this is achieved in the global element method.

Choice of trial functions

Each element is mapped onto a standard shape; four cornered elements onto a square with cartesian coordinates (s,t), three cornered elements onto a sector of a circle with polar coordinates (r,θ) which are then mapped linearly into [-1,1], so that the final mapped element is also the square $\{s,t: [-1,1] \times [-1,1]\}$. In this mapped square the trial functions (6.24) are used.

Although systematic, this procedure leaves considerable flexibility. The expansion set is uniquely determined by the map used, which has two stages. First is a blending function map (Gordon and Hall, 1973) from the curvilinear element to a square or sector. This map has the advantage of treating curved sides exactly while retaining the smoothness properties of the original element. This can be followed by a nonlinear map to deal with any singularities.

It is possible in this way to deal effectively with a variety of awkward regions. In particular, infinite or semi-infinite elements can be handled effectively, as can singularities such as occur at re-entrant corners.

Evaluation of the stiffness matrix

In two space dimensions the Galerkin equations appropriate to the global element method contain both double and single integrals. The form these take is discussed in detail in Delves and Phillips (1977) and some of the integrals for a specific problem are displayed in full in Hendry and Delves (1978). The major cost in computing comes from the double integrals, a typical double integral not involving derivatives has the form

$$I_{i,j,k,\ell} = \int_{-1}^{1} \int_{-1}^{1} p_i(s)\, p_j(t) B(s,t) p_k(s) p_\ell(t) \; ds dt \qquad (6.25)$$

where $B(s,t)$ is a known function and $p_i(s)$ is a trial function from (6.24).

To evaluate (6.25) numerically with a product p-point quadrature rule takes $O(p^2)$ operations. The $MN^2 \times MN^2$ matrix for an M-element region can be viewed as an $M \times M$ block matrix. Only the diagonal blocks involve double integrals and the (m,n) block is in fact null unless elements m and n have a common side or are coupled through periodic or other boundary conditions. Hence, with $p \approx N$, it takes $O(MN^6)$ operations to accumulate the stiffness matrix numerically, if each double integral is computed individually. This daunting count can however be reduced to $O(MN^4)$ as follows. For simplicity, consider $i,j,k,\ell \geq 3$. Then from (6.24) and the product relation

$$T_p(x)T_q(x) = \tfrac{1}{4}[T_{p+q}(x) + T_{|p-q|}(x)] \tag{6.26}$$

we find

$$I_{i,j,k,\ell} = \frac{\pi^2}{16}[\bar{B}_{i+k,j+\ell} + \bar{B}_{|i-k|,j+\ell} + \bar{B}_{i+k,|j-\ell|} + \bar{B}_{|i-k|,|j-\ell|}] \tag{6.27}$$

where

$$\bar{B}_{pq} = \frac{4}{\pi^2}\int_{-1}^{1}\int_{-1}^{1}(1-s^2)(1-t^2)T_{p-3}(s)T_{q-3}(t)\,ds\,dt\,. \tag{6.28}$$

Equation (6.28) identifies $\bar{B}_{pq}$ as a coefficient in the series

$$(1-s^2)^{3/2}(1-t^2)^{3/2}B(s,t) = \sum_{r=0}^{\infty}{}'\sum_{q=0}^{\infty}{}'\bar{B}_{p+3,q+3}\,T_p(s)T_q(t) \tag{6.29}$$

(substitute (6.29) in (6.28) and use the orthogonality relations for the Chebyshev polynomials). If we can compute this series efficiently, then (6.27) implies that we can construct the stiffness matrix in $O(MN^4)$ operations (that is $O(N^4)$ for each diagonal block), provided that we can repeat the trick for the integral involving derivatives. We can (see Delves and Phillips, 1977); indeed handling the derivative terms neatly is the major reason for the detailed choice of trial functions in (6.24) rather than the simpler choice $p_i(s) = T_i(s)$.

It remains to construct (6.29) numerically. We can do this straightforwardly enough using a product Gauss-Legendre rule to evaluate (6.28) directly; however, it is neater and quicker, and leads to computable error estimates, if we first construct the expansion

$$B(s,t) = \sum_{p=0}^{\infty}{}' \sum_{q=0}^{\infty}{}' B_{pq} T_p(s) T_q(t) . \tag{6.30}$$

This can be produced accurately in $O(N^2 \log N)$ operations using fast Fourier transform techniques, and then to generate $\bar{B}_{pq}$ using the known Chebyshev expansion of the additional factor $(1-s^2)^{3/2}(1-t^2)^{3/2}$. The simpler alternative of producing $\bar{B}_{pq}$ directly leads to large numerical errors due to the slow convergence of (6.29).

Solution of the equations

Finally, we consider briefly the numerical solution of the Galerkin equations. Viewed as a block $M \times M$ matrix, and there are only $O(M)$ non-zero blocks. This sparseness can be taken account of if we solve by block Gauss elimination. However, in practice M is not likely to be large enough for this sparseness to be very significant. The non-null blocks are in general full, so that, even allowing for the sparsity, block elimination takes $O(M N^6)$ operations. However as with the matrix set-up times, this can be reduced by taking appropriate account of the structure of the matrix. This structure is discussed in detail in Delves and Phillips (1977).

We write the diagonal blocks as $A_{mm} = V_{mm} + S_{mm}$ where V contains all the double integrals and S the line integrals. Then we find that V has large diagonal entries stemming from the second derivative terms. As we move away from the diagonal, the elements of V decrease rapidly. This follows from (6.27) and the observation that the coefficients $\bar{B}_{pq}$ decrease with increasing p,q. Viewed as an $N \times N$ block matrix, V is block asymptotically diagonal (see Musa and Delves, 1978; Delves, 1977). This implies that a convergent iterative scheme can be constructed for the solution of equations

$$V\underset{\sim}{\alpha} = \underset{\sim}{b} \tag{6.31}$$

with $O(N^4)$ operations. Since S_{mm} is of low rank the scheme for (6.31) can be extended with the same operations count to equations of the form

$$(V+S)\underset{\sim}{\alpha} = \underset{\sim}{b}. \tag{6.32}$$

Finally the off-diagonal blocks A_{mn} involve only line integrals and are also of low rank. The iterative scheme can then therefore be extended to cover a block M × M system.

The resulting scheme is not simple; but then neither are the corresponding special schemes for finite element matrices discussed in chapter 15. In both cases it is sensible to take account of the matrix structure to obtain large gains in efficiency.

6.5. GLOBAL METHODS FOR PROBLEMS WITH SINGULARITIES

Within a global framework, singularities or any other feature of the solution which are hard to reproduce using standard trial functions, are conventionally and conveniently treated by adding one or a few special terms. The complete expansion then has the form

$$u \approx U_N = \sum_{i=1}^{Q} \beta_i \psi_i + \sum_{i=1}^{N} \alpha_i \varphi_i \equiv \sum_{i=1}^{N+Q} \alpha_i' \varphi_i' \; . \tag{6.33}$$

Here, the functions φ_i represent the first N terms from a complete set of functions as in (6.2), the functions ψ_i are picked to reflect the expected analytic structure of the solution in the neighbourhood of the anticipated difficulty. Features which may be treated in this way include:

(1) Interface problems in which the coefficients in the differential equation are discontinuous or have a discontinuous low-order derivative along a line representing, usually, some physical interface in the problem being modelled.

(2) Point singularities in which the presence of either a re-entrant corner in the region or a discontinuity in the differential equation or the boundary condition yields a point at which the behaviour of the solution is not smooth.

We illustrate the effect of adding judiciously chosen

terms, with two examples taken from Yates, Hendry and Delves (1978).

The equation

$$\nabla^2 u(r,\theta) = 2r^{\sqrt{3}-1}(1 + \sqrt{3}\cos^2\theta) \tag{6.34}$$

inside the elliptic region

$$r < R(\theta) = 1/(1+0.5\cos\theta) , \tag{6.35}$$

subject to the mixed boundary conditions

$$\left(\frac{5}{4} + \cos\theta\right)^{\frac{1}{2}} \frac{\partial u}{\partial n} - \frac{1+\sqrt{3}}{r^2} u = r^{\sqrt{3}} \cos\theta \sin^2\theta, \tag{6.36}$$

has the exact solution

$$u(r,\theta) = r^{\sqrt{3}+1}\cos^2\theta \tag{6.37}$$

and a singular point at $r = 0$. We choose the trial functions

$$\varphi_i(r,\theta) = \left(1 - \frac{r}{R(\theta)}\right)^j \begin{Bmatrix}\cos k\theta \\ \sin k\theta\end{Bmatrix}, \quad \begin{cases} j = 0,1,\ldots,N-1 \\ k = 0,1,\ldots,N-1 \end{cases}. \tag{6.38}$$

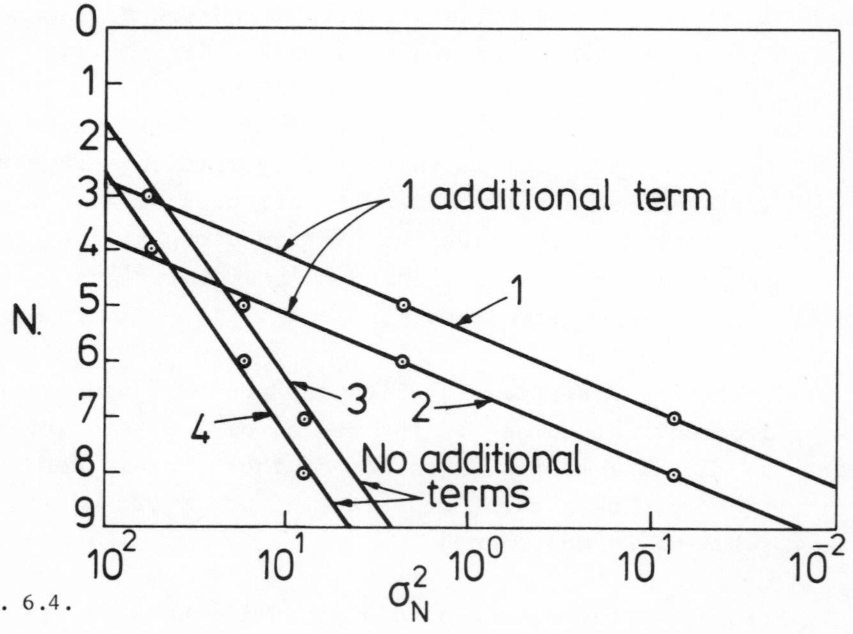

Fig. 6.4.

Figure 6.4 shows the rather slow convergence obtained with the trial functions, and the great improvement obtained by adding the single core term

$$\psi_1(r,\theta) = r^{\sqrt{3}+1} . \qquad (6.39)$$

Even the results with the term (6.39) are less good than for the previous, simpler problems. This slower convergence shows up more clearly in the second example.

The second order equation

$$[x^2 \frac{d^2}{dx^2} + x \frac{d}{dx} + 1]\, u(x) = 0, \qquad x \in [10^{-3},1] \qquad (6.40)$$

with boundary conditions

$$u(10^{-3}) = \cos(-3 \log 10) + \sin(-3 \log 10),$$

$$(6.41)$$

$$u(1) = 1$$

has the exact solution

$$u(x) = \cos(\log x) + \sin(\log x) . \qquad (6.42)$$

We choose as trial functions the Legendre polynomials scaled to the interval [0,1]

$$\varphi_i(x) = P_{i-1}\,(2x-1) \qquad i = 1,2,\ldots,N$$

and take the additional terms

$$\psi_i(x) = (\log x)^i \qquad i = 1,2,3 .$$

The results with and without the additional terms are shown in Figure 6.5. Without core terms, no convergence is obtained. After including the three terms, the error σ_N^2 is asymptotically well fitted by the form

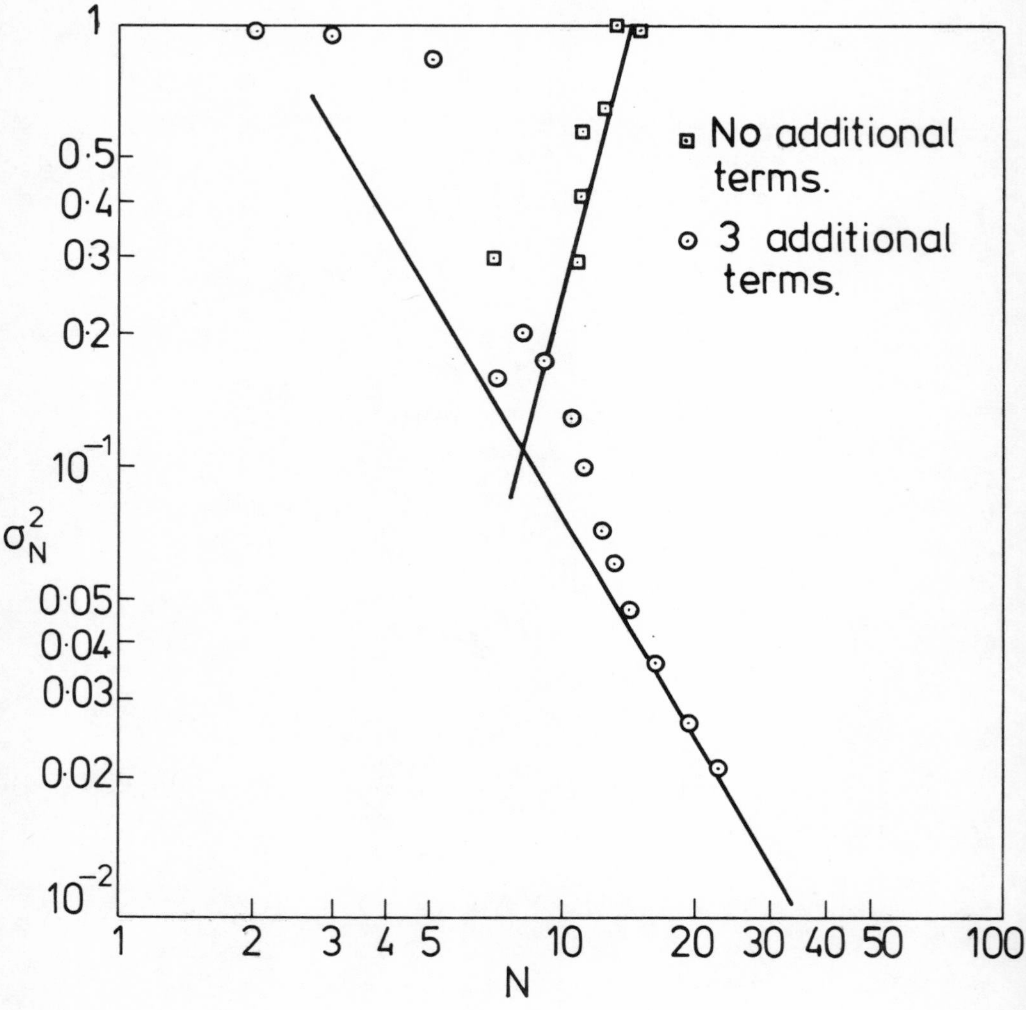
1
0·5
0·4
0·3
0·2
10^{-1}
σ_N^2
0·05
0·04
0·03
0·02
10^{-2}
1
2
3
4
5
10
20
30
40
50
100
N
No additional terms.
3 additional terms.

Fig. 6.5.

$$\sigma_N^2 \sim c\ N^{-1.7}$$

representing a respectable but still rather slow rate of convergence. We can increase the convergence rate by adding more terms, but to achieve exponential convergence the number of additional terms Q must increase at the same rate as the number of regular terms N and adding more than a few terms leads to very illconditioned sets of equations.

Both difficulties would seem to be overcome if we drop the 'regular' trial functions φ_i in (6.33) and use only the singular set ψ_i. But these may not be appropriate over the whole region; and if we have several distinct singularities, each with its own terms, dropping all except one set is bound to fail.

6.6. REGIONAL METHODS FOR PROBLEMS WITH SINGULARITIES

We recall from section 6.3 that a regional method splits the given region Ω up into a small number of subregions, the global element method then uses a global expansion over each subregion. Now if we choose the subregions so that each contains at most one difficult feature, we can choose trial functions within each subregion to be tailored to the form of that feature.

A single expansion is used within a given element and so we avoid the interference between the terms ψ_i and the terms φ_i in (6.33), and the illconditioning which stems from this. We automatically obtain a high order treatment of the singularity and hence rapid convergence may be achieved.

These apparent advantages are in fact realised, as the following examples illustrate. A global element solution of the Motz problem (see section 3.1) was obtained (Hendry and Delves, 1978) in which we isolated the singular points at O in a semicircle of radius 1/4, and divided the remainder of the region as shown in Figure 6.6. Within the semicircle, we used the known expansion of the form (4.11) of section 4.3. In the remaining regions we used a blending function map and a set of product orthogonal polynomials in the mapped elements. Solving the global element equations, we obtained the results shown in Figure 6.7 which give a plot of the error across the element interfaces, and at the boundary of the region. As the degree N of the

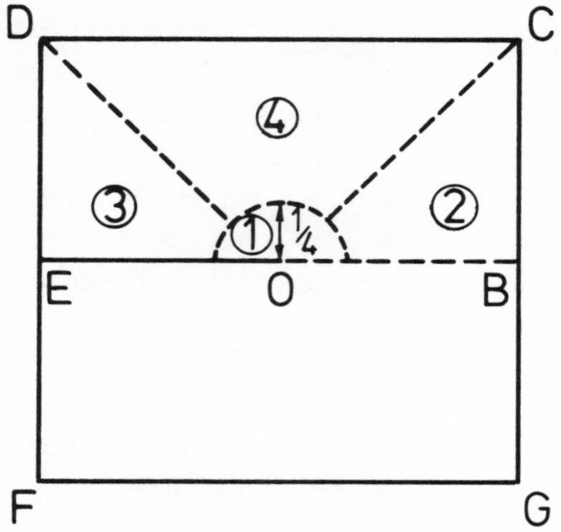
D
C
4
3
2
1
1/4
E
O
B
F
G

Fig. 6.6.

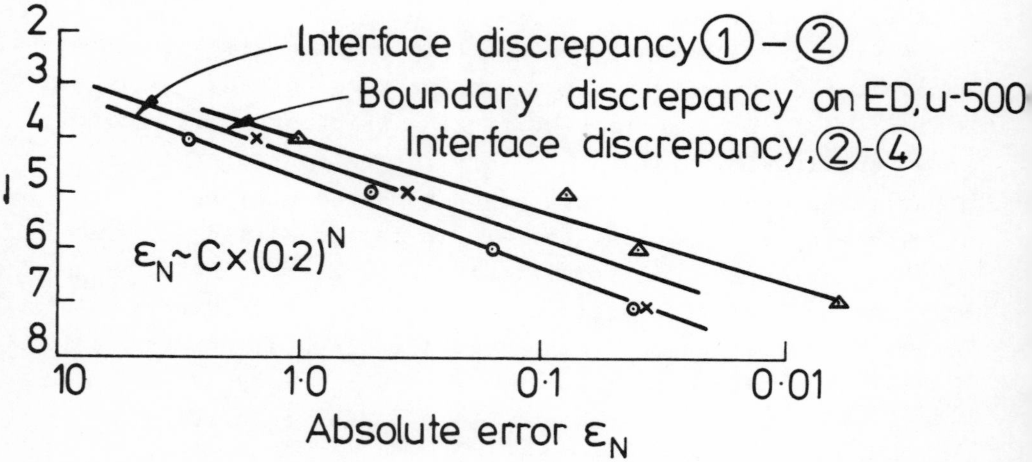
Interface discrepancy 1 – 2
Boundary discrepancy on ED, u-500
Interface discrepancy, 2-4
$\varepsilon_N \sim C \times (0{\cdot}2)^N$
2
3
4
5
6
7
8
10
1·0
0·1
0·01
Absolute error ε_N

Fig. 6.7.

approximating polynomial is increased there are $3N^2 + N$ parameters in the approximation. The results show an exponential decrease; that is, convergence is as rapid as is obtained in section 6.2 for non-singular problems.

Table 6.2 shows some of the calculated parameters α_i in equation (4.11) for the calculation with N = 9. These are

TABLE 6.2

Calculated value of the singular coefficient α_i

	α_1	α_2	α_3	α_4
GEM (N=9)	401.162	87.6558	17.2381	-8.07045
Rosser and Papamichael (1973)	401.1625	87.65592	17.23792	-8.071215
Morley (1973)	401.41	88.61	-20.07	-26.78

in excellent agreement with the exact values of Rosser and Papamichael (1973) given in section 4.3. Also shown are the parameters calculated by Morley (1973) using a set of singular core functions adjoined to a finite element basis. As noted by Morley, the accuracy of these parameters is less than that of his overall calculation; this is symptomatic of the illconditioning encountered when using adjoined special functions.

An alternative problem defined over the same domain as in Figure 6.6, and first suggested by Fix, Gulati and Wakoff (1973) is

$$\nabla^2 u = 0$$

subject to

$$u = 0 \quad \text{on} \quad EO$$

$$u = 1/8 \quad \text{on} \quad DC$$

and

$$\frac{\partial u}{\partial n} = 0 \quad \text{on} \quad OB,\ BC,\ ED.$$

This problem also has an $r^{\frac{1}{2}}$-singularity and we may use the same elements and expansions as for the previous problem. Figure 6.8 shows some of the results obtained with a semi-circle of radius 0.5 for region 1 in Figure 6.6. The stress intensity factor σ_0 is essentially the first coefficient α_1 and σ_{0N} is the global element approximation. In Figure 6.8 results are compared with those obtained by Fix *et al.* (1973) using respectively cubic spline Lagrange (SS_h^{SL}) and Hermite (SS_h^H) bases augmented by four additional terms. These have been plotted at the value of N corresponding most closely to the total number of variational parameters in use. As with the Motz problem, the global element results are superior; also as before in the finite element solution, point accuracy is higher than that to which σ_0 is determined.

6.7. DISCUSSION

The results of this chapter show that high order regional methods such as the global element method, are capable of producing very rapid convergence and high accuracy even for singular problems. Based on these results, we identify classes of problem for which such methods may prove advantageous. For problems with smooth solutions and smooth or simple regions the global element method will converge rapidly. In singular problems, especially where the detailed behaviour near the singular point is of interest, again, rapid convergence is possible, together with a very clean representation of the singular part of the solution.

Not all problems are of these two types; for example, where the region involved is so irregular that a large number of global elements would be indicated, then finite elements appear to be the more rational choice. Even in such cases, however, it may be possible to identify a portion of the region, (as in Figure 6.9 for example) which could be treated using global elements. This could be achieved in principle by including a global element option within the finite element library.

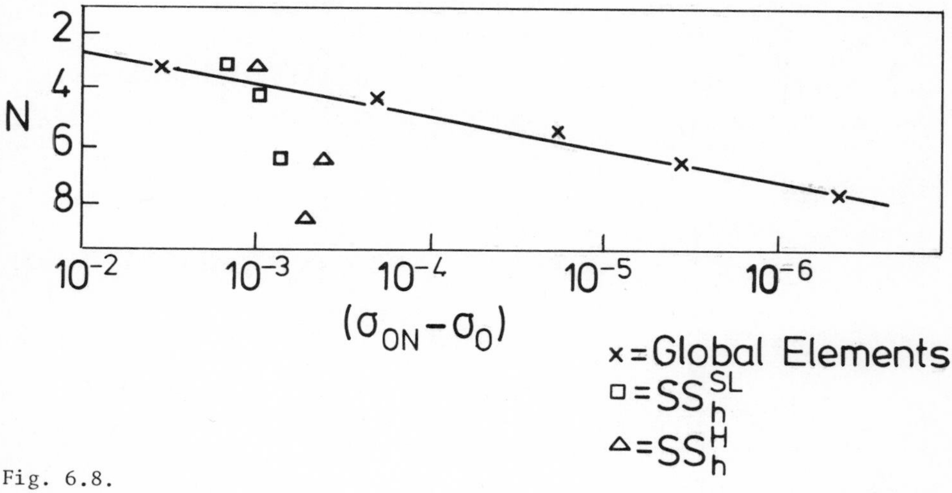

Fig. 6.8.

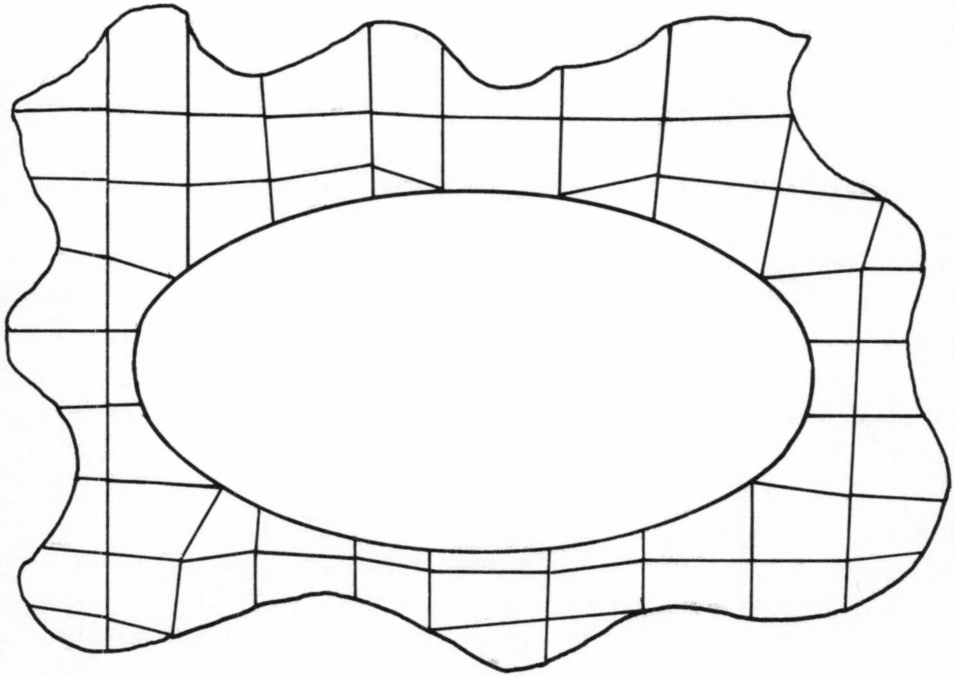

Fig. 6.9.

7

TWO DIMENSIONAL BIHARMONIC PROBLEMS

7.1. INTRODUCTION

Let $\Omega \subset \mathbb{R}^2$ be an open bounded domain with boundary Γ and define $\bar{\Omega} \equiv \Omega \cup \Gamma$. In $\bar{\Omega}$ the *first type* boundary value problem for a partial differential equation involving the biharmonic operator ∇^4 acting on the function u(x,y) is

$$\nabla^4 u(x,y) \equiv \frac{\partial^4 u}{\partial x^4} + 2\frac{\partial^4 u}{\partial x^2 \partial y^2} + \frac{\partial^4 u}{\partial y^4} = f(x,y), \qquad (x,y) \in \Omega, \quad (7.1)$$

$$u(x,y) = g_1(x,y), \qquad (x,y) \in \Gamma \quad (7.2)$$

and

$$\frac{\partial u(x,y)}{\partial n} = g_2(x,y), \qquad (x,y) \in \Gamma, \quad (7.3)$$

where $\partial/\partial n$ is the derivative in the direction of the outward normal to the boundary and f, g_1 and g_2 are given functions defined respectively on Ω and Γ. We note that for fourth order problems, unlike the mixed boundary value problems for Poisson's equation considered in earlier chapters, two boundary conditions are required at each point of Γ.

Important applications of problems involving equations of the type (7.1) occur in continuum mechanics in both linear elasticity and in fluid flow. In particular such problems occur in two dimensional elasticity where in plane stress and plane strain situations the unknown function is the Airy stress function and in thin plate flexure theory where the unknown function is the transverse displacement of an isotropic plate bent by a load. For plane stress and plane strain situations the functions g_1 and g_2 in (7.2) and (7.3) are derived from the applied boundary tractions and, in the absence of body forces, f in (7.1) is zero. For the bent plate g_1 and g_2 are the edge conditions, $g_1 = g_2 = 0$ on Γ for the clamped plate, and f is related to the applied load. In two dimensional creeping flow of a viscous fluid u(x,y) is the stream function. Details of the occurrence of the biharmonic operator in elasticity can be found in standard texts such as Love (1944), Sokolnikoff (1956), Timoshenko and Goodier (1951) and Wang (1953). Equations of this type are also considered in more recent texts on finite element methods such as Gallagher (1975) and Zienkiewicz (1977). An excellent discussion is given in Fox and Southwell (1946). The situation in fluids is presented in von Mises and Friedrichs (1971) and Zienkiewicz (1971).

The present paper is concerned with methods for the numerical solution of problems of the type (7.1) - (7.3). Some theoretical aspects of biharmonic functions are first stated briefly in section 7.2, after which, using the notation of chapter 2, weak formulations of fourth order problems are defined in a Sobolev space setting. Finite element and finite-difference methods are described in sections 7.3 and 7.4 respectively, after which applications to a biharmonic problem from fracture are given in section 7.5.

7.2. SOME THEORETICAL ASPECTS

The homogeneous form of (7.1) is the *biharmonic equation*

$$\nabla^4 u(x,y) = 0 \ . \tag{7.4}$$

Although the purpose of this section is to reformulate problems of the type (7.1) - (7.3) in weak form, we first discuss briefly two interesting properties of biharmonic functions which are useful in numerical techniques.

The relation of the biharmonic operator ∇^4 to the Laplacian operator ∇^2 is clear, since $\nabla^4 \equiv \nabla^2\nabla^2$. Jaswon and Symm (1977) show that the biharmonic function $u(x,y)$ can be expressed in terms of two harmonic functions ϕ and ψ as $u = r^2\phi + \psi$, where $r^2 \equiv x^2 + y^2$. These same authors use this splitting for the numerical solution of some plate problems by integral equation techniques.

It is well known that a function, harmonic in a bounded domain Ω and continuous on $\overline{\Omega}$, attains its maximum and minimum values on Γ. This maximum principle is most useful for bounding errors in numerical solutions to harmonic problems. Unfortunately, as might be expected since a biharmonic problem involves two conditions at each boundary point, majorisation formulae for biharmonic functions involve both function and derivative values on the boundary. The well known formula here is due to Miranda (1948-49), who considers the case where in (7.1) and (7.2) respectively $f = 0$ and $g_1 = 0$. The Miranda result has recently been extended by Rosser (1978), for Ω a circle or a rectangle, to the case where in (7.1) $f = 0$ whilst the boundary conditions in (7.1) and (7.3) are non-homogeneous. In this context it has been suggested by Collatz (1977) that majorisation formulae of this type may be useful in the approximating of u, if a problem similar to (7.1) - (7.3) for which the solution is known explicitly can be set up and a bound on the difference between u and this solution can be found with the majorisation formula.

As for second order equations we write the weak form as

$$a[v,v] = (f,v)_0 \ .$$

The biharmonic operator has two distinct Dirichlet forms, see Agmon (1965),

$$a_1[u,v] \equiv \int_\Omega \nabla^2 u \, \nabla^2 v \; dxdy, \tag{7.4}$$

$$a_2[u,v] \equiv \iint_\Omega \left(\left(\frac{\partial^2 u}{\partial x^2} - \frac{\partial^2 u}{\partial y^2}\right)\left(\frac{\partial^2 v}{\partial x^2} - \frac{\partial^2 v}{\partial y^2}\right) + 4\,\frac{\partial^2 u}{\partial x \partial y}\,\frac{\partial^2 v}{\partial x \partial y}\right) dxdy \tag{7.5}$$

from which it follows that there are infinitely many distinct Dirichlet forms of the type:

$$a[u,v] = t\, a_1[u,v] + (1-t)\, a_2[u,v] \ . \tag{7.6}$$

The bilinear form $a_2[u,v]$ satisfies on $\mathcal{H}_2$ a coercivity condition that there exist constants $\alpha > 0$ and $\beta < \infty$ such that

$$\alpha \| v \|^2_{\mathcal{H}_2(\Omega)} \leqq a_2[v,v] + \beta \| v \|^2_{\mathcal{L}_2(\Omega)} \qquad \text{for all} \quad v \in \mathcal{H}_2 \ . \tag{7.7}$$

The form $a_1[u,v]$ does not satisfy this condition, whilst for $-3 < t < 1$ $a[u,v]$ in (7.6) does, (Agmon, 1965).

The form $a_1[u,v]$ can be obtained from $\nabla^4 u$ by forming $(\nabla^4 u,v)_0$ and using Green's formula so that

$$(\nabla^4 u,v)_0 = (\nabla^2 u,\nabla^2 v)_0 + \int_\Gamma \frac{\partial(\nabla^2 u)}{\partial n} v \, ds - \int_\Gamma \nabla^2 u \frac{\partial v}{\partial n} ds \ . \qquad (7.8)$$

In view of the nonuniqueness of the Dirichlet form for the biharmonic operator, many weak problems can be obtained via $(\nabla^4 u,v)_0$. Since $|\nabla^2 v| = |v|_2$ over $\mathcal{H}_2^0$, $|\nabla^2 v|$ is a norm over $\mathcal{H}_2^0$ and $a_1[u,v]$ is $\mathcal{H}_2^0$ elliptic. By the Lax-Milgram lemma it is well known, see Lions and Stampacchia (1967), that there exists a unique function $u \in \mathcal{H}_2^0$ which minimises the functional

$$I[v] = \int_\Omega |\nabla^2 v|^2 \, dxdy - 2\int fv \, dxdy$$

and which is also the solution of

$$a_1[u,v] = (f,v)_0 \quad \text{for all} \quad v \in \mathcal{H}_2^0 \ . \qquad (7.9)$$

It follows from (7.8) that the solution of (7.9) formally satisfies

$$\begin{aligned} \nabla^4 u(x,y) &= f(x,y) \ , && (x,y) \in \Omega \ , \\ u(x,y) &= \frac{\partial u(x,y)}{\partial n} = 0 \ , && (x,y) \in \Gamma \ . \end{aligned} \qquad (7.10)$$

Problem (7.10) is the Stokes problem for the creeping flow of an incompressible fluid, where $u(x,y)$ is stream function and f is a source term.

For the plate problem, where $u(x,y)$ is the displacement, the strain energy of the bent plate is defined as $a_3[v,v]$ where

$$a_3[v,w] = \int [\nabla^2 v \, \nabla^2 w - (1-\nu)\left(\frac{\partial^2 v}{\partial x^2}\frac{\partial^2 w}{\partial y^2} - 2\frac{\partial^2 v}{\partial x \partial y}\frac{\partial^2 w}{\partial x \partial y} + \frac{\partial^2 v}{\partial y^2}\frac{\partial^2 w}{\partial x^2}\right)] dxdy,$$

$$v,w \in \mathcal{H}_2 \ , \qquad (7.11)$$

where ν, $0 < \nu < \frac{1}{2}$, is the Poisson's ratio; see Landau and Lifschitz (1967) and Washizu (1968). The bilinear form $a_3[v,w]$ is in fact coercive over $\mathcal{H}_2$ for $-7 < \nu < 1$, see Scott (1976).

In the case of the clamped plate $u \in \mathcal{H}_2^0$ solves

$$a_3[u,v] = (f,v)_0 \quad \text{for all } v \in \mathcal{H}_2^0 \quad , \qquad (7.12)$$

where $a_3[u,v]$ is $\mathcal{H}_2^0$-elliptic. The solution of (7.12) formally satisfies (7.10).

Two different weak forms thus lead to the same fourth order problem (7.10). It is shown by Ciarlet (1978, p. 30) that for the case of homogeneous boundary conditions, $u = \partial u/\partial n = 0$ on Γ, the contribution of the second term in the integral of (7.11) is zero. The weak form of the clamped plate problem can thus be written as (7.9).

7.3. FINITE ELEMENT METHODS

The finite element method is now described briefly for the clamped plate problem. Greater detail can be found in Whiteman (1976a). For the weak form (7.12) the solution $u \in \mathcal{H}_2^0$ is approximated by $U \in \mathcal{H}_h$, where $\mathcal{H}_h \subset \mathcal{H}_2^0$ is a finite dimensional space and U is the solution of the problem

$$a_3[U,V] = (f,V)_0 \quad \text{for all } V \in \mathcal{H}_h \; . \qquad (7.13)$$

It follows from (7.12) and (7.13) that

$$a_3[u-U,V] = 0$$

and hence from the continuity and $\mathcal{H}_2^0$-ellipticity of the bilinear form that Cea's Lemma holds in the form

$$|u - U|_2 \le \frac{\alpha}{\gamma}|u - V|_2 \quad \text{for all } V \in \mathcal{H}_h \; . \qquad (7.14)$$

When the finite element method is applied, the region Ω is partitioned into elements E and as before these are usually triangles or rectangles, the size of which is indicated by a mesh parameter h. In any element there are k nodes, with each of which is associated a basis function $\varphi_E^i(x,y)$. An interpolant to point evaluations of u and certain of its derivatives at the nodes is defined and from this the approximating function $U_E(x,y)$ is derived in each element. If the φ_E^i in each element are chosen so that globally $\varphi_i \in \mathcal{H}_2^0$, then the inclusion $\mathcal{H}_h \subset \mathcal{H}_2^0$ is satisfied and (7.14) holds. The approximation U(x,y) is a linear combination of the φ_i in which the (unknown) coefficients are the point evaluations of φ_i and its appropriate derivatives. These coefficients, and hence U are calculated by substituting in (7.13) and solving the resulting global stiffness linear system, see Whiteman (1976a, p.30) or chapter 15 of this volume.

Inequality (7.14) holds when V is the $\mathcal{H}_h$-interpolant to u. The problem of bounding $\| u-U \|$ as in chapter 2 is thus one of approximation theory. The basis functions $\varphi_i(x,y)$ are usually piecewise polynomial functions over $\mathcal{H}$, which in each element are polynomial of order p. Many bounds analagous to those in section 2.10 have been derived for this type of approximation; for example with triangular elements by Ciarlet and Raviart (1972). In this case using (7.14) we find that

$$|u-U|_2 \leq Ch^{\mu}|u|_k \qquad (7.15)$$

with

$$\mu = \min(p-1,\ k-2) \ ,$$

where C is a constant independent of h and μ, and $|\ |_k$ is the k-order seminorm, and $|u|_2$ is a norm in $\mathcal{H}_2^0$. The inequality (7.15) will be relevant only if the solution u of (7.12) satisfies certain smoothness conditions. For example for $p \geq 2$ in order to have $\mu = 1$ we must have $k = 3$; that is the third order derivatives of u must be in $\mathcal{L}_2$. If Ω is a convex polygon it is known that $u \in \mathcal{H}_3 \cap \mathcal{H}_2^0$ so that an O(h) bound can be obtained.

Inequality (7.14), and hence (7.15), demands that $\mathcal{H}_h \subset \mathcal{H}_2^0$ so that *conforming trial functions* are used (see chapter 5). In the present case the necessary and sufficient condition for a piecewise polynomial function to belong to a subspace $\mathcal{H}_h \subset \mathcal{H}_2^0$ is that it is continuously differentiable on $\bar{\Omega}$ and that it satisfies the essential boundary conditions of the differential problem in the classical sense. The disadvantage of a piecewise polynomial trial function possessing $C^1(\bar{\Omega})$ continuity is that in each element it has a large number of degrees of freedom, so that the resulting global stiffness matrix is large. For triangular elements examples of $C^1(\bar{\Omega})$ functions can be produced with the Zlámal 21 degree of freedom quintic local function, with the 18 degrees of freedom reduced version of this and with the 12 degrees of freedom Clough-Tocher function defined on a three part macro-triangle (see Mitchell and Wait, 1977). As in chapter 5, if it is desired that the number of degrees of freedom be reduced, one has to use finite dimensional spaces of non-conforming functions which have the property that $\mathcal{H}_h \not\subset \mathcal{H}_2^0$. In this case although the global continuity of the first derivative of the approximating function is lost, following section 5.2 error inequalities can be derived for functions based, for example, on the Morley 6 degree of freedom triangle or the Adini 12 degree of freedom rectangle. The application of these to the clamped plate problem is discussed by Ciarlet (1974). A non-conforming finite element method based on the Adini rectangle has been applied by Janovsky and Prochazka (1976) to the problem of a clamped plate with ribs.

7.4. FINITE-DIFFERENCE METHODS

When the finite-difference method is applied to any elliptic boundary value problem, the differential equation is approximated at each point of a mesh by a difference equation. After taking boundary conditions into account the difference equations for all the mesh points together form a linear system of the form

$$A\underset{\sim}{U} = \underset{\sim}{b} \qquad (7.16)$$

in which the $\underset{\sim}{U}$ is the vector of approximations to $u(x,y)$ at the mesh points. The matrix A in (7.16) is in general sparse and has a banded structure, so that the system is often solved using an iterative method (see chapter 14). Thus in choosing a finite-difference replacement one frequently takes into account not only its accuracy, but also the form of the matrix that it will produce.

Considering now the biharmonic operator of (7.1), we put a square mesh over Ω and use the notation $x_j = \bar{x} + jh$, $y_k = \bar{y} + kh$,

$u(x_j,y_k) = u_{jk}$, where $(\bar{x},\bar{y})$ is some point of $\mathbb{R}^2$. At a mesh point (x_j,y_k) for which the appropriate neighbours are contained in $\bar{\Omega}$, it can be shown using Taylor's series that

$$\Delta^4 u\Big|_{j,k} \approx \frac{1}{h^4}\Big\{u_{j-2,k} + u_{j,k-2} + u_{j,k+2} + u_{j+2,k} + \\ + 2(u_{j-1,k-1} + u_{j-1,k+1} + u_{j+1,k-1} + u_{j+1,k+1}) - \\ - 8(u_{j-1,k} + u_{j,k-1} + u_{j,k+1} + u_{j+1,k}) + 20\, u_{j,k}\Big\}, \tag{7.17}$$

where, provided that the sixth order derivatives of u are bounded over Ω, the local discretisation error is $O(h^2)$. The thirteen point finite-difference replacement at (x_j,y_k) to the differential equation (7.1) derived from (7.17) is therefore

$$\Delta_{13}\, u_h\Big|_{jk} = h^4 f_{jk} , \tag{7.18}$$

where Δ_{13} is the difference star of Figure 7.1.

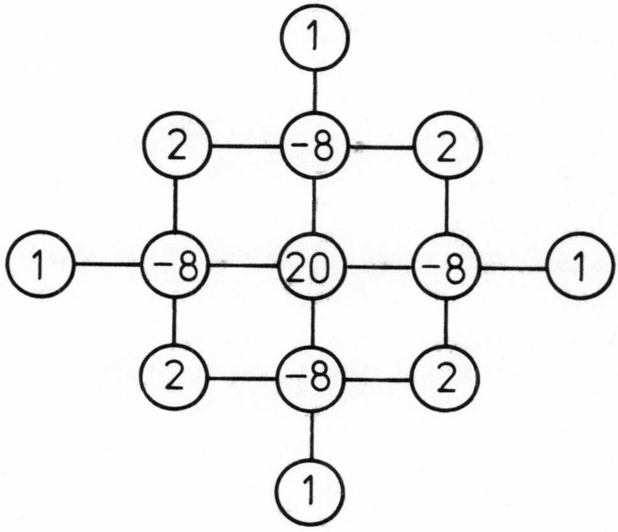

Fig. 7.1.

The spread of the 13 point star means that even for the case where Γ is rectangular, modifications have to be made for many mesh points in Ω to take the boundary conditions into account. The usual approach is to remove exterior "imaginary" mesh points by the use of the normal derivative boundary condition (7.3). Techniques for achieving this for the cases both of rectangular and of curved boundaries are discussed in detail by Fox (1950), who applies the finite-difference method to several fourth order problems.

If a system of the form (7.16) is to be solved iteratively using S.O.R. it is desirable from the point of view of determining the optimal relaxation factor ω_b that the matrix A should possess Young's property A. The technical details of S.O.R. and of other iterative methods can be found in chapter 14. Tee (1963) gives a case with the 13 point star where this property is not achieved, but then derives a 17 point finite-difference replacement for $\nabla^4 u$ for which the resulting matrix does possess property A.

The identity $\nabla^4 \equiv \nabla^2\nabla^2$ was mentioned in section 7.2. This allows the equation (7.1) to be split into a coupled pair of second order equations of the form $\nabla^2 u = v$, $\nabla^2 v = f$. Smith (1977) approximates both of these Poisson equations using finite differences and defines an iterative technique for solving the coupled discrete system. The coupled equation approach is also used in the finite element context by Glowinski (1973) and Bourgat (1976).

7.5. A BIHARMONIC PROBLEM CONTAINING A BOUNDARY SINGULARITY

In an effort to illustrate the numerical methods already discussed, we now give as an example a biharmonic problem from linear elastic fracture mechanics. This is a plane strain situation, in which a two dimensional rectangular elastic solid containing a crack is subject to in-plane loading, and has the added interest that a stress singularity is present at the crack tip. This problem, defined in the cracked rectangle of Figure 7.2, with loading σ normal to BC and EF, is of type (7.1) - (7.3) where $u(x,y)$ is the Airy stress function.

By virtue of symmetry about AOD it is necessary to consider only the part of the problem in the rectangle $\Omega \equiv$ OABCDO with boundary Γ so that

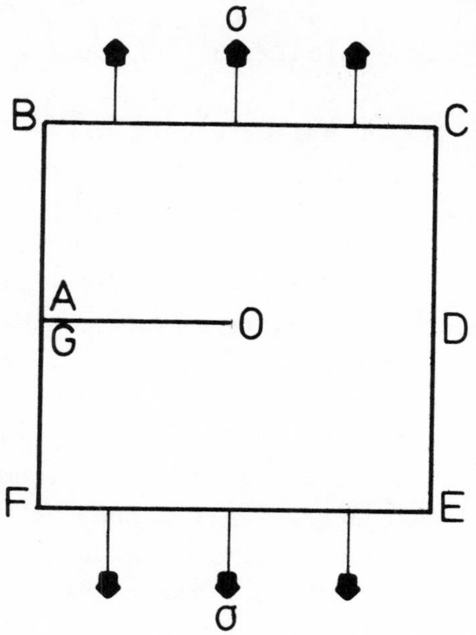

Fig. 7.2.

$$\begin{aligned} &\nabla^4 u(x,y) = 0 , && (x,y) \in \Omega , \\ &u = \frac{\partial u}{\partial n} = 0 , && \text{on OA and AB} , \\ &u = \sigma\left(\frac{x^2}{2} + ax + \frac{a^2}{2}\right), \ \frac{\partial u}{\partial n} = 0, && \text{on BC} , \\ &u = 2\sigma a^2, \ \frac{\partial u}{\partial n} = 2\sigma a , && \text{on CD} , \end{aligned} \tag{7.19}$$

where BC = 2a and OA = a.

The stress function u of (7.19) can be expanded as a series involving unknown coefficients of the form

$$u(r,\theta) = \sum_{n=1}^{\infty} (-1)^{n-1}\alpha_{2n-1}\, r^{n+\frac{1}{2}}\left\{-\cos\left(n-\frac{3}{2}\right)\theta + \frac{2n-3}{2n+1}\cos\left(n+\frac{1}{2}\right)\theta\right\} + \\ + (-1)^n \alpha_{2n}\, r^{n+1}\{-\cos(n-1)\theta + \cos(n+1)\theta\} , \tag{7.20}$$

where (r,θ) are polar coordinates about the crack tip with OD as zero angle and the coefficient α_1 is related to the stress intensity factor, see Whiteman (1978). For the case OA = a = 0.4, AB = 0.7 and $\sigma = 10^4$ a finite-difference solution based on the 13 point star of Figure 7.1 for a mesh of length $h = 0.1$ has been calculated. On the same mesh but with each square subdivided by a diagonal into two triangles a finite element solution has also been calculated using in each element the 21 degrees of freedom Zlǎmal C^1 element. Finally a collocation technique based on the series (7.20) truncated to N terms has been used to approximate u. This series automatically satisfies the boundary conditions on OA and $\overline{OD}$ and is matched at M points on the part ABCD of the boundary to the appropriate boundary values, so that an over-determined system of 2M equations in the N unknowns $\alpha_1, \alpha_2, \ldots, \alpha_N$ is obtained. This is solved using a linear programming technique for the α_i and hence an approximation to $u(x,y)$ is obtained. The discussion of majorisation formulae in section 7.2 was made with boundary matching techniques such as the above in mind. The aim with such techniques is to use these formulae to bound errors throughout Ω with known or calculable errors on Γ. Approximate values for $u(x,y)$ obtained with collocation taking $N = 23$ and $M = 47$ are given in Table 7.1. These values are found to be of uniform accuracy throughout the region OABCDO. Finite-difference and finite element results obtained with the standard methods and uniform meshes are also given in Table 7.1. As can be expected these are found to be inaccurate in the neighbourhood of O due to the presence of the singularity. Although no effort is made to rectify this here, many variations of standard finite-difference and finite element techniques have been proposed for dealing with singularities of this type, particularly in the context of fracture; in particular see Bernal and Whiteman (1970), Gregory, Fishelov, Schiff and Whiteman (1978), Schiff, Whiteman and Fishelov (1974) and the references contained in these papers.

TABLE 7.1

B			50		200		450		800		1250		1800		2450	C
0	1.1		49		201		449		801		1251		1798		2453	3197
0			48		191		432		778		1231		1789		2447	3200
	1.1	48	48	192	191	433	433	779	780	1231	1232	1789	1789	2447	2449	3201
0			40		164		386		721		1178		1756		2437	3200
	-1.1	40	39	165	164	387	387	723	723	1182	1181	1759	1758	2438	2437	3199
0			27		122		312		628		1091		1699		2418	3200
	-0.6	28	27	133	123	317	315	634	632	1102	1097	1709	1703	2422	2419	3200
0			12		71		215		496		963		1615		2389	3200
	1.1	14	13	76	73	223	219	509	504	988	973	1638	1622	2399	2393	3201
0			0		23		105		325		789		1501		2351	3200
	-0.3	1	0.2	27	24	116	110	343	337	842	807	1550	1516	2367	2356	3199
0			-3		-2		19		133		586		1383		2313	3200
	-1.0	-3	-3.4	-2	-1.8	24	20	149	147	685	617	1461	1406	2341	2322	3199
	0		0		0		0									3200
									0	592	508	1417	1355	2328	2307	3201
A									O							D

At any point P the results have the following significance

		Finite elements
Finite differences	P	Linear Programming

Numerical results for problem with crack

8

INTEGRAL EQUATIONS

8.1. INTRODUCTION

Some acquaintance with integral equations is to be presumed, given familiarity with problems involving partial differential equations (see for example, Lonseth, 1977). The theoretical and practical investigation of many such problems can be undertaken by integral equation techniques of a variety of forms. Thus, perhaps one third of the entries in the bibliography of Noble (1971) relate to applications connected with partial differential equations.

However (to the best of our knowledge), there is no work which compares and contrasts *all* the various techniques for obtaining integral formulations of differential equations. Although such a paper would be a valuable contribution to the literature, it is not our intention to undertake such an exposition here. We shall refer to Colton (1976), Courant and Hilbert (1962, p. 362), Henrici (1957), Bergman and Schiffer (1953). The book of Green (1969) is also of interest, providing a coherent introduction to the application of integral equations.

There also exist books devoted to the study of integral equations arising in a particular context (Jaswon and Symm, 1977) and the numerical solution of integral equations (Anselone, 1971; Atkinson, 1976; Baker, 1977; Delves and Walsh, 1974).

8.2. INTEGRAL EQUATIONS

It is our purpose, for the present, to outline some very basic results in the theory of integral equations, illustrating the results by examples with applications in differential equations.

We commence by defining the concept of an integral equation: this is a functional equation in which the unknown function occurs as an "essential" part of an integrand. Since derivative terms may also be present, an "integro-differential equation" is an integral equation. We require that the contribution to the integrand be an "essential" one in order to exclude from the class equations of the form, say, $(d/dx)\int_0^x f(y)\,dy = \gamma(x)$ where the integral is removable to give $f(x) = \gamma(x)$. However, many (but not all) integral equations can be reformulated as differential equations. (If $f(x) - \int_0^x f(y)\,dy = g(x)$ then $f'(x) - f(x) = g'(x)$ with $f(0) = g(0)$.) The original equation is nevertheless to be regarded as a genuine integral equation even though it is *equivalent*, under manipulation, to a differential equation.

Examples of integral equations are (Fredholm type)

$$\int_0^1 k(x,y)f(y)\,dy = g(x), \qquad 0 \le x \le 1, \tag{8.1}$$

$$f(x) - \int_0^1 k(x,y)f(y)\,dy = g(x), \qquad 0 \le x \le 1, \tag{8.2}$$

$$\kappa_r f_r(x) = \int_0^1 k(x,y)f_r(y)\,dy, \qquad 0 \le x \le 1, \tag{8.3}$$

where, for example, $k(x,y) = \log|x-y|$, and (Volterra type)

$$\int_0^x k(x,y)f(y)\,dy = g(x), \qquad x \ge 0, \tag{8.4}$$

$$f(x) - \int_0^x k(x,y)f(y)\,dy = g(x), \qquad x \ge 0, \tag{8.5}$$

where $k(x,y) = (x^2-y^2)^{-\frac{1}{2}}$, for example.

We are largely concerned, here, with integral equation formulations of problems which can be recast as (partial) differential equations. The general theory is however relevant. In many texts on integral equations the theory is presented for equations in which the integral is taken over an interval $[a,b]$ of the real line - taken as $[0,1]$ in equations (8.1) - (8.3) - but the theory extends to many situations where the range of integration is a closed or open curve or a surface or volume. (In the latter cases the distinction between equations of Fredholm type and of Volterra type may blur or disappear.) For general statements of results, refer to Zabreyko *et al.* (1975). Complications can occur when the range of integration is unbounded or when the integrand contains a strong form of singularity. The abstract formulation of an integral equation permits the essential features to be isolated. Thus, a *kernel* $k(x,y)$ defines an integral operator on a suitable normed, linear, function space $\mathscr{X}$ by setting

$$K\phi(x) = \int k(x,y)\phi(y)\,dy \tag{8.6}$$

(with integration over a suitable domain) and the linear integral equations

$$\int k(x,y)f(y)\,dy = g(x) \tag{8.7}$$

(referred to as an equation of the first kind) and

$$\mu f(x) - \int k(x,y)f(y)\,dy = g(x) \tag{8.8}$$

(an equation of the second kind parametrized† by the assigned value $\mu \neq 0$) appear as

$$Kf = g \tag{8.9}$$

and

$$(\mu I-K)f = g \tag{8.10}$$

where $g \in \mathscr{X}$ and the solution is sought in $\mathscr{X}$.

In the case where (8.10) is uniquely solvable for every $g \in \mathscr{X}$, we can define the operator $(\mu I-K)^{-1}$ on $\mathscr{X}$ such that $f = (\mu I-K)^{-1}g$. The sensitivity of (7.10) to perturbation (which governs in part the accuracy obtainable in an approximate method) is then measured by the condition number:

$$\sigma(\mu I-K) = \|\mu I-K\|\,\|(\mu I-K)^{-1}\|. \tag{8.11}$$

† The equation of second kind is often parametrized by $\lambda = \mu^{-1}$ in $f(x) - \lambda\int k(x,y)f(y)\,dy = \gamma(x)$, $\gamma(x) \Rightarrow \lambda g(x)$.

The abstract formulation allows us to develop a theory based on the properties of K and allows us to dispense, for example, with the traditional preoccupation with the condition $\iint|k(x,y)|^2dxdy < \infty$.

The homogeneous version of (8.8) is associated with the eigenvalue problem, consisting of finding non-trivial functions $f_r(x)$ with $Kf_r = \kappa_r f_r$, namely

$$\int k(x,y)f_r(y)\,dy = \kappa_r f_r(x) \quad . \qquad (8.12)$$

Such values κ_r and corresponding functions $f_r(x)$ will be called eigenvalues and eigenfunctions. Associated with (8.12) is the analogous equation

$$\int \overline{k(y,x)}\,\phi_r(y)\,dy = \bar{\kappa}_r\phi_r(x) \quad ; \qquad (8.13)$$

if κ_r is simple then its condition number is defined as $\sigma(\kappa_r) = |(f_r,\phi_r)|/\|f_r\|\,\|\phi_r\|$ with function norms in $\mathscr{X}$ and its dual space. (The space $\mathcal{L}_2$ of square-integrable functions, being self-dual, is commonly taken.) If $\sigma(\kappa_r)$ is small relative to unity the eigenvalue κ_r is *illconditioned.*

8.3. FREDHOLM THEORY

The solvability of (8.10) depends upon the choice of space $\mathscr{X}$; for a general abstract result see Zabreyko *et al.* (1975, p.122). Let us suppose $\mathscr{X} = C[a,b]$ equipped with uniform norm, and $|ab| < \infty$.

Theorem 8.1.

Assume either (i) $k(x,y)$ *is continuous for* $a \le x,\ y \le b$ *or (ii)* $k(x,y) = h(x,y)|x-y|^{-\alpha}$ *where* $0 < \alpha < 1$ *and where* $h(x,y)$ *is continuous for* $a \le x,\ y \le b$. *(In the latter case* $k(x,y)$ *is said to be weakly-singular.) Then the equation*

$$\mu f(x) - \int_b^a k(x,y)f(y)\,dy = g(x) \qquad a \le x \le b \qquad (8.14)$$

has a continuous solution if and only if

$$\int_a^b g(y)\overline{\phi(y)}\,dy = 0$$

for every continuous function $\phi(x)$ *satisfying*

$$\int_a^b \overline{k(y,x)}\phi(y)\,dy = \bar{\mu}\ \phi(x) \qquad a \le x \le b\ .$$

(*The number of linearly independent functions* $\phi(x)$ *satisfying this equation is finite.*)

The latter result is associated with the name of Fredholm. The Fredholm theory is valid whenever K is a compact regular (integral) operator on $\mathscr{X}$, f, g $\in \mathscr{X}$. (The theory in the space $\mathcal{L}_2[a,b]$, for kernels satisfying $\int_a^b\int_a^b |k(x,y)|^2 < \infty$ is discussed by Smithies (1958).)

The number of linearly independent functions $\phi(x)$ satisfying the last equation is equal to the number of linearly independent eigenfunctions satisfying the homogeneous version of (8.14), namely (cf. equation (8.12))

$$\mu f(x) = \int_a^b k(x,y)f(y)\,dy\ . \tag{8.15}$$

Thus Theorem 8.1 states a result which might be conjectured from the theory of linear algebraic equations, that the solution of (8.14) is unique unless μ is an eigenvalue (satisfying (8.15) for some non-trivial f(x)). If μ is an eigenvalue ("of k(x,y)") then solutions exist only for a restricted class of functions g(x), and a particular solution requires the determination of p parameters where p is the geometric multiplicity of μ.

Our concentration on this result deserves some justification: in practice, Fredholm integral equations occurring in the study of partial differential equations frequently have non-unique solutions.

<u>Examples of non-unique solutions</u>. Suppose Ω is the domain exterior to a simple closed curve Γ in the plane. Associated with Γ is the equation of the second kind

$$\pi f(p) - \int_\Gamma g(p,q)'f(q)\,ds = -\phi(p), \quad p \in \Gamma\ , \tag{8.16}$$

with $g(p,q) = \log|p-q|$ and where, in the notation of Jaswon and Symm, $g(p,q)' = \log|p-q|' \equiv \log'|q-p| = (\partial/\partial n)\log|q-p|$ (the derivative being taken along the normal into Ω, p being held fixed), and ds indicates an element of arc length. The solution f(p) permits the determination of a harmonic function

$$u(P) = \int_\Gamma f(q)\log'|q-P|ds, \quad P \in \Omega \cup \Gamma \tag{8.17}$$

satisfying the Dirichlet conditions $u(p) = \phi(p)$ on Γ with $u(P) = O(1/|P|)$ as $P \to \infty$. However, the general theory indicates that (8.16) has a solution only if $\int_\Gamma \phi(q)\lambda(q)ds = 0$ where $\lambda(p)$ is a solution of $\pi\lambda(p) = \int_\Gamma \log|q-p|' \lambda(q)ds$ (the kernel of the latter operator being the adjoint of $g(p,q)'$). The existence of such a function $\lambda(p)$ may be deduced from the fact that $\int_\Gamma \log'|q-p|ds = \pi$. The condition on ϕ is satisfied, given the validity of (8.17), and indeed the solution of (8.16) contains an arbitrary constant *whilst the resulting function* u(P) *is uniquely defined.*

The preceding treatment follows classical lines, but the comparative difficulty of computing $\log'|q-p|$ and other considerations have prompted the use of alternative formulations (Jaswon and Symm, 1977; Symm, 1964, p.18).

In the preceding example we considered the Dirichlet problem for Laplace's equation. The exterior Neumann problem for the Helmholtz equation, in the form

$$\left.\begin{aligned} &(\nabla^2 + k^2)u(P) = 0\,, && P \in \Omega\,, \\ &u'(p) = \phi(p)\,, && p \in \Gamma\,, \\ &|P|^{\frac{1}{2}}\left(\frac{\partial u}{\partial|P|} - iku\right) \to 0 \quad \text{as} && |P| \to \infty \end{aligned}\right\} \tag{8.18}$$

has also received some attention (Ursell, 1973). Writing $g(P,Q) = \frac{1}{2}i\pi\, H_0^{(1)}(k|P-Q|)$, the solution may be sought in the form

$$u(P) = \int_\Gamma g(P,q)f(q)ds, \qquad P \in \Omega\,, \tag{8.19}$$

where

$$\pi f(p) - \int_\Gamma g'(p,q)f(q)ds = -\phi(p), \qquad p \in \Gamma \tag{8.20}$$

but the solution $f(p)$ *is not uniquely determined* if k^2 is an eigenvalue of the interior Dirichlet problem. Some attention has therefore been given to revising the representation of $u(P)$ to give

$$u(P) = \int_\Gamma \gamma(P,q)f_*(q)ds , \qquad P \in \Omega \tag{8.21}$$

where the equation

$$\pi f_*(p) - \int_\Gamma \gamma'(p,q)f_*(q)ds = -\phi(p) \tag{8.22}$$

is uniquely solvable and $\gamma(P,Q)$ differs from $g(P,Q)$ by an analytic wave function. This can be achieved either for all real k or for a range of real k, depending on the complexity of $\gamma(P,Q)$.

Note that even though an equation of the second kind is uniquely solvable it can be illconditioned.

Further Theory

Two standard analytical techniques may be used to analyze (8.12). The first applies where $k(x,y)$ is "degenerate" and will not be pursued here. In the second it is conventional to write

$$f_{r+1}(x) = \lambda \int_a^b k(x,y)f_r(y)dy + \gamma(x) \tag{8.23}$$

where $\lambda = 1/\mu$ and $\gamma(x) = \lambda g(x)$ (see Theorem 8.1). The iteration converges in a space $\mathscr{X}$ if $|\lambda|^n\|K^n\| \to 0$, or $|\lambda|\rho(K) < 1$. (The quantity $\lim_{n\to\infty}\|K^n\|^{1/n}$ is the spectral radius $\rho(K)$.) In particular the iteration converges for a compact Volterra integral operator (Zabreyko *et al.* 1975). If we set $f_0(x) = \gamma(x)$, and the iteration converges, then

$$f(x) = \gamma(x) + \lambda K\gamma(x) + \lambda^2K^2\gamma(x) + \ldots \tag{8.24}$$

where K^r, the r-th power of K, is an integral operator whose kernel is the "r-th iterated kernel" of $k(x,y)$. Observe that the iteration (8.23) converges even when $\mu = \lambda^{-1}$ is an eigenvalue of $k(x,y)$ provided $|\lambda|\rho(K) = 1$, given the choice $f_0(x) = \gamma(x)$, and the existence of a solution (not unique).

The iteration (8.23) is mentioned quite often in integral equation formulations of differential equations but its convergence rate can be slow and in our view other iterative techniques sometimes seem preferable to this one.

When the integral equation (8.14) has a non-unique solution, a related equation which has a unique solution can be obtained from which the general solution of (8.14) can be deduced. For, the kernel $k(x,y)$ can be written as $k_{\mu,1}(x,y) + k_{\mu,2}(x,y)$ where (Zabreyko *et al.* 1975, p.118) the eigenvalues and eigenfunctions of $k_{\mu,1}(x,y)$ are precisely μ and the associated eigenfunctions $f_1(x),\ldots,f_p(x)$ of $k(x,y)$, and $k_{\mu,2}(x,y)$ has the remaining eigenvalues and eigenfunctions, then $f(x) = f_*(x) + \Sigma_1^p \alpha_r f_r(x)$ where

$$\mu f_*(x) - \int_a^b k_{\mu,2}(x,y) f_*(y)\,dy = g(x). \qquad (8.25)$$

We can think of $k_{\mu,2}(x,y)$ as a 'deflated' kernel. The construction of $k_{\mu,1}(x,y)$ involves a knowledge of the eigenfunctions and generalized eigenfunctions of $k(x,y)$ and of $k(y,x)$ corresponding to the eigenvalue μ; it is simplified if $k(x,y) = \overline{k(y,x)}$ or when μ is a simple eigenvalue.

8.4. EQUATIONS OF THE FIRST KIND

The theory of integral equations of the first kind is much less satisfactory than that of equations of the second kind. It is well-known that when K is a compact operator on $\mathscr{X}$ then the equation $Kf = g$ in $\mathscr{X}$ is ill-posed (see Baker, 1977, Miller, 1974).

To investigate (8.7) in the space of square-integrable functions we seek the singular system $\{u_n(x),\ v_n(x);\ \mu_n^{-1}\}_{n\ge 1}$ satisfying

$$\left.\begin{aligned} \int k(x,y)u_n(y) &= \mu_n v_n(x)\ , \\ \int \overline{k(y,x)}\ v_n(y)\,dy &= \mu_n u_n(x) \end{aligned}\right\} \quad \mu_n > 0\ , \qquad (8.26)$$

where $u_n(x)$, $v_n(x)$ are square-integrable. A square-integrable solution of (8.7) then exists if $\Sigma_{n\geq 1}\ \mu_n^{-2}|(g,v_n)|^2 < \infty$ and is given by

$$f(x) = \sum_{n\geq 1} \mu_n^{-1}(g,v_n)u_n(x) \ , \qquad (8.27)$$

almost everywhere. This theory gives little or no general insight into the existence of *smooth* solutions.

However, in the space of square integrable functions it can be seen on examining (8.27) that the degree of ill-posedness depends upon the rate of increase of the values μ_n as $n \to \infty$: a faster decrease corresponds to worse conditioning. (If $\mu_n \sim n^{\alpha}$ as $n \to \infty$ we could suggest α as a measure of illconditioning. For a weakly-singular kernel this measure generally indicates a relatively well-conditioned problem, compared with the continuous case.)

Observe that if there exists $\phi \neq 0$ such that $K\phi = 0$ then any solution of $Kf = g$ is not unique and vice-versa.

Because (8.27) relates only to square-integrable solutions, it may be theoretically more satisfactory (when possible) to reduce the first-kind equation to one of the second kind. A direct approach using complex variable theory may also assist.

A general technique is the following. We consider the equation

$$Mf = g \qquad (8.28)$$

in the space $\mathscr{X}$. We seek a linear operator L whose domain includes the range of M such that LM can be written in the form $\mu I-T$ where T is compact (and linear) on $\mathscr{X}$, for some constant $\mu \neq 0$.

Then, if g is in the range of M, any solution of (8.28) satisfies

$$(\mu I-T)f = Lg \qquad (8.29)$$

to which the Fredholm theory, for equations of the second kind, applies. Further, if $L\phi = 0$ implies $\phi = 0$ the equations (8.28) and (8.29) are precisely equivalent, and L is called a left-regularizor for M. In (8.9) M is the compact operator K.

In the spirit of the above, consider the following. Let Ω be a domain in two dimensions bounded by a simple closed smooth curve Γ and consider the equation $\nabla^2 u(P) = 0$ for $P \in \Omega$ where $u(p)$ assumes compatible values on Γ. We seek the solution $u(P)$ in the form $\int_\Gamma \log|P-q|\; f(q)ds$. The boundary condition gives the equation of the first kind for $f(q)$:

$$\int_\Gamma \log|p-q|\; f(q)ds = u(p) \;, \qquad p \in \Gamma \;. \tag{8.30}$$

We suppose there is no non-trivial solution of (8.30) corresponding to the case $u(p) \equiv 0$, $p \in \Gamma$; that is, suppose Γ is not a "Γ-contour" (Jaswon and Symm, 1977). Then equation (8.30) may be shown to have a unique solution when $u(p) \equiv 1$ (Muskhelishvili, 1953). It also follows that any solution of (8.30) is unique, for general $u(p)$.

Let n denote the inward normal to Γ and denote by L the operator which assigns to a function ϕ the function $\phi' = (\partial/\partial n)\phi$ on Γ; we take for the domain of L the functions defined on Γ which have an extension to $\Omega \cup \Gamma$ satisfying $\nabla^2\phi(P) = 0$. Thus the equation $L\phi = \psi$ is associated with the Neumann problem, solvable if $\int_\Gamma \psi(q)ds = 0$. We have (8.30) and apply L to obtain

$$\pi f(p) + \int_\Gamma \log_i^{\prime}|p-q|f(q)ds = u'(p) \tag{8.31}$$

which is a Fredholm equation of the second kind. The homogeneous equation has a solution $\lambda(p)$ (that is, $-\pi$ is an eigenvalue) since the adjoint kernel has eigenvalue $-\pi$ and eigenfunction 1, and the condition for solvability is the Gauss condition $\int_\Gamma u'(q)ds = 0$. If $f_0(p)$ is a particular solution orthogonal to $\lambda(p)$ the general solution is $f_0(p) + \alpha\lambda(p)$.

Of course the equation $L\phi = 0$ has non-unique solution, above. To ensure uniqueness, restrict the domain of L to functions $\phi(p)$ such that $\phi(p_0) = u(p_0)$ for some fixed $p_0 \in \Gamma$. Then L can be applied legitimately to (8.30). Of the functions $f_0(p) + \alpha\lambda(p)$ only one satisfies (8.30) at $p_0 \in \Gamma$, this is the unique solution of (8.30). (Now, L is a left-regularizor for the particular problem.)

Regularization can be used to reduce certain singular integral equations (in which the integral must be interpreted as a Cauchy principal value) to equations for which the Fredholm theory is applicable.

If $a^2(p) - b^2(p) \neq 0$ for $p \in \Gamma$, a closed contour with continuous curvature in the complex plane, the equation

$$a(p)f(p) + \frac{b(p)}{\pi i} \int_\Gamma \frac{f(q)}{q-p} ds = g(p), \quad p \in \Gamma, \tag{8.32}$$

can be regularized. Regularization is, however, frequently a theoretical (rather than practical) device.

As with systems of algebraic equations, nonlinear theory is more complicated than the linear case. For nonlinear equations

$$f(x) - \int_b^a H(x,y,f(y))dy = g(x) \tag{8.33}$$

results are stated by Zabreyko *et al.* (1975, p.390). The corresponding Volterra equations can be studied in their own right, frequently with greater ease. Analysis of an iterative technique (based for example on Newton's method) is frequently helpful. Thus the iteration

$$f_{n+1}(x) = \int_a^b H(x,y,f_n(y))dy + g(x) \tag{8.34}$$

yields useful insight particularly in the Volterra case.

The Theodorsen integral equation arising in conformal mapping assumes the form

$$f(x) = x - \frac{1}{2\pi} \int_0^\pi \log\left(\rho \frac{f(x+y)}{f(x-y)}\right) \cot \frac{y}{2} dy \tag{8.35}$$

where the contour Γ has polar representation $r = \rho(\theta)$. The iteration of the form (8.34) obtained on identifying (8.35) with (8.33) is proposed in Gram (1962).

8.5. ALTERNATIVE FORMULATIONS

The intimate connection between the Dirichlet problem

$$\nabla^2 u(P) = 0 , \qquad P \in \Omega ,$$
$$u(p) = \phi(p) , \qquad p \in \Gamma , \tag{8.36}$$

and the conformal mapping problem referred to in the above example (that of finding an analytic function $z = f(w)$ mapping $\{w = x + iy \mid (x,y) \in \Omega\}$ onto $|z| \le 1$ with $f(w_0) = 0$ for some interior point w_0) is discussed elsewhere. We refer in particular to Gram (1962, p.168) and note that the two problems are equivalent in the sense that, if one problem has been solved, the solution of the other can be deduced.

From our point of view it is of interest to observe that the problem of conformal mapping may be reduced to the solution of an integral equation in a wide variety of ways. Gram (1962) gives details of (a) the Neumann-Poincaré integral equation (the Potential Method), (b) the Gerschgorin integral equation, (c) the Theodorsen integral equation, (d) the Friberg integro-differential equation, and (e) the Royden integral equation and also Schwarz's and von Neumann's alternating methods. More recently, integral equations of the first kind have been proposed for the solution of conformal mapping (Symm, 1966) and the numerical performance of methods using this formulation have been contrasted with those based on second kind equations (Hayes, Kahaner, and Kellner, 1975).

The mapping $z = f(w)$ discussed above can be obtained from the representation $f(w) = (w-w_0) \exp\{g(w)+h(w)\}$ where g,h are real-valued harmonic conjugates, $\nabla^2 g = 0$, $g = -\log|w-w_0|$, $w \in \Gamma$; thus, for some function $\hat{f}(w)$ to be found,

$$g(w) = \int_\Gamma \log|w-\omega|\hat{f}(\omega)|d\omega| \qquad w \in \Omega \cup \Gamma. \tag{8.37}$$

(It is necessary to scale, if the transfinite diameter of Ω is 1.) Taking $w \in \Gamma$, (8.37) becomes an integral equation for $f(w)$. If we write $w_0 = 0$ for convenience and $\psi(w) = i \log|w/f(w)| = f(w) + i\phi(w)$ where f,ϕ are real-valued we obtain

$$f(w) = \frac{1}{\pi} \int_\Gamma \frac{\partial}{\partial n} \log|w - \omega| f(\omega)|d\omega| + \Phi(w), \qquad w \in \Gamma, \tag{8.38}$$

for an appropriate function $\Phi(w)$. This is an integral equation of the second kind (the Lichtenstein-Gerschgorin equation). Here, $f(w) = \arg \{f(w) - w\}$; $w \in \Gamma$. Let Γ be parametrized by arclength s, so $w = w(s)$, on Γ, and $\arg \{f(w(s))\}$ may be denoted $\theta(s)$. Gram (1962) gives the equivalent equation for $\theta(s)$, and an ALGOL 60 procedure for solving the integral equation approximately. Hayes, Kahaner and Kellner, in the paper referred to above, conclude that the formulation (8.37) is competitive in practice with (8.38).

A general conclusion relating to the solution of partial differential equations may be inferred: not only is it necessary to choose between direct methods (finite difference methods and variational methods, etc.) for the *differential equation* and corresponding methods for related *integral equation* formulations, it is also necessary to select one of a number of integral equation formulations.

Lest it be thought that integral equations arise only from elliptic problems we should assure the reader that this is not the case (see for example Roach, 1970). Frequently, of course, time-dependent parabolic problems yield elliptic problems when the time-variable is discretized, and integral equation techniques can be applied (see for example Gerdes, 1978). Also, Burton (1974) notes that a certain time-dependent problem reduces to the solution of Helmholtz' equation, for which he considers integral equation methods. Finally, we note that moving boundary problems can sometimes be recast as integral equations. (Hansen and Hougaard, 1974, derive a Volterra integral equation of the second kind for a moving boundary problem in diffusion theory.)

A common feature of the integral equation formulations which have been chosen to illustrate our discussion, in preceding Examples, has been the reduction of the dimension of the problem. Thus the solution of Helmholtz' or Laplace's equation† in two dimensions can be reduced to the solution of an integral equation in one dimension (the integral along a contour being reducible to an integral with respect to arc length). Similarly, three dimensional problems can be reduced

†For more general equations see, for example, Gilbert (1970).

to an integral equation for a function of two variables (see, for example, Jaswon and Symm, 1978; Symm, 1974, pp. 312-320). However, the solution u(P) of the original problem is required over the domain Ω, in the original higher dimensional space, and must be *recovered* from the solution of the integral equation. (Exceptionally, this may not be required for the conformal mapping problem. Sometimes the solution of the integral equation has a physical significance and is actually required.) The general technique typified in the Examples depends upon finding a representation for u(P) of the form, say,

$$u(P) = \int_\Gamma K_1(P,q)f(q)ds + \int_\Gamma K_2(P,q)f'(q)ds + \nu(P) \qquad (8.39)$$

$$P \in \Omega$$

where the prime denotes the normal derivative, and deriving an integral equation for f(p). We observed in one of the examples in section 19.3 that there is some arbitrariness in the choice of representation (8.39) but it should also be clear that complete freedom in the choice of $K_1(P,q)$, $K_2(P,q)$ and $\nu(P)$ is out of the question!

We shall refer to a general expression of the form (8.39) in the next section but in the interim we give a preliminary example.

In solving Laplace's equation in two (three) dimensions we may seek a solution in the form of a simpler layer potential $u(P) = \int_\Gamma g(P,q)f(q)ds$ or a double layer potential $u(P) = \int_\Gamma g(P,q)'f_*(q)ds$ where, in two dimensions, $g(P,q) = \log|P-q|$ and, in three dimensions, $g(P,q) = 1/|P-q|$. The appropriateness of such representations depends upon the behaviour of u(P); for example $\int_\Gamma 1/|P-q|'f_*(q)ds$ is $O(|P|^{-2})$ in an exterior region as $|P| \to \infty$ and is unsuitable as a representation of u(P) unless the latter function has this behaviour. In his discussion (1974) of Helmholtz' equation, Burton observes the possible choices (i) $K_1(P,q) = G_k(P,q)$, $K_2(P,q) = 0$, (ii) $K_1(P,q) = 0$, $K_2(P,q) = G_k(P,q)'$ where $G_k(P,q) = \exp(ik|P-q|)/(4\pi|P-q|)$ in three dimensions and

$G_k(P,q) = (i/4)H_0^{(2)}(k|P-q|)$ in two dimensions.[†] Burton also notes (see Kuprazde, 1965) that for general functions $u(P)$ satisfying the equation, the representation with $K_1(P,q) = G_k(P,q)$, $K_2(P,q) = G_k'(P,q)$ is required.

Cole (1977), considering acoustic problems for an open arc, which result in Helmholtz' equation, considers alternative representations (and also, by way of introduction, surveys a number of previous integral equation techniques). The connection between harmonic functions and analytic function theory is indicated by the conformal mapping problem. Gakhov (1966, p. 294) discusses conditions under which a function analytic in a bounded domain Ω can be represented as $(1/2\pi 1)\int_\Gamma \{\mu(\omega)/(w-\omega)\}d\omega$, for real μ. Kershaw (1974) indicates briefly the derivation of integral equations for the Dirichlet problem for Laplace's equation for such a representation.

8.6. COMPUTATIONAL EFFORT AND COMPLEXITY

The solution of integral equations of the type under discussion yields a function $f(p)$ for $p \in \Gamma$ from which $u(P)$, $P \in \Omega$ can be deduced. The received wisdom on whether such formulations will yield a saving of computational effort, compared with methods for the corresponding differential equation, is that they may do so when (i) the domain Ω is unbounded (exterior problems) and (ii) when the problem is a higher order (e.g. biharmonic) problem. The difficulty with integral equation formulations arises from the fact that although the number of dimensions is reduced, the approximate methods generally lead to systems of equations which are not sparse, as in the case of the differential equations. This disadvantage (which arises from the global nature of an integral rather than the local nature of a derivative) requires a compensating advantage. Such an advantage is apparent in the exterior problem since the difficulty of dealing with an infinite region is eliminated; at the same time, numerical questions relating to the imposition of the boundary condition at infinity are superseded by the analytical question of the appropriate representation (8.39) for the solution

[†]Other writers use different scale factors, on occasions.

u(P) in terms of a boundary function f(p).

The relative advantage of integral equations is also apparent in biharmonic problems, to the extent that such problems lead to less sparse equations than arise, say, in Laplace's equation.

We note, in passing, that multiply-connected regions appear to pose little difficulty for integral equation methods, in general.

The complexity of the kernel in an integral equation has a substantial effect on the efficiency of an integral equation method. No general conclusion regarding computational effort is at present possible, but it may be mentioned that Terry (1967), comparing integral and differential equation techniques for an *interior* Neumann problem (close to the hearts† of all of us), appears to conclude that relative efficiency of his methods depends upon the geometry of Γ. In general, Terry appeared to favour differential equations.

For the casual reader, the utility of a particular formulation may well depend upon the nature and extent of library programs. Let us draw to the reader's attention the codes of Symm and Pitfield (1974), of Hayes (1968), and of Blue (1977) for Laplace's equation.

The integral equation formulation adopted by Symm for the treatment of Laplace's equation in two dimensions is somewhat different from the formulations of earlier examples: it applies to fairly general boundary conditions. The method applies to singly- or multiply-connected, bounded or unbounded domains, the restriction imposed being that the boundary Γ be bounded, and consist of closed contours free from cuts or cusps and the boundary data should be piecewise continuous (jumps in normal derivatives are permitted at corners of Γ).

The starting point for the integral equation formulation in Symm's method is Green's 'third identity'

$$\int_\Gamma \log|P-q|u'(q)ds - \int_\Gamma \log|P-q|'u(q)ds = \eta(P)u(P), \quad P\in\Gamma\cup\Omega. \tag{8.40}$$

† The equations studied by Terry arise from problems in electrocardiology.

Derivatives are taken along the normal into Ω, and $\log|P-q|'$ denotes the derivative for varying q, fixed P. If $P \in \Omega$, $\eta(P) = 2\pi$ whilst, if $P = p \in \Gamma$, $\eta(p)$ is the angle in Ω between tangents to Γ at either side of p. We thus obtain the integral equation

$$\int_\Gamma \log|p-q|u'(q)ds - \int_\Gamma \log|p-q|'u(q)ds = \eta(p)u(p), \quad p\in\Gamma \qquad (8.41)$$

from which we determine $u(q)$, $u'(q)$; $q \in \Gamma$. The latter values generate $u(P)$ from (8.40). To be precise, (8.41) may have to be recast as a set of integral equations. Let $u(p) = \phi(p)$ be prescribed for $p \in \Gamma_1$, $u'(p) = \psi(p)$ be prescribed for $p \in \Gamma_2$ where $\Gamma = \Gamma_1 \cup \Gamma_2$. Then

$$\int_{\Gamma_2} \log|p-q|'u(q)ds -$$
$$- \int_{\Gamma_1} \log|p-q|u'(q)ds = \eta(p)\phi(p) + \int_{\Gamma_1} \log|p-q|'\phi(q)ds +$$
$$+ \int_{\Gamma_2} \log|p-q|\psi(q)ds \qquad (8.42)$$

for $p \in \Gamma_1$ with a related equation for $p \in \Gamma_2$. For the interior Neumann problem, a restriction $\int_{\Gamma_0} u(q)ds = \alpha$ is imposed where Γ_0 encloses $\Gamma - \Gamma_0$.

A difficulty which arises when the inner-boundary of an exterior domain, or the outer boundary of an interior domain, is a 'Γ-contour' (as for (8.30)) must be avoided by rescaling. Finally,† the formula (8.40) for an exterior domain yields $u(P)$ such that

$$u(P) = \{\frac{1}{2\pi} \int_\Gamma u'(q)ds\} \log|P| + O(|P|^{-1}) \qquad (8.43)$$

as $|P| \to \infty$ and if a solution bounded at infinity is sought then a term $2\pi c$ is included in the left of the representation (8.40); in the Dirichlet problem c is determined such that $\int_\Gamma u'(q)ds = 0$.

Blue (1977) deals with two- and certain three-dimensional problems. The technique of Blue is based on (8.40) but he

†For further details see Pitfield and Symm (1974).

supplements this equation with the identity

$$\pi u'(p) = \int_\Gamma \log'|p-q|u'(q)ds -$$

$$- \int_\Gamma \{u(q)-u(p)\}\log'|p-q|'ds, \qquad p \in \Gamma \qquad (8.44)$$

for use at certain $p \in \Gamma$ (when $p \in \Gamma$ is not a corner point). We understand the aim of the modification is to improve conditioning, but we have not ourselves investigated this method, to date.

For the treatment of the biharmonic equation $\nabla^4\chi = 0$ on Ω, with χ and χ' prescribed on Γ, we may write $\chi = r^2u + \nu$ where $\chi \equiv \chi(x,y)$, $r = \sqrt{x^2+y^2}$ and u, ν are *harmonic*, if Ω is bounded, simply-connected, and contains the origin. (We take $\chi = r^2u + \nu + c$ where c is a constant if Ω is an exterior domain, containing the origin.) Now u, ν can be expressed as

$$u(P) = \int_\Gamma \log|P-q|f(q)ds\ ,$$

$$\nu(P) = \int_\Gamma \log|P-q|\phi(q)ds$$

for $P \in \Omega \cup \Gamma$. The prescribed values of $\chi(p)$, $p \in \Gamma$, yield

$$\chi(p) = r^2(p)\int_\Gamma \log|P-q|f(q)ds + \int_\Gamma \log|P-q|\phi(q)ds$$

$$(+c)$$

which constitutes an integral equation for f and ϕ. The latter equation is complemented by an equation relating $\chi'(p)$, $p \in \Gamma$, to $f(p)$ and $\phi(p)$. (We have $\chi'(p) = r^2u'(p) + 2\{x(p)x'(p)+y(p)y'(p)\} \times u(p) + \nu'(p)$ and the usual expressions in terms of $f(p)$, $\phi(p)$ for $u(p)$, $u'(p)$ and $\nu'(p)$ can be substituted.) Thus we obtain coupled integral equations.

For the exterior problem, c is determined by the condition $\int_\Gamma f(q)ds = 0$ which ensures uniqueness of χ. For details see Jaswon and Symm 1977.

8.7. CONDITIONING

Whilst we have broached the question of computational effort, we have not yet discussed stability and conditioning. The subject is one which merits further investigation, but some immediate observations are in order.

(i) We tend to sympathize with any distrust of first kind integral equation formulations, given their ill-posed nature. Smoothing methods can be applied to give well-posed problems but these should surely be considered only as a palliative. We have noted, however, that there are degrees of ill-posedness and we have to recognize that first kind formulations, in which the kernel is weakly-singular, have found a considerable following in practice. Thus they have been recommended by Symm, and by Hayes Kahaner and Kellner (1975). Also, as a random example, Allami and Thomas (1975) report success with a two-dimensional Fredholm integral equation of the first kind.

(ii) Study of the relative merits of second kind equations reveals the role of the condition number introduced in (8.11), and those who advocate second kind equations in preference to first kind equations must not overlook the possibility of ill-conditioning for second kind equations. Where the integral equation has a non-unique solution the corresponding discretized equations may be singular, inconsistent, or ill-conditioned, and special action must be taken.† Certain discretization techniques may preserve solvability; then a particular solution should be sought.

(iii) It is quite possible that mild ill-conditioning of some integral equation formulations is acceptable, when the solution of the integral equation is an intermediate stage in the computation of the solution u(P) of a differential equation. Thus if

$$\int_{\Gamma} \log|p-q|f(q)ds = u(p), \qquad p \in \Gamma \tag{8.45a}$$

and

$$u(P) = \int_{\Gamma} \log|P-q|f(q)ds, \qquad p \in \Omega \cup \Gamma \tag{8.45b}$$

†Atkinson (1966).

and *the two integrals are discretized in the same manner for* $p \in \tilde{\Gamma} \subset \Gamma$, when solving the first and evaluating the second, it is quite clear that the computed approximation $\tilde{u}(P)$ will assume the values $u(p)$ for $p \in \tilde{\Gamma}$.

(iv) In the latter context we note that for Dirichlet's problem for Laplace's equation in a bounded domain Ω if $u(P)$ and $\tilde{u}(P)$ are harmonic in Ω then $\sup_{\Omega}|u(P)-\tilde{u}(P)| \leq \sup_{\Gamma}|u(p)-\tilde{u}(p)|$. (A similar maximum principle can be cited for mixed boundary value problems; Jaswon and Symm, 1977, p. 141). There is therefore some theoretical advantage in ensuring that the discretization results in an approximation $\tilde{u}(P)$ which is harmonic.

8.8. A SURVEY OF NUMERICAL METHODS

Basic numerical methods for integral equation can be cast in a general framework applicable to the majority of functional equations. Methods for integral equations, specifically, are discussed fairly exhaustively in the works of Anselone (1971), Atkinson (1976), Baker (1977), Delves and Walsh (1974), and the second part of Jaswon and Symm (1977) along with Ivanov (1976), and will only be surveyed here. The categories of numerical method are the following:

(1) Approximation methods involving simplification of the functional equation, e.g. modification of the range of integration or of the kernel.

(2) Discretization methods.

(3) Projection methods and generalizations thereof.

(4) Variational methods.

Appearing under (1) are those methods of which replacement of the kernel $k(x,y)$ of (8.8) by an approximation $\tilde{k}(x,y) = \sum_1^N X_i(x)Y_i(y)$ is illustrative. When the kernel has the latter form the equation can be solved analytically once certain quadratures are performed. The discretization methods of (2) involve the application of quadrature formulae, or product-integration formulae, to approximate integrals (and difference formulae to approximate derivatives). The projection methods (3), applied to a functional equation $Tf = 0$, where T is an (in general) non-linear operator acting in a space $\mathscr{X}$, involve the choice of a projection operator P_n

mapping $\mathscr{X}$ *onto* a subspace $\mathscr{X}_n$ which is spanned by a basis $\varphi_0(x), \ldots, \varphi_n(x)$. Replacement of the original equation by the equation $P_n T\tilde{f} = 0$, for $\tilde{f} \equiv P_n\tilde{f} \in \mathscr{X}_n$ then follows. Extensions of this technique are possible, but within this framework can be set Galerkin and collocation methods. Variational methods arise, for example, when the solution of a functional equation in $\mathscr{X}$ is characterized as a stationary point of a real-valued functional $J\{\varphi\}$ over $\mathscr{X}$; an approximate solution in a subspace $\mathscr{X}_n$ is determined as a staionary point of $J\{\varphi\}$ over $\mathscr{X}_n$. In various instances methods from one category may also be regarded as members of another category. The methods indicated are discussed in more specific and more practical terms than those indicated here by Baker (1977) and Delves and Walsh (1974), for example, and a number of salient features are admirably summarized by Noble (1977), whilst Jaswon and Symm (1977) should also be referred to.

The treatment of Baker (1977) is largely devoted to integral equations over an interval of the real line, though some attention is also paid to integral equations on a contour. Parametrization (e.g. by arc length) reduces the latter problem to the former but generalizations have to be made when this parametization is unsuitable, or for multidimensional equations. In such circumstances it is frequently convenient to replace the contour Γ by a more convenient contour Γ^* in the integral equation. Typically, Γ^* may be polygonal, or piecewise circular, in one dimension, or piecewise planar etc. in two dimensions. (This illustrates stage (1) of the above summary.) The basic techniques are then applied to the approximating equation.

Some initial mathematical analysis of the integral equation is useful to establish whether (i) the kernel and (ii) the solution are weakly-singular, continuous, or differentiable of high-order. The numerical techniques may have to be determined by the answers to these questions. Thus, a method of treating singularities in the solution is indicated by Jaswon and Symm (1977, p. 205) and illustrated by examples. Whilst singularities in the kernel $k(x,y)$ of (8.7) and (8.8) can frequently be removed by analytical techniques, direct methods† can be applied to equations with weakly- or intrinsically-singular kernels.

There is little doubt that discretization and collocation methods are the simplest in practice. The discretization techniques applied to equations (8.7) and (8.8) yield respectively the formulae

$$\sum \nu_{ij}\tilde{f}(y_j) = g(z_i) \tag{8.47}$$

(with due warning that the latter equations may be ill-conditioned and unreliable) and

$$\mu\tilde{f}(y_i) - \sum \upsilon_{ij}\tilde{f}(y_j) = g(y_i). \tag{8.48}$$

Such equations arise on applying integration formulae

$$\int k(z_i,y)\varphi(y)dy \simeq \sum \nu_{ij}\varphi(y_j), \tag{8.49}$$

or, more particularly

$$\int k(y_i,y)\varphi(y)dy \simeq \sum \upsilon_{ij}\varphi(y_j). \tag{8.50}$$

(The choice of points z_i in (8.47) need not generally be the same as the quadrature points y_i but ν_{ij} and υ_{ij} are otherwise analogous.) For continuous kernels and solutions, (8.49), (8.50) may be derived by applying a quadrature rule

$$\int \Psi(y)dy \simeq \sum \Omega_j\Psi(y_j) \tag{8.51}$$

(for periodic integrands the trapezium rule has good features) so that $\nu_{ij} = \Omega_j K(z_i,y_j)$, $\upsilon_{ij} = \Omega_j K(y_i,y_j)$. Possible failure of this technique for (8.7) is discussed by Baker (1977). For a weakly-singular kernel, and a continuous solution, formulae (8.49) and (8.50) can be derived by approximating $\varphi(x)$ by an interpolating function $\tilde{\varphi}(x)$, which is (say) piecewise constant, piecewise linear, piecewise polynomial, or spline, and determining coefficients such that

†For first kind equations, direct treatment to a singular equation is probably *preferable*.

$$\int k(z_i,y)\tilde{\varphi}(y)dy = \sum \nu_{ij}\varphi(y_j) \tag{8.52}$$

with the analogue for (8.50). The simplest illustration is provided by partitioning the range of integration Γ into sub-intervals Γ_i with $y_i \in \Gamma_i$ and setting

$$\tilde{\varphi}(x) = \varphi(y_i), \qquad x \in \Gamma_i \tag{8.53}$$

so that

$$\nu_{ij} = \int_{\Gamma_j} k(z_i,y)dy \tag{8.54}$$

in (8.49). The resulting equations (8.47) (and analogously, (8.48)) can then be seen to be collocation equations. In practical cases, the geometry of the contour Γ may make it difficult to evaluate (8.54), and the values ν_{ij} may be approximated by using a quadrature rule (8.51) - see Jaswon and Symm (1977, p. 153). (The resulting approximations would be the same if Γ were replaced by a nearby curve Γ^*.) In the presence of a weak singularity, such a procedure is unsatisfactory if $z_i \in \Gamma_j$ and Jaswon and Symm propose as one remedy the replacement of the offending curve Γ_j by a piecewise linear curve (or circular arc) Γ_j^* which permits the evaluation of $\int_{\Gamma_j^*} k(z_i,y)dy$.

In the case where an auxiliary equation such as $\int f(y)dy = \alpha$ (see section 8.6) has to be applied to ensure a unique solution, the integral is discretized in the same manner. If this equation is added to the original equations the resulting system will be over-determined and may be inconsistent. It would appear that in general the discretization may be devised (see Jaswon and Symm p.178) so that inconsistency does not arise and so that one of the 'collocation' equations can be dropped to yield a satisfactory problem. Atkinson (1966) has proposed a method for solving second kind equations with non-unique solutions but a general technique for computing a (pseudo-) solution of a linear system in the presence of ill-conditioning, singularity, or over- or under-determined situations, involves recourse to minimal least-squares solutions.

Lynn and Timlake (1967) consider the Neumann problem in three dimensions, and the resulting two-dimensional integral equations. The discretization procedure is similar to that indicated above, and the coefficient matrix is singular. The discussion of the report indicates how a deflation may be employed to yield a non-singular coefficient matrix, and how deflations may also be employed to yield fast convergence in iterative schemes. For further discussion, see for example, Ikebe, Lynn and Timlake (1968).

The discussion immediately preceding has given emphasis to problems arising from elliptic problems. For a further summary of such numerical techniques see Colton (1976, pp. 136-143). We also refer to the work of Cryer (1970a), Elvius (1971) and of Atkinson and Gilbert in Gilbert (1970), whilst methods for general integral equations are discussed in the literature cited earlier in this chapter.

8.9. FURTHER COMMENTS

We consider that this is not the appropriate place to supply an exhaustive list of further reading on integral equations and their role in the study and treatment of differential equations. Our list of references includes a number of papers which perhaps fortuitously have caught our attention; the books cited, in particular, will be helpful to those who wish to enagage on a literature search.

PART II

NUMERICAL METHODS FOR PARABOLIC EQUATIONS

This part contains a description of discretization techniques for parabolic partial differential equations, their stability and their convergence properties. This is followed by a discussion of techniques for solving some of the difficult problems which arise frequently in practice; for example, the diffusion-convection equation, boundary layer problems and "stiff" kinetics problems are all discussed.

The material of this part is intended to be self-contained except where specifically stated. The reader may feel some need to consult part I for a discussion of finite elements and singularities and part III for a discussion for finite differences and solution techniques for the discretized equations. A reader who finds the material of this section too advanced is advised to consult Smith (1977) or Ames (1969) first.

Contents

Chapter	Title	Contributor
9	Parabolic Equations and their Discretization.	J.M. Watt
10	Semi-discretization of Parabolic Equations and the Numerical Solution of Semi-discretized Equations.	I.Gladwell
11	The Diffusion-Convection Equation.	I.M.Smith
12	Boundary Layer and Similarity Solutions for Parabolic Equations.	A.Prothero
13	Strongly Nonlinear Parabolic Equations-Examples.	A.Prothero

9

PARABOLIC EQUATIONS AND THEIR DISCRETIZATION

9.1. INTRODUCTION

Both parabolic and hyperbolic partial differential equations usually arise from mathematical descriptions of time-dependent or evolutionary processes; the solution of such equations can be thought of as *evolving* as time increases from a given initial state under the influence of certain boundary conditions.

If the state of the system, or solution of the equation, is a function $u(\underset{\sim}{x},t)$ where $\underset{\sim}{x}$ is a space variable of one or more dimensions and t is time, then parabolic equations can be distinguished from hyperbolic equations by the fact that in the parabolic case the speed of propagation of disturbances is infinite, whereas for the hyperbolic case it is finite.

Parabolic equations are used as a model of diffusion in an isotropic medium, heat conduction, fluid flow through porous media, boundary layer flow over a flat plate, and of many others. We discuss hyperbolic equations in chapters 21-23.

A particular parabolic equation, the equation describing the conduction of heat in a medium with conductivity a, is

$$u_t - a\,\nabla^2 u = f(\underset{\sim}{x},t)$$

where $u(\underset{\sim}{x},t)$ is the temperature and $f(\underset{\sim}{x},t)$ is the rate of supply of heat per unit volume, for instance due to radioactivity or microwave heating.

We concentrate on the more general equation

$$u_t - Lu = 0 \qquad (9.1)$$

where $u \equiv u(\underset{\sim}{x},t) \equiv u(x_1,x_2,\ldots,x_n,t)$ and $u_t = \partial u/\partial t$, $u_i = \partial u/\partial x_i$, $u_{ij} = \partial^2 u/\partial x_i \partial x_j$, etc. . (In one space dimension we use u_x, u_{xx} for u_1 and u_{11}.) We also assume that

$$Lu = \sum_{i,j=1}^{n} a_{ij}\, u_{ij} + \sum_{i=1}^{n} b_i\, u_i + cu\ , \tag{9.2}$$

is an elliptic operator; in other words that the matrix $A = \{a_{ij}\}$ satisfies, for some $\gamma > 0$,

$$\underset{\sim}{v}^T A \underset{\sim}{v} \geq \gamma\, \underset{\sim}{v}^T \underset{\sim}{v} \tag{9.3}$$

for all vectors $\underset{\sim}{v}$.

We can distinguish three cases for the coefficients a_{ij}, b_i and c:

(i) They are constant.

(ii) They are variable, depending on $\underset{\sim}{x}$ and t.

(iii) They are variable, depending on $\underset{\sim}{x}$, t, u and u_i.

In cases (i) and (ii) the equations are linear, whereas in case (iii) they are said to be quasilinear.

Fully nonlinear equations can also occur, the term cu often being replaced by a function of u. The treatment of nonlinear equations is usually made to depend on a linearized or variational equation. Elliptic operators containing derivatives of order up to 2m can also be studied with little extra complication.

The differential equation (9.1) does not yet determine u. In order to do this, initial and boundary conditions are needed. If (9.1) holds for $\underset{\sim}{x} \in \underset{\sim}{R}^n$ and for $0 \leq t \leq T$, the solution is determined if $u(\underset{\sim}{x},0)$, the initial value of u, is given for all $\underset{\sim}{x} \in \underset{\sim}{R}^n$, and we have a *pure initial value problem* or *Cauchy problem*. If however only a limited open region $\Omega \subset \underset{\sim}{R}^n$ is being considered *boundary conditions* must also be specified. That is some condition on u and its derivatives must be given for all $\underset{\sim}{x} \in \Gamma$, the boundary of Ω, and for all t, $0 < t \leq T$. In this situation we have a *mixed* or *initial-boundary* problem.

In the second order case which we are considering, the main types of boundary condition are:

(i) Dirichlet condition

$$u(\underset{\sim}{x},t) = g(\underset{\sim}{x},t), \forall \underset{\sim}{x} \in \Gamma, \quad 0 < t \leq T.$$

In the heat conduction case this means that the temperature u on the surface of a body Ω is specified at all time.

(ii) Neumann condition

$$u_n(\underset{\sim}{x},t) = g_1(\underset{\sim}{x},t), \quad \forall \underset{\sim}{x} \in \Gamma, \quad 0 < t \leq T .$$

(The notation u_n indicates differentiation along the outward normal to Γ .) Physically this means that the rate of flow of heat from the surface is specified.

(iii) A combination of the above two types of boundary condition may occur in the form

$$\alpha(\underset{\sim}{x},t)\; u(\underset{\sim}{x},t) + \beta(\underset{\sim}{x},t)\; u_n(\underset{\sim}{x},t) = g_2(\underset{\sim}{x},t) ,$$

$$|\alpha| + |\beta| \neq 0 , \quad \forall \underset{\sim}{x} \in \Gamma, \quad 0 < t \leq T .$$

9.2. PROPERTIES OF THE SOLUTION

Consider first the simple Cauchy problem

$$u_t = au_{xx}, \quad x \in \underset{\sim}{R}, \ t > 0 ,$$

$$u(x,0) = f(x), \quad x \in \underset{\sim}{R} .$$

The solution can be written as

$$u(x,t) = \frac{1}{2(\pi t)^{\frac{1}{2}}} \int_{-\infty}^{\infty} \exp\left\{\frac{-a(x-\lambda)^2}{4t}\right\} f(\lambda)d\lambda$$

From this solution we can easily deduce the following results:

(i) The value of $u(x,t_0)$ depends on that of u at time $t = 0$, that is on $f(x)$ *for all values* of x (of course the magnitude of this dependence is very small if

$x - \lambda$ is large);

(ii) $u(x,t)$ is a continuous function of x and t, and in fact all its derivatives with respect to both x and t are continuous;

(iii) $u(x,t) \leq \max_{-\infty<x<\infty} f(x)$.

These results are generalised in Friedman (1964). A typical existence theorem, can be stated loosely as:

If L is parabolic and has coefficients which are continuous functions of $\underset{\sim}{x}$ and t and if $c(\underset{\sim}{x},t) \leq 0$ everywhere, the boundary Γ of Ω has no reentrant corners, and the boundary and initial conditions are jointly continuous, then there exists a unique solution to the Dirichlet problem (9.1) which has continuous second derivatives with respect to x.

Friedman also shows that if the coefficients a_{ij}, b_i and c of L are m times differentiable with respect to x, then in the interior of the region the solution is (m+2) times differentiable with respect to x, and the x-derivatives of order up to m are continuously differentiable with respect to t. These results can be extended up to the boundary provided that the boundary, and the initial and boundary conditions are sufficiently smooth.

The third property can be generalised by considering the parabolic equations (9.1) where L as defined by (9.2) has continuous coefficients and where $c(\underset{\sim}{x},t) \leq 0$ for $\underset{\sim}{x} \in \Omega$, $0 < t < T$. Then, if $u_t - Lu \leq 0$ for $\underset{\sim}{x} \in \Omega$, $0 < t \leq T$, and if u has a positive maximum which is attained at an interior point $(\underset{\sim}{\overline{x}},\overline{t})$ of $\Omega \times [0,T]$, then U is constant on $\Omega \times [0,T]$.

9.3. SIMPLE DISCRETIZATIONS

Consider the equation

$$u_t - au_{xx} = f(x,t), \quad a > 0 \; , \tag{9.4}$$

with initial condition

$$u(x,0) = g(x) \; , \tag{9.5}$$

and a mesh in the (x,t)-plane with interval Δx and Δt.

We use the notation $u_j^n = u(x_j,t_n)$, $f_j^n = f(x_j,t_n)$ etc., and use U_j^n to denote the calculated approximation to u_j^n. An obvious very simple difference replacement for (9.4) and (9.5) is

$$M(\Delta x,\Delta t)\ U_j^n \equiv (U_j^{n+1} - U_j^n)/\Delta t - a\ \delta_x^2\ U_j^n/(\Delta x)^2 = f_j^n\ ,$$
$$U_j^0 = g_j\ . \tag{9.6}$$

We define the *local truncation error* of the discretization $M(\Delta x,\Delta t)$ of (9.4) to be

$$\tau_j^n = M(\Delta x,\Delta t)u(x_j,t_n) - \{u_t(x_j,t_n) - au_{xx}(x_j,t_n)\},$$

and say that the discretization is *consistent* if τ_j^n tends to zero with Δx and Δt. The *global truncation error* is

$$U_j^n - u(x_j,t_n)$$

where U_j^n is given by (9.6). The discretization is convergent if the global truncation error tends to zero with Δx and Δt with $j\Delta x$ and $n\Delta t$ fixed. These definitions are extended to other discretizations in the obvious way.

We cannot immediately deduce that a consistent discretization is convergent. For this to be so the discretization must be *stable*, that is small perturbations in the right hand side of equations (9.6) must lead to small perturbations in the solution U_j^n. Stability is the subject of the next two sections.

The approximation U_j^{n+1} given by (9.6) depends on just three values of U at time t_n, and working backwards we find that U_j^n depends only on $g_{j-n-1}, g_{j-n}, \ldots, g_{j+n+1}$. Now we know that the exact solution $u(x_j,t_{n+1})$ depends on $g(x)$ for *all* values of x. Hence the approximate solution cannot tend in the limit to the exact solution unless, as Δx, $\Delta t \to 0$, $x_{j-n-1} \to -\infty$, $x_{j+n+1} \to \infty$ when $(x_j,t_{n+1}) = (\bar{x},\bar{t})$ is a fixed point, that is $j\Delta x$ and $(n+1)\Delta t$ are fixed. Now $n+1 = \bar{t}/\Delta t$ so we need $(n+1)\Delta x = \bar{t}\ \Delta x/\Delta t \to \infty$, as $\Delta x,\Delta t \to 0$ so

$$\Delta t = o(\Delta x) \ . \tag{9.7}$$

We see later that for stability the stronger condition

$$\Delta t \le (\Delta x)^2/2a \tag{9.8}$$

must be imposed.

By expanding in Taylor series, the local truncation error of (9.6) is easily seen to be of order $O(\Delta t) + O(\Delta x)^2$ and so if $\Delta t = k(\Delta x)^2/a$ with $k \le 1/2$, so that (9.8) is satisfied, the local discretization error is $O(\Delta x)^2$.

A condition of the form of (9.7) (or (9.8)) is clearly needed to ensure the convergence of any explicit method, and it has the consequence that explicit methods often need very small time steps to ensure stability and convergence and for this reason such simple methods are often too inefficient to use.

To overcome this problem we use a suitable *implicit* method, the simplest of which is the *Crank-Nicolson method*

$$(U_j^{n+1} - U_j^n)/\Delta t \ -\tfrac{1}{2}\, a(\delta_x^2\, U_j^{n+1} + \delta_x^2\, U_j^n)/(\Delta x)^2 = f_j^{n+\frac{1}{2}} . \tag{9.9}$$

This equation can be rearranged as

$$-pU_{j+1}^{n+1} + (1+2p)U_j^{n+1} - pU_{j-1}^{n+1} = pU_{j+1}^n + (1-2p)U_j^n + $$
$$+ pU_{j-1}^n + \Delta t f_j^{n+\frac{1}{2}} \tag{9.10}$$

where $p = a\Delta t/2(\Delta x)^2$. The local discretization error in this case is $O(\Delta t)^2 + O(\Delta x)^2$. From this latter form it is clear that the components of $\underset{\sim}{U}^{n+1}$ can be obtained by solving a tridiagonal set of linear equations. Thus each component of $\underset{\sim}{U}^{n+1}$ depends on *all* the values of $\underset{\sim}{U}^n$, in agreement with the situation in the analytic case, and the stability analysis below shows that the method is stable for all values of Δt and Δx. Thus larger steps Δt can be taken without destroying stability and this method is often more efficient than the explicit one.

As a further example we give a three time-step nine-point scheme which has local truncation error $O(\Delta t)^2 + O(\Delta x)^4$ and which is stable for all values of Δt and Δx:

$$\begin{aligned}\Delta t M(\Delta x,\Delta t)U_j^n \equiv \left(\tfrac{1}{8} - p\right) U_{j-1}^{n+1} + \left(\tfrac{5}{4} + 2p\right) U_j^{n+1} + \left(\tfrac{1}{8} - p\right) U_{j+1}^{n+1} - \\ - \tfrac{1}{6} U_{j-1}^n - \tfrac{5}{3} U_j^n - \tfrac{1}{6} U_{j+1}^n + \\ + \tfrac{1}{24} U_{j-1}^{n-1} + \tfrac{5}{12} U_j^{n-1} + \tfrac{1}{24} U_j^{n-1} = \Delta t f_j^{n+1}.\end{aligned}$$

Richtmyer and Morton (1967, p.890) give a table of fourteen different discretizations for (9.4). In the next chapter we shall see how to derive discretizations via semi-discretization techniques.

9.4. STABILITY

It is convenient (in the spirit of the previous section) to make the definitions

$$\underset{\sim}{U}^n = \{U_j^n : x_j \in \Omega\}, \quad \underset{\sim}{U} = \{\underset{\sim}{U}^n : 0 \le n \le T/\Delta t\} .$$

We assume that Δt is given as a function of Δx which for convenience we now denote by h; usually $\Delta t = r(\Delta x)^2 \equiv rh^2$, where r is a constant.

There are many definitions of stability used in particular circumstances, but the basic idea is to consider the complete set of equations originating from the discretization of the differential equation and its boundary conditions. These equations are solved for the approximate solution $\underset{\sim}{U}$, and we now write them as

$$M_h(\underset{\sim}{U}_h) = \underset{\sim}{\beta}_h , \tag{9.11}$$

the suffix h on all quantities indicating their dependence on h, it being important to remember that as h decreases the number of equations increases, as well as their coefficients changing. If we also have a perturbed set of equations with solution $\underset{\sim}{V}_h$

$$M_h(\underset{\sim}{V}_h) = \underset{\sim}{\gamma}_h ,$$

we say the discretization is *stable* in some norm $\|.\|$ if

$$\|\underset{\sim}{U}_h - \underset{\sim}{V}_h\| \le S\|\underset{\sim}{\beta}_h - \underset{\sim}{\gamma}_h\|$$

for constant S and for all values of h, $0 < h \le h_0$. Thus, loosely, a discretization is stable if small perturbations in the equations cause small perturbations in their solution, uniformly for all small h.

In an evolutionary system such as we are considering it is more convenient to consider each time-step of the calculation separately. We also simplify our discussion to the usual case of linear equations, and also to one-step methods. (A multistep method can always be written in one-step form (Richtmyer and Morton, 1967, ch. 7).)

We therefore suppose that (9.11) can be written in the form

$$\underset{\sim}{U}^{n+1} = C(h)\underset{\sim}{U}^n + \tau(h)\underset{\sim}{\psi}_h(\underset{\sim}{\beta}_h) \qquad (9.12)$$

where

$$\tau(h) = \Delta t$$

and $\underset{\sim}{U}^0$ is given. (This does not preclude the approximate equations being implicit.) For (9.11), written in the form (9.12), to be stable we need $\|C(h)^n\| \le K$ for $n\tau(h) \le T$ and, a requirement usually trivially satisfied,

$$\|\underset{\sim}{\psi}_h(\underset{\sim}{\beta}) - \underset{\sim}{\psi}_h(\underset{\sim}{\gamma})\| \le K_1\|\underset{\sim}{\beta} - \underset{\sim}{\gamma}\|$$

see Godunov and Riabenkii (1964, ch. V).

As an example consider again $u_t - au_{xx} = f(x,t)$ with $u(0,t) = 0$, $u(1,t) = 0$, for $t \in (0,T]$, and $u(x,0)$ given for $x \in \Omega = [0,1]$. Take $\Delta x = h$, $\Delta t = rh^2$, $J = 1/h$ and use the explicit difference replacement (9.6) for $0 < j < J$. The equations can be written as

$$\begin{bmatrix} U_1^{n+1} \\ U_2^{n+1} \\ \vdots \\ U_{J-1}^{n+1} \end{bmatrix} = \begin{bmatrix} 1-2ra & ra & & & \\ ra & 1-2ra & ra & & \\ & \ddots & \ddots & \ddots & \\ & & \ddots & \ddots & ra \\ & & & ra & 1-2ra \end{bmatrix} \begin{bmatrix} U_1^n \\ U_2^n \\ \vdots \\ U_{J-1}^n \end{bmatrix} + rh^2 \begin{bmatrix} f_1^n \\ f_2^n \\ \vdots \\ f_{J-1}^n \end{bmatrix}. \quad (9.13)$$

Now if $0 < 2ra \le 1$, and $C(h)$ is the matrix in the above equation, we have

$$\| C(h) \|_\infty = 1$$

and hence

$$\| C(h)^n \|_\infty \le \| C(h) \|_\infty^n = 1$$

and the stability criterion is satisfied. (It is in fact satisfied if and only if $0 \le 2ra \le 1$.) So for stability

$$\Delta t \le (\Delta x)^2/2a ,$$

which is a very serious restriction on the allowed values of Δt.

On the other hand using the Crank-Nicolson method (9.9) we get, with $p = ra/2$,

$$C(h) = \begin{bmatrix} 1+2p & -p & & & \\ -p & 1+2p & -p & & \\ & \ddots & \ddots & \ddots & \\ & & \ddots & \ddots & -p \\ & & & -p & 1+2p \end{bmatrix}^{-1} \begin{bmatrix} 1-2p & p & & & \\ p & 1-2p & p & & \\ & \ddots & \ddots & \ddots & \\ & & \ddots & \ddots & p \\ & & & p & 1-2p \end{bmatrix}$$

and it can be shown that $\| C(h) \| \le 1$ for all positive values of p, that is the method is stable, without restriction on the time step Δt; but the analysis in this case is more complicated, see Mitchell (1969).

However in this simple case where the differential equation has constant coefficients the analysis is simplified by a Fourier transformation of the variation in the x-direction, provided we have a Cauchy problem with periodic initial data; see Isaacson and Keller (1966, p.523) for a full treatment. We seek a trial solution of the form

$$U_j^n = z^n e^{ij\xi} \qquad |\xi| \leq \pi .$$

For the explicit case substituting U_j^n into (9.6) we obtain

$$z^{n+1}e^{ij\xi} = z^n e^{ij\xi}\{1-2ra + ra(e^{i\xi} + e^{-i\xi})\}$$

and cancelling

$$z = 1 + 2ra(\cos\xi - 1) \ .$$

We have $|z| \leq 1$ and the trial solution non-increasing for all ξ if and only if

$$ra \leq \tfrac{1}{2} \ ,$$

which is the restriction we quoted previously.

For the Crank-Nicolson method (9.9) we similarly get

$$z = \frac{1+ra(\cos\xi - 1)}{1-ra(\cos\xi - 1)} \ ,$$

and $|z| \leq 1$ for all r and ξ, so the method is unconditionally stable.

If instead of a one-step method we have a k-step method, z takes the value of the eigenvalues of a matrix of order k called *amplification matrix* and denoted by $G(\Delta t,\xi)$. Isaacson and Keller (1966, p.523) show that the relationship between $C(\Delta t)$ which is a matrix of order kJ and $G(\Delta t,\xi)$ which is a matrix of order k is, when spectral norms are used,

$$\| C(\Delta t)^n \| = \max_{\xi:|\xi|\leq\pi} \| G(\Delta t,\xi)^n \| \ . \qquad (9.14)$$

Here G is of much smaller order than C (G is often of order 1) but we do have to maximise over ξ. For example, $G = [z]$ in the examples above.

For any matrix A, if $R(A)$ denotes the spectral radius, we have

$$R(A)^n = R(A^n) \le \|A^n\| \le \|A\|^n.$$

For stability we require, for some constant K,

$$\|C(\Delta t)^n\| \le K, \quad \text{for } n\Delta t \le T \, .$$

So a necessary condition for stability is that

$$R(C(\Delta t))^n \le K \, ,$$

or equivalently,

$$R(C(\Delta t)) \le 1 + K_1 \, \Delta t \, . \tag{9.15}$$

A sufficient condition is that

$$\|C(\Delta t)\| \le 1 + K_2 \, \Delta t \, . \tag{9.16}$$

If we apply (9.15) instead to G, and make use of (9.14), we get the Von Neumann necessary condition for stability

$$R(G(\Delta t,\xi)) \le 1 + K_1 \, \Delta t \, , \; |\xi| \le \pi \tag{9.17}$$

Remember that in the one-step case G is of order one, so that

$$R(G) = |G| \, .$$

When a system of q equations is being considered the orders of G and C are increased by a factor q but the analysis is otherwise unchanged.

Many conditions for stability have been devised, (some of them necessary and some sufficient, but rarely both) with the object of providing an easy algebraic way of checking stability using the amplification matrix G. It is not sufficient merely to check that the eigenvalues z_i of G satisfy $|z_i(\Delta t)|^n \leq K$ because the eigenvectors may tend to linear dependence as $\Delta t \to 0$. A discussion of this work is given in Richtmyer and Morton (1967, ch. 4).

9.5. STABILITY IN MORE COMPLICATED SITUATIONS

For more complicated systems involving nonlinear equations, variable coefficients and boundary effects the *energy method* can be used (see Richtmyer and Morton (1967, ch. 6)). But especially for parabolic equations there has been some important recent work by Widlund, Osher and Varah. The aim of this work has been to make the investigation of stability in the variable coefficient, boundary value case depend on the Fourier analysis of the constant coefficient case.

We make the following definitions. We say that a *difference scheme is parabolic* if there is a constant $\delta > 0$ independent of $\underset{\sim}{x}$, t and ξ such that the eigenvalues z_s of the amplification matrix satisfy

$$|z_s| \leq 1 - \delta\, |\xi|^2$$

for all x, t and all $|\xi| \leq \pi$.

For k-step methods with $k > 1$ we also need the *root condition* that when $\xi = 0$ the k eigenvalues z_s of the amplification matrix have modulus less than or equal to unity, and the eigenvalues of modulus unity are simple. The *strong root condition* is satisfied if when $\xi = 0$ the only eigenvalue of modulus unity is $z_1 = 1$. A parabolic difference scheme which satisfies the strong root condition is *strongly parabolic*.

The main theorem of Widlund (1966) tells us that, for the Cauchy (or pure initial value) problem for linear equations with variable coefficients, the difference scheme is stable if it is parabolic and satisfies the root condition, and if each problem obtained from the original by *freezing* the coefficients at $(\bar{\underset{\sim}{x}},\bar{t})$, that is by replacing $a_{ij}(\underset{\sim}{x},t)$, $b_i(\underset{\sim}{x},t)$ and $c(\underset{\sim}{x},t)$ by $a_{ij}(\bar{\underset{\sim}{x}},\bar{t})$, $b_i(\bar{\underset{\sim}{x}},\bar{t})$ and $c(\bar{\underset{\sim}{x}},\bar{t})$), is stable in the Fourier sense for all $\bar{\underset{\sim}{x}}$ and for all $0 < \bar{t} \leq T$. Widlund was not able to prove the necessity of all these conditions, but shows that they are likely to be necessary.

Osher (1972) and Varah (1971) have extended this analysis to deal with boundary conditions. Varah considers a system of equations of the form

$$\underset{\sim}{u}_t - A(x)\, \underset{\sim}{u}_{xx} = \underset{\sim}{0}$$

in the quarter plane $x \geq 0$, $t \geq 0$ where $\underset{\sim}{u}$ has q components; whereas Osher considers the single equation

$$u_t - a(x,t)u_{xx} - b(x,t)u_x - c(x,t)u = f(x,t) \qquad (9.18)$$

in a region $0 \leq x \leq 1$, $t \geq 0$. Both consider inhomogeneous initial and boundary data.

The particular conditions assumed and the technical details of the proof are complicated and we concentrate on the general ideas of Osher's paper. He shows that if the coefficients in (9.18) are continuous and the boundary conditions are either Dirichlet or mixed then a consistent difference approximation is stable provided it is strongly parabolic, is Fourier stable for frozen coefficients as for Widlund's theory, and some extra conditions involving the approximations to the boundary conditions are satisfied.

We take the boundary conditions at $x = 0$, and investigate for each time t_n, infinite sets of equations consisting of these conditions, together with the set of difference approximations to the differential equation, with coefficients 'frozen' at $\bar{x} = 0$, $t = t_n$ and with range $[0,\infty)$ for x.

The stability requirement is that:

(i) Of all the solutions of this set of equations at time t_n there is just one which is bounded; that is one for which

$$\sum_{i=0}^{\infty} (U_i^n)^2 < \infty.$$

(ii) (This condition is a check on the normal modes.) Suppose that $U_i^n = z^n V_i$ is a solution of the infinite set of equations for $n = 0,1,\ldots$; $i = 0,1,\ldots$, then for $|z| > 1$ the unique bounded solution is zero.

An analogous condition is required to hold at $x = 1$. We can get an intuitive idea of the meaning of these conditions by an argument using the ideas (but not details) of Godunov and Ryabenkii's stability analysis (Richtmyer and Morton, ch. 6). If h is very small and there are a very large number of steps, the coefficients of the differential equation, and its discretization, will be almost constant over a large number of steps near any point $(\bar{x},\bar{t})$. So, near $(\bar{x},\bar{t})$, stability will be governed by the stability of the corresponding frozen coefficient case, and the normal modes will be the same near $(\bar{x},\bar{t})$ as for frozen coefficients. Now consider as a candidate for a 'normal mode' of the variable coefficient equation a mode which is identical to the frozen mode near $(\bar{x},\bar{t})$, but which tapers gradually off to zero away from this point. If the tapering off is slow, this mode will approximately satisfy the equation satisfied by the normal modes of the variable coefficient equation. The condition that the equation with frozen coefficients satisfies the Fourier stability test then ensures that the frozen problems are stable in the interior of the interval, and the special boundary conditions ensure this is also true at the boundaries.

The general nature of the results of Varah and Osher, suggest that they should be applicable to more general systems of equations. These extensions do not seem to have been made. Other recent papers on this subject are Widlund (1970), Gustafsson and Sundström (1978), Gottlieb and Gustafsson (1976), and Verrit (1973).

10

SEMI-DISCRETIZATION OF PARABOLIC EQUATIONS AND THE NUMERICAL SOLUTION OF SEMI-DISCRETIZED EQUATIONS

In the first part of this chapter we outline semi-distretization techniques for reducing parabolic partial differential equations to systems of ordinary differential equations. These techniques include the well-known method of lines. Later, we discuss the solution of the systems of ordinary differential equations arising from such semi-discretizations.

10.1. SEMI-DISCRETIZATION TECHNIQUES

In chapter 9 we saw how finite-difference techniques may be used to discretise a parabolic equation in both space and time. Here, we will consider space semi-discretizations independently of time discretizations. We consider a linear parabolic equation

$$u_t = Lu + f \tag{10.1}$$

where L is an elliptic operator and u and f are functions defined on a domain $\Omega \times [0,\infty)$ with Dirichlet boundary conditions given on the boundary, Γ, of Ω for all $t > 0$, and with initial conditions given on Ω at time $t = 0$. We will not discuss more general boundary conditions here, but refer the interested reader to Mitchell and Wait (1977). We will briefly discuss nonlinear problems later.

The aim of a semi-discretization is to reduce equation (10.1), the boundary conditions and the initial conditions to a system of ordinary differential equations

$$A\underset{\sim}{a}' = B\underset{\sim}{a} + \underset{\sim}{c} \tag{10.2a}$$

with initial conditions

$$\underset{\sim}{a}(0) = \underset{\sim}{a}_0 \ . \tag{10.2b}$$

Here A and B are square matrices and $\underset{\sim}{c}$ is a vector and all may depend on the independent variable, time, if L and/or f in equation (10.1) or the boundary conditions do. If (10.1) is a constant coefficient equation and the boundary conditions are time-independent these terms in (10.2) are time-independent in general. The vector of unknowns $\underset{\sim}{a} \equiv \underset{\sim}{a}(t)$ may represent an approximation to the solution u at chosen discretization points (in the method of lines) or may describe coefficients in a series expansion for u.

For example, let us consider in detail the finite-difference approximations of the last chapter for the simple problem

$$u_t = u_{xx} \tag{10.3}$$

with boundary conditions u = 0 on both x = 0 and x = 1 for all time, and initial condition u = g(x) at t = 0. If we use the simple central-difference approximation of the last chapter we obtain equation (10.2a) with A an identity matrix, $\underset{\sim}{c} = \underset{\sim}{0}$,

$$B = \frac{1}{h^2}\begin{pmatrix} -2 & 1 & & & \\ 1 & -2 & 1 & & \\ & \ddots & \ddots & \ddots & \\ & & 1 & -2 & 1 \\ & & & 1 & -2 \end{pmatrix} \tag{10.4}$$

where h = 1/(n+1) is the discretization step in the space direction,

$$\underset{\sim}{a} \equiv \underset{\sim}{U} = [U_1, U_2, \ldots, U_n]^T$$

where $U_i \approx u(ih,t)$, i = 1,2,...,n and with (8.2b) taking the form

$$\underset{\sim}{a}(0) = [g(h), g(2h), \ldots, g(nh)]^T .$$

This reduction to the form (10.2) is essentially the first step in the method of lines, and the method would usually be completed by solving equations (10.2) using one of the methods of the next section.

Similarly if we consider the problem

$$u_t = \nabla^2 u \tag{10.5}$$

defined on a region $\Omega \times [0,\infty)$ where Ω is a region in two dimensions, we require boundary conditions for u on Γ for all $t > 0$, and initial conditions for u on Ω at $t = 0$. If we use the usual five-point central-difference approximation to $\nabla^2 u$, we obtain

$$\underset{\sim}{U}' = B\underset{\sim}{U} \tag{10.6}$$

with appropriate initial conditions for $\underset{\sim}{U}$ on Ω at $t = 0$, where the components of $\underset{\sim}{U}$ are approximations to the solution at the mesh points and B is the usual matrix corresponding to the five-point operator, see chapter 14.

The simplicity of equation (10.6) compared with the more general equation (10.2) may be of no practical value as its only advantage, namely that explicit methods can be used to integrate (10.6), often cannot be exploited due to stability and accuracy considerations, see the next section and chapter 9.

In what follows, we discuss equations in one space dimension mainly, as they will suffice to illustrate the ideas involved. However it must be borne in mind that, in practice, two and three space dimensions are of more interest and that the systems (10.2) arising in these cases can often only be solved implicitly and then by using special techniques, see chapters 14, 15 and 17.

As a final example of the use of a finite-difference method, we consider Numerov's method applied to problem (10.3), see Hall and Watt (1976) for its application to ordinary differential equations. We obtain equation (10.2a) with

$$A = \frac{1}{12}\begin{bmatrix} 10 & 1 & & & \\ 1 & 10 & 1 & & \\ & \ddots & \ddots & \ddots & \\ & & 1 & 10 & 1 \\ & & & 1 & 10 \end{bmatrix} \tag{10.7}$$

B as in (10.4), and $\underset{\sim}{c} = \underset{\sim}{0}$. This system of ordinary differential equations has a very similar form to that arising from the finite element approach below.

Let us consider an approximation to the solution u of (10.1) of the form

$$U(P,t) = \sum_{i=1}^{m} a_i(t)\varphi_i(P), \quad P \in \Omega, \ t \geq 0 , \tag{10.8}$$

where the functions $\varphi_i(P)$ are twice continuously differentiable functions which satisfy homogeneous Dirichlet boundary conditions on Γ; for example when $\Omega = [0,1]$, we have $\varphi_i(0) = \varphi_i(1) = 0$, $i = 1,2,\ldots,m$. We may then consider the Galerkin method for the case when L is a second order elliptic operator

$$(U_t,\varphi_j) = (LU,\varphi_j) + (f,\varphi_j), \quad j = 1,2,\ldots,m, \tag{10.9}$$

where $(.,.)$ is an inner product; for example on $[0,1]$ we might choose

$$(\varphi,\psi) = \int_0^1 \varphi(x)\psi(x)\,dx \ . \tag{10.10}$$

Equation (10.9) can be written in the form (10.2) with

$$A_{ij} = (\varphi_i,\varphi_j), \ B_{ij} = (L\phi_i,\phi_j) \ , \tag{10.11}$$

$$c_i = (f,\varphi_i), \ a_i = a_i(t) \ .$$

If L is self-adjoint, that is if

$$(L\varphi,\psi) = (\varphi,L\psi) \tag{10.12}$$

for twice continuously differentiable functions ϕ and ψ satisfying the boundary conditions, then, by integration by parts, we may write

$$(L\varphi,\psi) = a[\varphi,\psi] \qquad (10.13)$$

where $a[\varphi,\psi]$ is a bilinear form in φ and ψ and their first derivatives. For example, if we consider one space dimensional problems with $L\varphi = \varphi_{xx}$, integration by parts gives

$$(L\varphi,\psi) = -(\varphi_x,\psi_x) \equiv a[\varphi,\psi] \; . \qquad (10.14)$$

Using this definition we can rewrite (10.9) as

$$(U_t,\varphi_j) = a[U,\varphi_j] + (f,\varphi_j), \quad j = 1,2,\ldots,m, \qquad (10.15)$$

which can be written in the form (10.2) with the definitions in (10.11) except that

$$B_{ij} = a[\varphi_i,\varphi_j] \; . \qquad (10.16)$$

However it is clear that the values B_{ij} defined in (10.11) and (10.16) are mathematically identical. The equation (10.15) is known as the Ritz formulation of (10.9) and corresponds to a "weak" formulation of the differential equation, see Fairweather (1978), Mitchell and Wait (1976) and Strang and Fix (1973). To use the forms (10.9) or (10.15) we may choose the functions φ_i as global or regional elements or, more conventionally, as finite elements or splines. If we choose finite elements or splines then we expect to obtain a large system (10.2) and so we must choose basis functions φ_i so that the matrices A and B are sparse and structured.

In one space dimension we consider equally-spaced knots $x_i = ih$, $i = 0,1,\ldots,n+1$, $h = 1/(n+1)$. We can choose the functions φ_i as a basis for the spline space S_k^r where

$$S_k^r = \{f \mid f \text{ is a polynomial of degree } r \text{ on each interval } [x_i,x_{i+1}];$$
$$f \text{ is } k \text{ times continuously differentiable on } [0,1]\}. \qquad (10.17)$$

For example, the chapeau functions

$$\varphi_i(x) = \left\{\begin{array}{ll} (x-x_{i-1})/h, & x \in [x_{i-1},x_i) \\ (x_{i+1}-x)/h, & x \in [x_i,x_{i+1}] \\ 0, & x \notin [x_{i-1},x_{i+1}] \end{array}\right\} \qquad (10.18)$$

belong to S_0^1, whereas the usual smooth cubic splines belong to S_2^3 and the cubic Hermite splines belong to S_1^3, see Schultz (1973). We can choose bases for the spaces S_k^r which give sparse systems (10.2) by choosing B-splines. The B-spline basis functions are designed so they have minimal support, that is they are chosen to have non-zero values on the smallest possible number of subintervals $[x_i,x_{i+1}]$; see Cox (1978), and the references therein, for a full discussion of B-spline bases and their evaluation. For example, the functions $\varphi_i(x)$, $i = 1,2,\ldots,m$ in (10.18) are a B-spline basis for the subspace of S_0^1 consisting of functions satisfying the boundary conditions $\varphi_i(0) = \varphi_i(1) = 0$, $i = 1,2,\ldots,m$ (in this case $m=n$). They have a support of two intervals, and hence there are two functions φ_i taking non-zero values on each subinterval. The B-spline basis for the space of smooth cubic splines S_2^3 satisfying the boundary conditions consists of $m(= n+4)$ functions each with a support of four intervals and hence these are four basis functions taking non-zero values on each subinterval. Similarly the cubic Hermite space S_1^3 has a B-spline basis consisting of m $(=2n+2)$ functions each with support of two intervals and there are four non-zero basis functions on each subinterval. In general, the greater the order of continuity then the greater the number of intervals of support; and the lower the order of continuity the greater the number of basis functions.

If we use any of the subspaces S_k^r in the Galerkin method (10.9) with $L\varphi = \varphi_{xx}$ then we see that we must require the basis functions to have integrable second derivatives. On the other hand, to use (10.15) the basis functions need only have integrable first derivatives (the forms (10.9) and (10.15) are not equivalent in this latter case as integration by parts is not valid). We prefer to use (10.15) where possible as less restrictions are placed on the basis functions.

Now let us consider again the simple problem (10.3). If we use the roof basis functions (10.18) then we obtain equation (10.2a) with B as in (10.4),

$$A = \frac{1}{6}\begin{pmatrix} 4 & 1 & & & \\ 1 & 4 & 1 & & \\ & \ddots & \ddots & \ddots & \\ & & 1 & 4 & 1 \\ & & & 1 & 4 \end{pmatrix} \tag{10.19}$$

and $\underset{\sim}{c} = \underset{\sim}{0}$. The matrices A and B are of order n. The most natural way to choose the initial conditions (10.2b) is to set

$$a_i(0) = (g,\varphi_i),\ i = 1,2,\ldots,n. \tag{10.20}$$

Using this Galerkin method, we can expect the same order of convergence as with simple central differences, that is $O(h^2)$.

To obtain a higher order of convergence, we must choose a space of spline functions of higher polynomial degree such as S_1^3 or S_2^3. If we use S_2^3, the matrices A and B are seven-diagonal and of order n+4; whereas if we use S_1^3, the matrices are also seven-diagonal and of order 2n+2, but with some zero elements in the band which can be exploited in the solution technique. Both these choices give $O(h^4)$ convergence. Note that the space S_0^3, which would lead to a sparser matrix, cannot be used because the bilinear functional $a[\phi,\psi]$ in (10.16) does not exist for this choice. It is possible to achieve *superconvergence* using spline Galerkin techniques. For example, using the space S_2^3, it can be shown that (super)convergence of $O(h^6)$ is attained at the knots $\{x_i\}$, see Thomée (1972).

In fact, there are many spline spaces which lead to the same order of convergence. The choice of which splines to use will normally be made by comparing the amount of work involved. Here studying the one-dimensional case can be very misleading if the conclusions are to be applied in higher dimensions. In general, when working in higher dimensions, most practitioners prefer to use finite element spaces with very compact support and low order continuity. Spaces with non-compact support lead to denser systems of algebraic equations and difficulties in treating boundary conditions and irregular boundaries.

Instead of the Galerkin method, we can attempt to solve the differential equation by collocating at a number of points $P_i \in \Omega$; for the general problem (10.1), we choose U in (10.8) to satisfy

$$U_t(P_i,t) = LU(P_i,t) + f(P_i), \quad i = 1,2,\ldots,m, \qquad (10.21)$$

which gives a system of equations (10.2a) with

$$A_{ij} = \varphi_j(P_i), \; B_{ij} = L\varphi_j(P_i), \quad c_i = f(P_i). \qquad (10.22)$$

The initial conditions can be obtained by interpolation to the initial data at the points P_i.

In the one dimensional case, we can use the B-spline bases for S_k^r in (10.21) as long as S_k^r is sufficiently differentiable so that the elements in (10.22) exist. In general, the matrices thus obtained have smaller bandwidths than for the same bases in the Galerkin method. Hence, for example, using the B-spline basis for S_2^3 and collocating at the knots we obtain a five-diagonal matrix of the same size as the matrix in the Galerkin method, if we treat the endpoints carefully. In the case of S_1^3 collocating at two points between each knot, we obtain a five-diagonal matrix with some exploitable sparsity within the band; this choice of collocation points gives the correct number of equations to determine the solution. It is necessary to choose interior points since we must avoid using the knots due to the lack of differentiability there. Note that the roof functions S_0^1 cannot be used in the collocation method since they are not sufficiently differentiable.

In most cases, the same order of convergence can be obtained with collocation methods as with Galerkin methods using the same space of functions. Indeed, working with the cubic Hermite splines S_1^3, a superconvergence result at the knots can be obtained for collocation at the Gaussian points (zeros of the Legendre polynomial of degree 2), see Douglas (1972), and similar results have been obtained in higher dimensions. The fact that the same order of convergence can be obtained with collocation as with Galerkin, and at less computational expense in setting up the problem (10.2) and in solving it, has lead to many workers preferring the collocation method. However, it must be borne in mind that the collocation equations are always unsymmetric whilst the Galerkin equations derived from using (10.15) will be symmetric. Also, the numerical stability of the collocation process depends critically on the choice of collocation points; for example, collocation at Gaussian points for S_1^3 is stable.

Finally we discuss the method of moments (often known as the generalised Galerkin method, see chapter 3) where we replace equation (10.9) by

$$(U_t,\psi_j) = (LU,\psi_j) + (f,\psi_j), \quad j = 1,2,\ldots,m \qquad (10.23)$$

where the test functions ψ_j may be independent of the basis (trial) functions φ_j in (10.8). This method includes the Galerkin method (by setting $\psi_j = \varphi_j$) and the collocation method (by setting ψ_j to a Dirac δ-function). The test functions ψ_j may be chosen in many ways but they are usually chosen to reduce the amount of computation involved in the Galerkin method whilst preserving the order of convergence associated with the trial functions φ_j. One form of the method of moments is the H^1-Galerkin method where we choose, for example, $\phi_j \in S_k^r$ and $\psi_j \in S_{k-2}^{r-2}$. Hence, we could use smooth cubic splines for the trial functions φ_j and roof functions for the test functions ψ_j. Similarly we could use cubic Hermite splines for the trial functions with discontinuous linear functions for the test functions. These choices preserve the rate of convergence associated with the Galerkin method but lead to equations (10.2) with a matrix of smaller bandwidth which is less costly to set up than in the Galerkin method. See Mitchell and Griffiths (1978) and Fairweather (1978) for other examples of the method of moments.

10.2. NUMERICAL INTEGRATION TECHNIQUES

In this section we briefly discuss the numerical solution of the differential problem (10.2) as it arises from the techniques described above. First we remark that if equation (10.1) is nonlinear then clearly so will be equation (10.2a). The way the nonlinearity enters the equation clearly depends on the form of (10.1). If f depends nonlinearly on the solution u in (10.1) then clearly $\underset{\sim}{c}$ will depend on $\underset{\sim}{a}$ in (10.2a) and the form of the dependence will be given by the technique used to approximate (10.1) by (10.2a). Similarly, if we have

$$Lu = (k(u)u_x)_x \tag{10.24}$$

as often arises in diffusion processes then B in (10.2a) will depend on $\underset{\sim}{a}$ but the form of the equation will be unchanged. However, more general forms of nonlinearity may lead to more complicated nonlinear terms. Clearly, the technique to be used for solving the problem (10.2) will depend on the specific form of the nonlinearity and on the sparsity and structure of A and B.

In fact, when A and B are constant matrices and $\underset{\sim}{c}$ depends only on the independent variable, t, we can write down an explicit solution of (10.2) as follows

$$\underset{\sim}{a}(t) = e^{A^{-1}Bt}\underset{\sim}{a}_0 + \int_0^t e^{A^{-1}B\,(t-s)}\underset{\sim}{c}(s)\,ds\ . \tag{10.25}$$

It is clear that this solution will not be useful in practice when A and B are large and sparse, but it can be useful in analysing techniques for solving (10.2). Indeed, there is also an explicit, but complicated solution for (10.2) in the case when A and B depend on t.

In chapter 9, we saw some techniques for solving equation (10.1) though they were presented in a different notation to that appropriate here. Hence, for example, the Explicit method for problem (10.1) in the notation of problem (10.2) when A and B are constant, is

$$A\underset{\sim}{a}^{n+1} = (A + \Delta t\ B)\underset{\sim}{a}^n + \Delta t\underset{\sim}{c}^n,\ \underset{\sim}{a}^0 = \underset{\sim}{a}_o; \tag{10.26}$$

and the Crank-Nicolson method is

$$\left(A - \frac{\Delta t}{2} B\right)\underset{\sim}{a}^{n+1} = \left(A + \frac{\Delta t}{2} B\right) \underset{\sim}{a}^{n} + \frac{\Delta t}{2} (\underset{\sim}{c}^{n+1}+\underset{\sim}{c}^{n}), \ \underset{\sim}{a}^{0}=\underset{\sim}{a}_0. \quad (10.27)$$

The last term in (10.27) is replaced by $\Delta t \ \underset{\sim}{c}^{n+\frac{1}{2}}$ in chapter 9 without loss of accuracy or stability. Note that the explicit method has no advantage over Crank-Nicolson in terms of simplicity unless A is diagonal as in chapter 9 or when using (10.4). There is further discussion of this points in chapter 11. Another widely used method is the Fully Implicit method which for problem (10.1) gives

$$(A - \Delta t B)\underset{\sim}{a}^{n+1} = A\underset{\sim}{a}^{n} + \Delta t\underset{\sim}{c}^{n+1}, \ \underset{\sim}{a}^{0} = \underset{\sim}{a}_0 \ . \quad (10.28)$$

These methods: Explicit, Crank-Nicolson and Fully Implicit correspond to well-known methods for solving systems of ordinary differential equations namely the Euler, Trapezium and Backward Euler methods respectively. We will discuss other ordinary differential equation methods for solving (10.2) and we will assume some knowledge of the methods involved. The reader is referred to Hall and Watt (1976) and Lambert (1973) for a survey of the appropriate material.

When solving (10.2) numerically, we must first note that the problem is almost always stiff for practical problems; that is, the eigenvalues of the algebraic eigenproblem

$$(A\lambda - B)\underset{\sim}{x} = \underset{\sim}{0} \quad (10.29)$$

have negative real parts of widely differing magnitudes. This means that to ensure stability the steplengths used in any numerical integration method for solving (10.2) must be short relative to the timescale of the problem unless the method used has been designed to solve stiff problems. For example, if the method is, at least, A-stable (see Lambert, 1973, Hall and Watt, 1976), it may be suitable for solving stiff problems and hence for solving parabolic problems. For example, Euler's method is not A-stable and hence there is a severe stability restriction as seen in chapter 9, but the Trapezium and Backward Euler methods are A-stable and have unconditional stability.

These stability considerations essentially rule out linear explicit methods for solving (10.2) as they cannot be A-stable. However, Verwer (1977a) has developed special explicit Runge-Kutta methods with very little storage requirement and with good stability properties when the eigenvalues of (10.29) are real and negative (as they often are for semi-discretizations of parabolic equations). Using these methods one cannot use the steplengths associated with implicit methods but the hope is that the low overhead of the explicit methods when A is diagonal will counterbalance the longer steplengths of the implicit methods. A Fortran subroutine implementing one of these methods has been written by Verwer (1977b) and some comparisons of various methods of this type have been made by Dekker, van der Houwen, Verwer and Wolkenfelt (1977). These methods can only be used economically when A is a diagonal matrix in (10.2), that is when simple finite differences are used or when "mass lumping" is used to diagonalise A (see chapter 11).

Amongst implicit methods for solving (10.2), two classes stand out as promising candidates. There has been much discussion of backward differentiation multistep methods since the appearance of Gear (1971). These methods have good stability properties when the eigenvalues of (10.29) are real and negative. They have been the subject of much development and there is good software available, see the packages GEAR in Hindmarsh (1974) and EPISODE in Hindmarsh and Byrne (1975). Of course, these packages have not been developed with the problem (10.2) in mind and to solve a practical problem in this form it would be necessary to modify them to economise on storage and exploit the sparsity and structure of A and B.

Many other subroutines from various classes of methods for stiff problems have been tested in Enright, Hull and Lindberg (1975). Most of these subroutines would not be useful in the context of solving (10.2). Amongst them is one member of the class of implicit Runge-Kutta routines which itself is not likely to be useful but in this class there are useful methods.

A full description of Runge-Kutta methods can be found in Hall and Watt (1976) and in Lambert (1973) and we will not discuss them in detail here. Let us instead consider the effect of using an implicit Runge-Kutta method on equation (10.2) with constant matrices A and B and with $\underset{\sim}{c} = \underset{\sim}{0}$ when the true solution is

$$\underset{\sim}{a} = e^{A^{-1}Bt}\underset{\sim}{a}_0 , \tag{10.30}$$

that is

$$\underset{\sim}{a}^{n+1} = e^{A^{-1}B\Delta t}\underset{\sim}{a}^n . \tag{10.31}$$

An implicit Runge-Kutta method with steplength Δt will replace (10.31) by

$$q(A^{-1}B\Delta t)\underset{\sim}{a}^{n+1} = p(A^{-1}B\Delta t)\underset{\sim}{a}^n \tag{10.32}$$

with $\underset{\sim}{a}^0 = \underset{\sim}{a}_0$, where p and q are polynomials. In fact $r(z) = p(z)/q(z)$ is exactly the rational function obtained when applying the implicit Runge-Kutta method to the test equation $y' = \lambda y$ to give $y^{n+1} = r(z)y^n$ where $z = \lambda\Delta t$, see Hall and Watt (1976, ch. 10) for further details. The practical implementation of methods such as (10.32) is discussed in Siemieniuch and Gladwell (1974) and in Smith, Siemieniuch and Gladwell (1977). Many fully implicit Runge-Kutta methods of maximum order of accuracy give rise to polynomials p and q in (10.32) such that r(z) is a Padé approximant to e^z, and for Padé approximants q has no real factors if q is of even degree. Such methods usually have good stability properties but high order methods of this type may be impractical for large problems (10.2) since they involve solving systems of Nm nonlinear equations per step where m is the number of components in $\underset{\sim}{a}$ and N is the degree of q. Even for linear problems, where (10.32) can be solved, fully implicit Runge-Kutta methods are impractical as there is no method for solving (10.32) for $\underset{\sim}{a}^{n+1}$ without forming A^{-1} unless one uses complex arithmetic.

These considerations have led to the development of semi-implicit Runge-Kutta methods, see Butcher (1964), Nørsett (1974) and Alexander (1977). In these methods one solves N sets of equations each of size m at each step. Furthermore each set of equations is very similar and the whole problem can be solved quite straightforwardly. In the case of (8.32), these methods correspond to the polynomial q having linear factors and, in the case of the methods of Alexander and Nørsett, the linear factors are identical (which implies that we solve a sequence of problems similar to (10.27) at each step). Such methods can be implemented without recourse to matrix multiplications and inversions unlike when using the fully implicit Runge-Kutta methods. The subroutine DIRK available from Alexander could be modified quite simply to solve the problem (8.2) economically as could a similar subroutine available from Nørsett. Some semi-implicit Runge-Kutta methods are compared with conventional methods in Smith, Siemieniuch and Gladwell (1977). See chapter 11 for further discussion.

11

THE DIFFUSION-CONVECTION EQUATION

11.1. INTRODUCTION

Parabolic equations containing first order spatial derivatives are called "diffusion-convection" equations because of the physical processes they describe. Typically, a suspended material (effluent, sediment etc.) is carried along ("convected") by a flow of fluid while at the same time its concentration is being attenuated ("diffused") within the flow. Numerically, the more interesting problems arise when the first order (convection) spatial derivatives are large in relation to the second order (diffusion) ones. That is, the solution behaves like the solution of the limiting hyperbolic (purely convective) case. In the converse situation, when diffusion is dominant, ordinary methods for parabolic equations can be used (see chapters 9 and 10).

A typical equation, written for brevity in its one-dimensional form is

$$\frac{\partial u}{\partial t} = D \frac{\partial^2 u}{\partial x^2} - v \frac{\partial u}{\partial x} \tag{11.1}$$

where u describes the concentration of a suspension convected with velocity v and diffusing according to the "diffusion coefficient" D. Of course equation (11.1) greatly oversimplifies the physical realities of turbulent flow and, in particular, practical difficulty attaches to the measurement of the number D. A useful measure of the degree of "convectiveness" in a problem is the dimensionless Peclet number $P_e = vx/D$ where x is some reference distance. In a numerical context a mesh step size of h is associated with a "cell Peclet number" $P_e = vh/D$.

A second source of the diffusion-convection equation lies in the Navier-Stokes equations, particularised for the case of incompressible flow. In that event

$$\frac{\partial u}{\partial t} = \nu \frac{\partial^2 u}{\partial x^2} - v \frac{\partial u}{\partial x} \tag{11.2}$$

where $u = -\partial^2\psi/\partial x^2$, $v = -\partial\psi/\partial x$. The coefficient of kinematic viscosity is ν and ψ is a stream function. The relevant dimensionless parameter is now the Reynolds number $R_e = vx/\nu$.

The past 20 years or so have seen various attempts at numerical solutions of equations of this kind. An exhaustive bibliography cannot be presented here but some key contributions are outlined in the next section, which shows the usual scope for rediscovery of known methods.

11.2. PREVIOUS WORK

Early finite-difference solutions were obtained by Peaceman and Rachford (1962), Stone and Brian (1963), Roberts and Weiss (1966) and many others. Price, Cavendish and Varga (1968) sum up the early experience and refer to the difficulties of oscillations and undue numerical diffusion in the solutions. To circumvent these problems, many authors developed rather specialised weighting techniques when forming the difference equations. The same ideas have more recently been proposed by Spalding (1972) under the title of "upwind" difference schemes.

"Classical" Galerkin techniques (in which the total region was spanned by the trial functions) were used by Cavendish, Price and Varga (1969) and by Culham and Varga (1971) (1971) recognising the near-hyperbolic nature of strongly convective problems.

Guymon (1970) obtained the first finite element solutions to diffusion-convection problems at the expense of transforming the space differential operator into its equivalent self-adjoint form by the substitution

$$c = u \exp(Pe/2) \quad . \tag{11.3}$$

Clearly because of the exponential this approach is numerically limited to small Peclet numbers, that is to essentially diffusive processes, whereas Price *et al.* (1968) had been working with Peclet numbers of the order of 10^4 to 10^5. Smith, Farraday and O'Connor (1973) produced Galerkin finite element solutions to equation (11.1). They used implicit integration in time (Crank-Nicolson) together with linear, parabolic or Hermitian cubic elements in one- and two-dimensional space. Somewhat earlier, Loziuk, Anderson and Belytschko (1972) solved a steady-state version of equation (11.1).

More recently, papers describing finite element solutions have proliferated, see, for example Christie, Griffiths, Mitchell and Zienkiewicz (1976), Heinrich, Huyakorn, Zienkiewicz and Mitchell (1977), Heinrich and and Zienkiewicz (1977), Huyakorn and Taylor (1977) and Mitchell and Griffiths (1978). One of the main areas of application is to solutions of water resources problems and a recent book describing the state of the art is edited by Gray, Pinder and Brebbia (1977). See chapter 10 for discussion of the techniques used to obtain finite element solutions.

11.3. PARTICULAR PROBLEMS ASSOCIATED WITH THIS EQUATION

Difficulties arise because we are interested in solving "large" engineering problems in two and three space dimensions. That is, we think in terms of 10^3 or 10^4 nodal unknowns or finite-difference points. The discretised system of ordinary differential equations is "stiff" and problems are often nonlinear or have non-constant coefficients. We cannot economically use arbitrarily fine meshes. The spatially semi-discretized approximation to equation (11.1) is of the form

$$A \frac{d\underset{\sim}{U}}{dt} = - B \underset{\sim}{U} \tag{11.4}$$

In all large system analysis there is a preference for explicit integration of these equations in time, because of computer storage limitations, and in finite element work the mass matrix A has often been *lumped*, that is diagonalised. Usually, lumping for a system (11.4) implies adding the elements of the rows of A and placing the result on the diagonal then setting all off-diagonal elements to zero; B is unchanged. We speak of a lumped mass matrix A in contrast to the *consistent* mass matrix A obtained from the Galerkin method. Algorithms of this kind usually have little or no intrinsic damping. At high Peclet or Reynolds numbers the spatial discretization alone introduces spurious oscillations into the numerical solution whereas the true solution is known on physical or analytical grounds to be non-oscillatory. As integration in time proceeds by a just-stable algorithm (for example, simple explicit) for a nonlinear system, additional oscillations are propagated by the time-discretization. Taken together, spatial and temporal oscillations can lead at best to physically unacceptable solutions and at worst to numerical overflows.

11.4. SOME SIMILAR COMPUTING PROBLEMS

Unwanted numerical oscillations arise in many aspects of the numerical solution of partial differential equations, and, in particular in the following areas.

(a) Incompressibility in solid mechanics.

Strong spatial oscillations are usually dealt with by smoothing locally or, in finite elements, by inexact "reduced" integration and by not evaluating at the nodes but at optimal sampling points.

(b) Elasto-plastic dynamics.

The appropriate semi-discretized differential equation is

$$M\frac{d^2\underset{\sim}{x}}{dt^2} + C\,\frac{d\underset{\sim}{x}}{dt} + K\underset{\sim}{x} = \underset{\sim}{P}(t) \tag{11.5}$$

in which M is the "mass matrix", C the "damping matrix" and K the "stiffness matrix". By making the substitutions:

$$\underset{\sim}{U} = \begin{pmatrix} \dot{\underset{\sim}{x}} \\ \underset{\sim}{x} \end{pmatrix}, \quad A = \begin{bmatrix} M & O \\ O & I \end{bmatrix}, \quad B = \begin{bmatrix} C & K \\ -I & O \end{bmatrix} \tag{11.6}$$

the form of equation (11.4) can be recovered. Problems of this kind are again large, stiff and usually nonlinear leading to a need to lump the matrix A and integrate explicitly in time in exactly the same way as is done in the diffusion-convection case. This has some benefits in that the lumped semi-discretized system is somewhat less stiff than the consistent one (11.3) but in the author's experience lumping introduces additional spurious oscillations, see Smith (1977). In this kind of problem quite elaborate artificial damping is often added in the solution algorithm to keep the spurious oscillations within bounds. Of course, the more physically acceptable results thus produced could be quite erroneous.

11.5. FINITE DIFFERENCES OR FINITE ELEMENTS?

In principle, all problems can be solved by either method. However, there are two practical advantages enjoyed by finite elements. These are (i) ease of modelling complex 2- and 3-dimensional geometries and (ii) computation strategies based on quadrature and matrix handling procedures which can be highly automated (and incidentally take on even greater potential significance in the context of array micro-processors). Maybe, in principle finite differences could be adapted to compete in these two areas, but this has just not been done.

Finite elements do have a (part real, part apparent) disadvantage when, as is usually done, they are formulated as a Galerkin or least squares weighted residual method (this would not apply in the case of collocation for example). The difficulty lies in the fully banded structure of the matrix A in equation (11.4). This fact has led some finite-difference enthusiasts to discount finite elements altogether (see, for example Weare, 1976, Leendertse, 1977) since, apparently, explicit methods of integration in time cannot succeed economically. However, as was pointed out above, elastodynamic problems have been solved explicitly by finite elements for many years by the simple expedient of lumping A to give a diagonal matrix. This can always be unambiguously done for simple elements (linear quadrilaterals for example) but can cause trouble in at least two ways. First, it can lead to anomalies (negative lumped mass!) for higher order elements such as quadratic quadrilaterals. This problem is fairly readily obviated by alternative nodal weighting schemes which preserve the element eigenvalues, see Smith (1977). Secondly, when, in the (generalised) Galerkin process, test and trial functions are made different, the banded matrix A

can become unsymmetric. Again the best approach is to use weighted diagonal lumping to preserve the true element eigenvalues. In the author's opinion, A should not in general simply be replaced by the identity matrix I as is sometimes done, see, for example, Mitchell and Griffiths (1978).

11.6. SPATIAL SEMI-DISCRETIZATION

Siemieniuch and Gladwell (1978) analysed a model diffusion-convection equation (11.1) which was spatially semi-discretized by various explicit finite-difference techniques to yield equation (11.4) with $A = I$. Then by examination of the eigenvalues of the matrix B it could be shown that for "large" mesh spacings oscillatory components would exist in the discretized solutions. Since, for the model problem chosen, an analytical treatment showed the true solution to be non-oscillatory, it could be asserted that whatever the time discretization, unwanted oscillations would appear in the finite-difference solutions due to space semi-discretization alone.

This type of phenomenon is shown in Figures 11.1 which

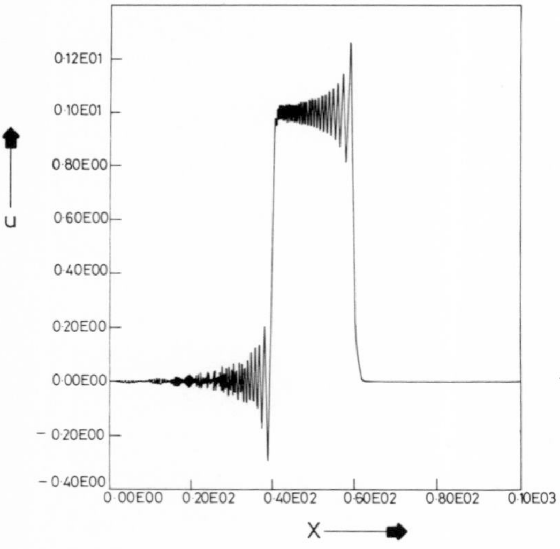

FIG.11·1 a.

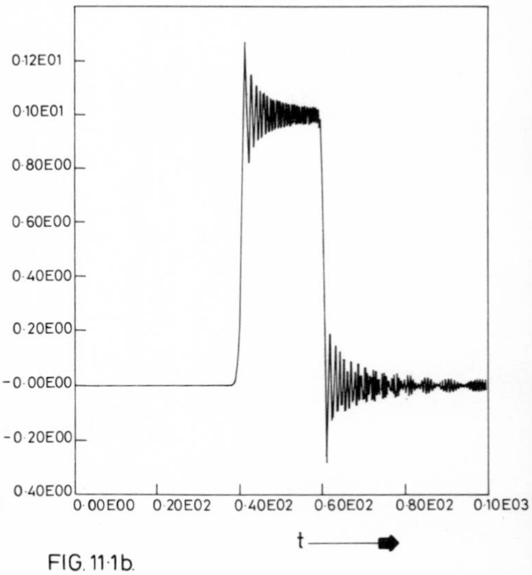

FIG. 11·1b.

relate to the problem (11.1) with boundary and initial conditions

$$\begin{aligned} u(x,0) &= 0 \qquad & 0 \le x \le 100 \\ u(0,t) &= 1 \qquad & 0 \le t \le 20 \\ u(0,t) &= 0 \qquad & t > 20 \end{aligned} \tag{11.7}$$

and with physical parameters u = 1, D = 0 (that is, a purely convective problem). The solution was obtained for linear (Chapeau) finite elements with distributed (not lumped) *A* matrix, $\Delta x = \Delta t = 0.1$, and a Crank-Nicolson implicit integration in time. Note that in the present discussion of spatial discretization, non-dissipative (undamped) time integration schemes are used. Figure 11.1a shows u as a function of x at t = 60, that is, when the centre of the pulse has been convected half way across the solution domain. The rather violent oscillations at the front of the pulse and in its wake are not relieved to any marked degree by decreasing the time-step. Figure 11.1b shows u as a function of t as it

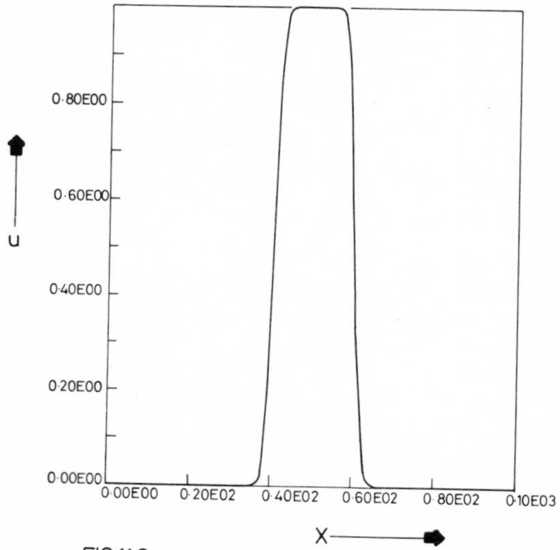

FIG.11·2 a.

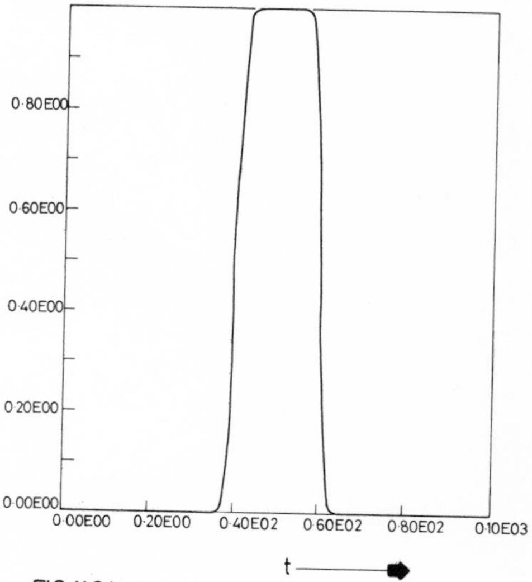

FIG.11·2 b.

would be seen by an observer stationed at x = 40. The oscillatory approximation to the true rectangular pulse is again evident.

Figures 11.2 relate to the same problem but this time a small diffusion factor has been introduced, that is, we set D = 0.02. The absence of oscillations shows that the space step is small enough in this instance to give a good approximation for this diffusion-convection problem. However, in general, such a space step will be uneconomic and for the remainder of the discussion we consider a space step of 1.0, that is, 10 times greater than before. For the purely convective case, the solution is plotted in Figures 11.3. It is more or less acceptable for engineering purposes, but bear in mind the facts that the *A* matrix is not lumped and that the time integration is implicit. It was pointed out earlier that difficulties are more likely to arise when *A* is lumped and explicit time integration is used. The effects of these changes are now considered in turn.

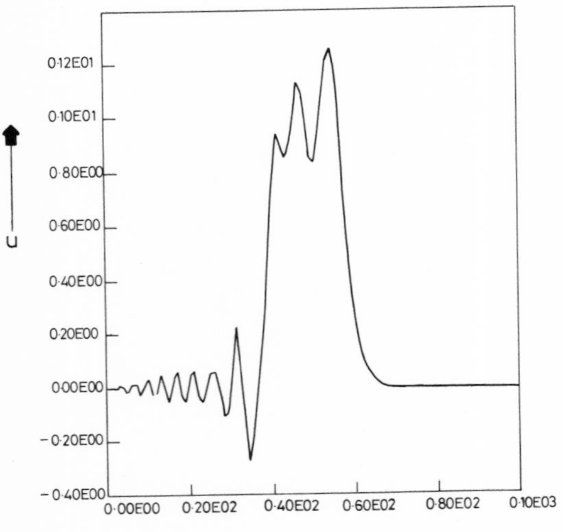

FIG. 11·3a

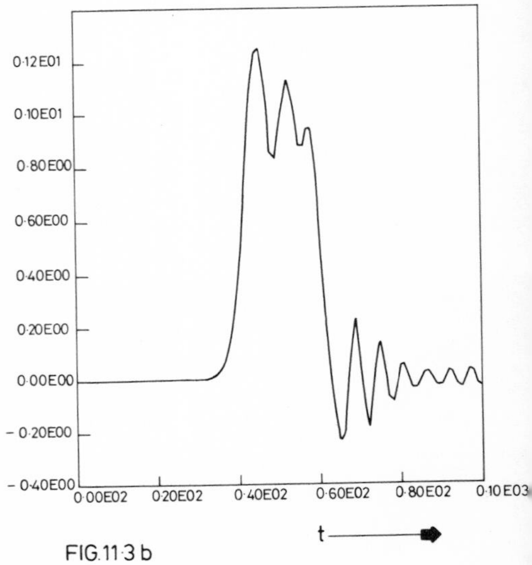

FIG. 11·3b

Figures 11.4 show the effect of lumping *A* while retaining implicit integration in time. There is no question that there is some deterioration in the solution, by comparison with Figures 11.3, in particular the front is less sharp and the oscillations in the wake as seen by the observer, Figure 11.4b, are much more pronounced. When explicit time integration is substituted for implicit integration, the solution, see Figures 11.5, is now completely unacceptable, particularly in respect of the oscillations in the wake, Figure 11.5b. The stable explicit timestep used in this case was 0.1, that is, one tenth of that in the implicit case.

On the basis of these results, it can be seen that lumping *A* together with integrating explicitly in time can be the source of difficulties in solving diffusion-convection problems. Methods are now described by which these difficulties can be at least partially alleviated.

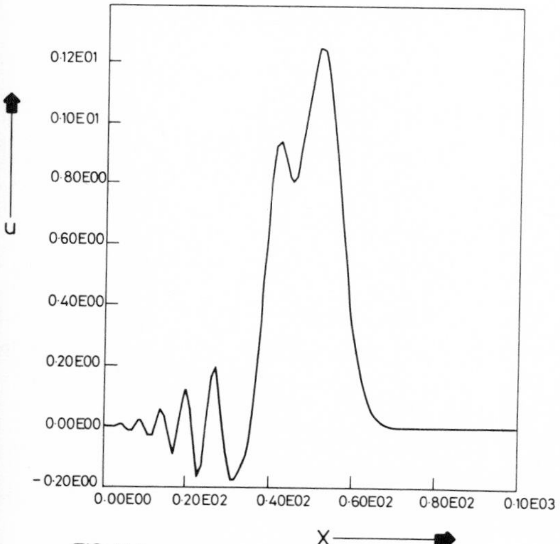

FIG. 11·4 a.

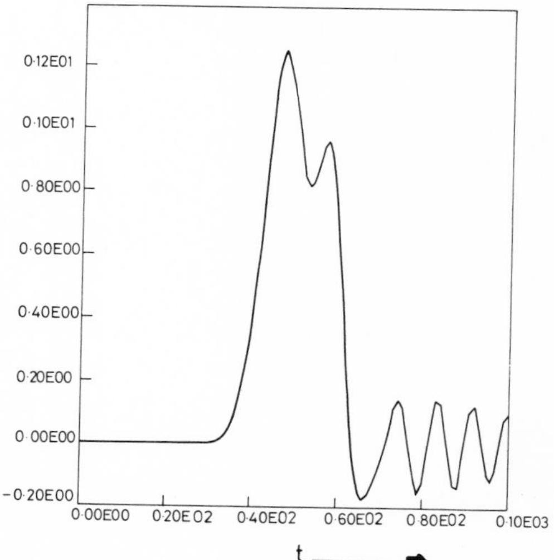

FIG.11·4 b.

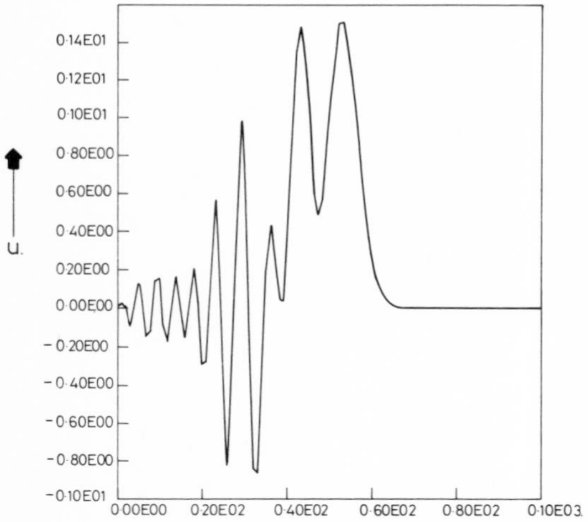

FIG. 11·5 a.

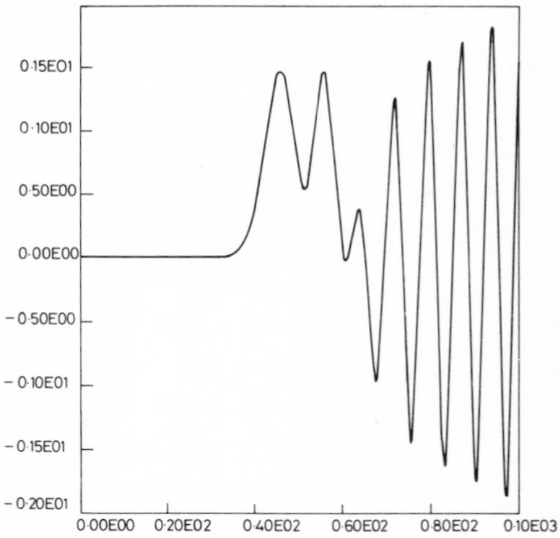

FIG. 11·5 b

11.7. IMPROVED SPATIAL DISCRETIZATION

An obvious way to improve spatial approximations is to use higher order interpolation in space. This can be done in essentially two ways, namely by retaining u as the only nodal unknown and using, for example, the isoparametric "quadratic" type of interpolation or by including derivatives of u in the interpolation. Both of these techniques were used in Smith *et al.* (1973) and it was concluded that the second technique gave the greater improvements. This seems intuitively justifiable on the grounds that the 'peaky' spatial oscillations shown in Figure 11.1a for example seem likely to be smoothed if derivative continuity is imposed. Results for the test problem of equations (11.7) are shown in Figures 11.6 where the smoothing is quite marginal. It must be remembered that the extra derivative unknowns increase the cost of solution unless coarser meshes are used (twice as coarse in the present results). In the case of the first technique of higher order interpolation, no striking improvements were found over linear

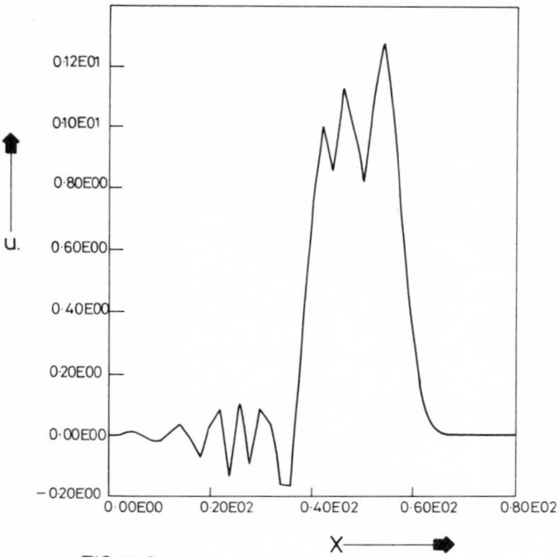

FIG. 11·6 a.

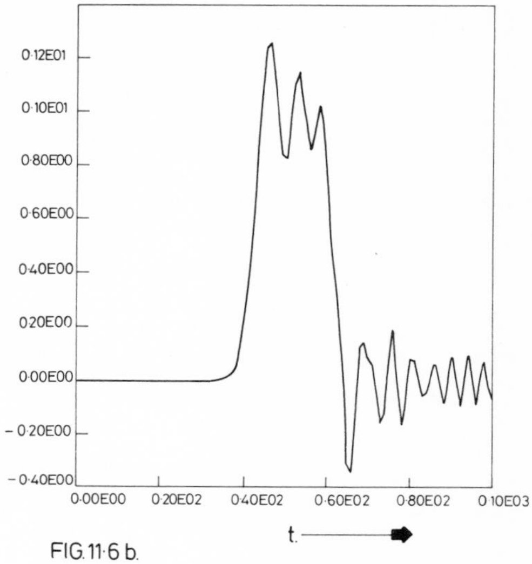

FIG. 11·6 b.

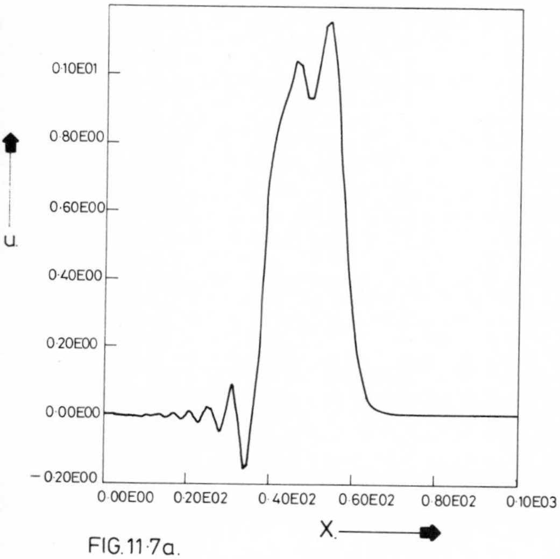

FIG. 11·7a.

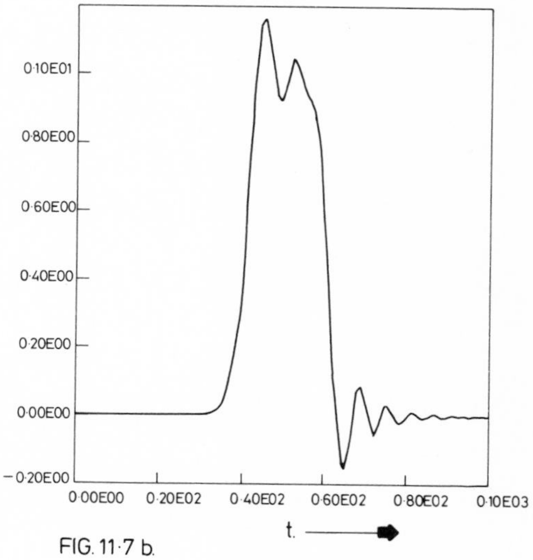

FIG. 11·7 b.

elements, and this finding has been confirmed by Heinrich *et al*. (1977).

The most popular technique for improving numerical solutions to the diffusion-convection equation without changing the order of spatial interpolation is nowadays called "upwinding". This is a variant of Stone and Brian's (1963) weighting system which was popularised for finite-difference methods by Spalding (1972) and adapted for finite elements (see chapter 3) by Christie *et al.* (1976) and the others referenced in the section on previous work above. In a finite element context consider the simple (linear) Chapeau shape functions N_i. Weighting or upwinding is achieved through using weighting functions W_i in the (generalized) Galerkin process which differ from the shape functions. Clearly there is a wide choice for the functions W_i given that they have to vanish at the nodes, but we consider only the parabolic case

$$W_i = N_i \pm \alpha F_i \tag{11.8}$$

where in local coordinates

$$F_i(\xi) = \frac{3}{2}(1 - \xi^2) \tag{11.9}$$

and in which α has been adjusted so that mass is conserved. After application of the Galerkin process, using the W_i as trial functions and the N_i as test functions, the element *A* matrix is found to be

$$A = L \begin{bmatrix} \frac{1}{3} - \frac{1}{4}\alpha & \frac{1}{6} - \frac{1}{4}\alpha \\ \frac{1}{6} + \frac{1}{4}\alpha & \frac{1}{3} + \frac{1}{4}\alpha \end{bmatrix} \tag{11.10}$$

where L is the element length. Similarly the element *B* matrix can best be written

$$B = -\frac{V}{2}\begin{bmatrix} (\alpha - 1) & -(\alpha - 1) \\ -(\alpha + 1) & (\alpha + 1) \end{bmatrix} + \frac{D}{L}\begin{bmatrix} 1 & -1 \\ -1 & 1 \end{bmatrix} \tag{11.11}$$

where the first part is associated with the first order (convective) term on the right hand side of equation (11.1) and the second part is associated with the second order (diffusive) term. It can be seen that the upwinding process is essentially applied to the convective term. The parameter α can range from 0 (standard method, no upwinding) to 1 (full upwinding).

To study the effectiveness of this process we return to our problem defined by equations (11.7) with $\Delta x = \Delta t = 1.0$, convective velocity $v = 1.0$ and diffusivity $D = 0.02$. Results for the "standard" solution (consistent A, implicit time integration) are shown in Figures 11.7. By comparison with Figures 11.3 it is seen that the small diffusion term causes quite a lot of smearing of the solution (peak concentration at $t = 60$ drops from 1.27 to 1.16 for example). With full upwinding, $\alpha = 1$, the results are modified as shown in Figures 11.8. The smoothing or smearing leading to an improved result is quite evident, with peak concentration dropping again, to 1.09.

The really dramatic influence of upwinding is found in the lumped, explicitly integrated solutions. Figures 11.9 show the solution with no upwinding which contains a lot of unwanted oscillations. As the upwinding parameter α is increased, to 0.1, 0.2, 0.5 and 1.0, Figures 11.10, smearing is evident in the solutions. For this problem a small value of α, say 0.2, does produce an improved solution but "full" upwinding, $\alpha = 1$, as is sometimes advocated, leads to unacceptable numerical diffusion. Therefore the drawbacks are that it is a smearing process and in novel problems could lead to non-conservative results; secondly, in complex two- or three-dimensional flow fields it could be difficult to define the direction of the "wind".

Other techniques for improved solutions may be mentioned briefly. One is the imposition of interior "penalty functions" which has proved useful in analogous problems such as incompressible elasticity. The technique is one of applying additional constraints through the Lagrange multiplier approach, and could no doubt be used on this equation with some success. However previously mentioned tricks such as smoothing and "reduced" integration have proved in other problems to be simpler to apply.

Gray and Pinder (1976) and van Genuchten and Gray (1978) have proposed various "dispersion-corrected" schemes for solving the (linear) diffusion-convection equation. These involve rather elaborate adjustments to the coefficients of

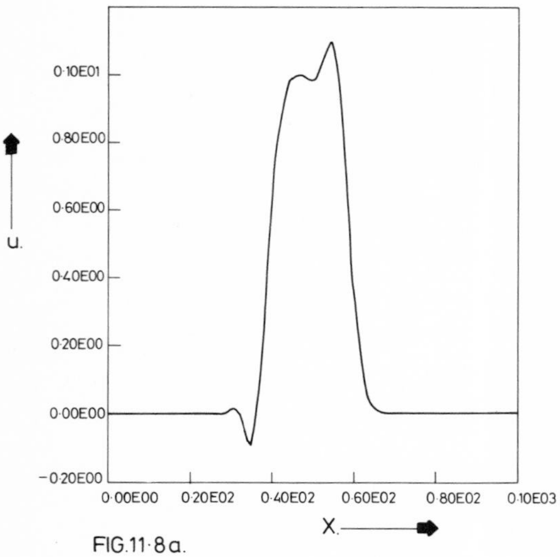

FIG.11·8a.

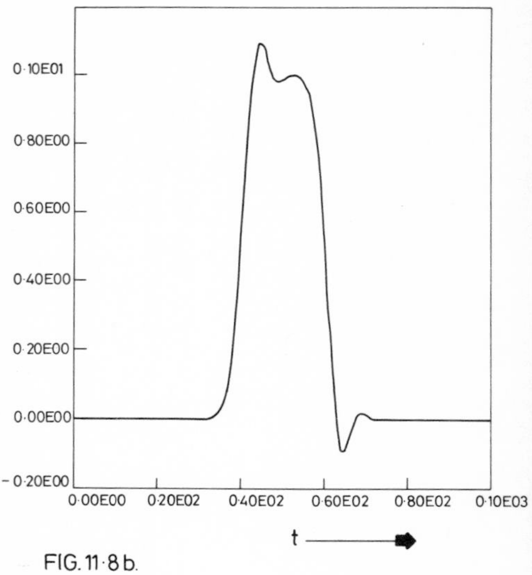

FIG. 11·8b.

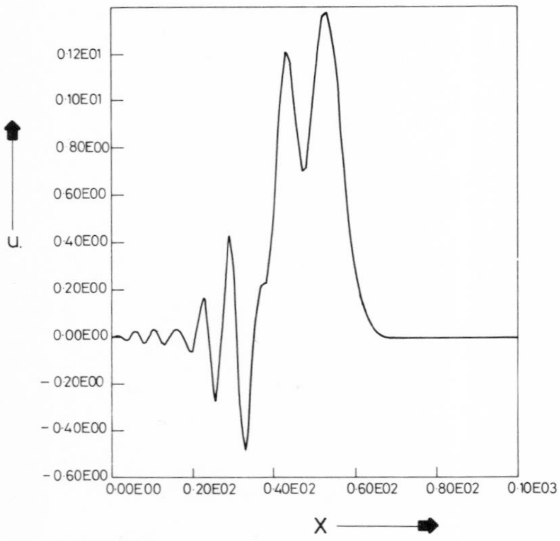

FIG. 11·9a.

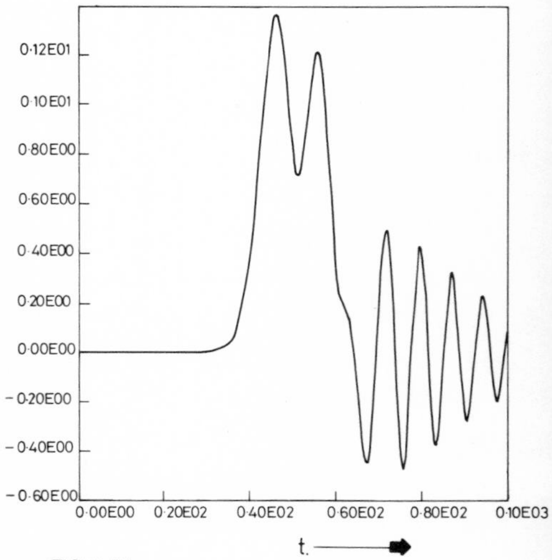

FIG.11·9b.

the discretized system in order, by comparison with Taylor series expansions, to make the schemes third order correct, fourth order correct and so on. For this writer, the benefits gained in the chosen problems seem somewhat marginal and the method not readily capable of extension to general domains or nonlinear problems. The authors state as their

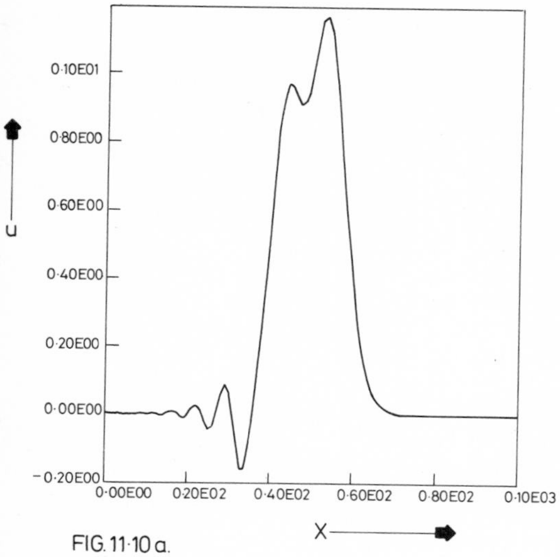

FIG. 11·10 a.

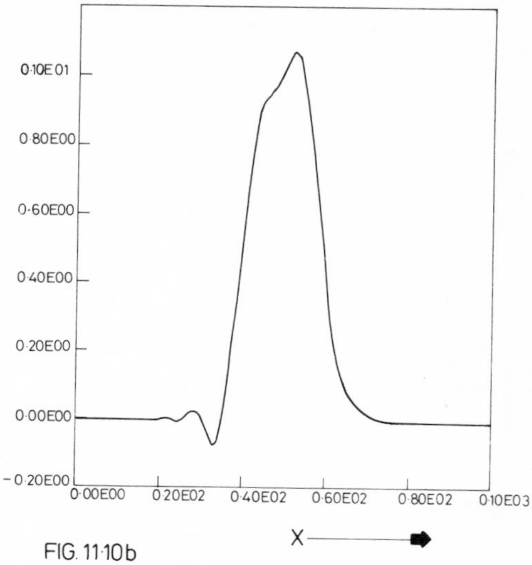

FIG. 11·10 b

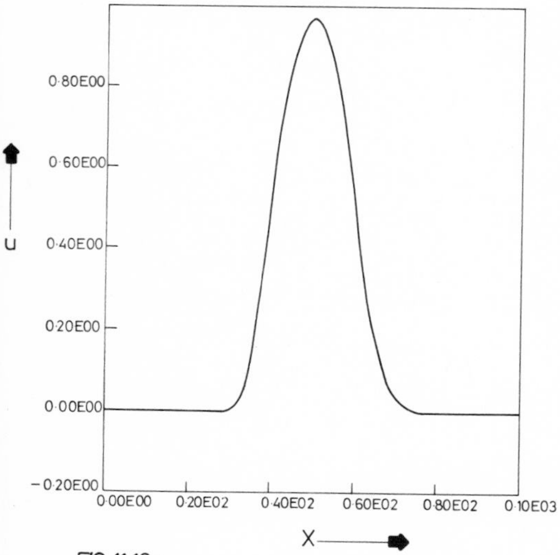

FIG. 11·10 c.

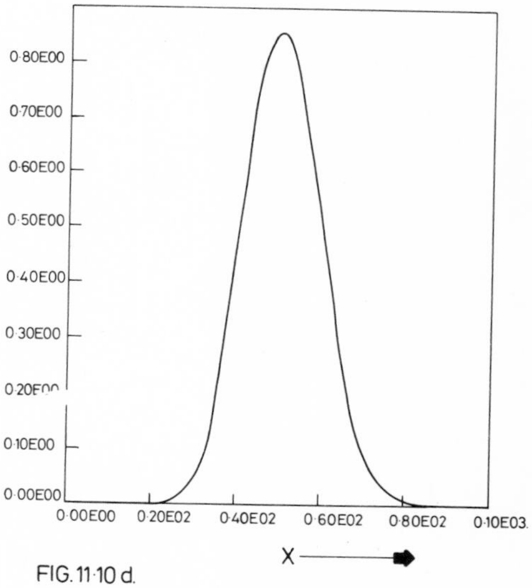

FIG. 11·10 d.

motivation for using dispersion-corrected schemes rather than upwinding their observation that "upstream weighting of the convective term generates unacceptable solutions for convection-dependent transport. The amplification curves show that this is due to excessive damping over a wide range of harmonics."

11.8. TIME DISCRETIZATION

The simplest non-dissipative algorithms for integrating with respect to time have been used in the above study. In the implicit calculations a centred Crank-Nicolson method was used and in the explicit calculations a forward Euler. In both implicit and explicit cases improved methods can be devised but it is recommended in the present application that these be as mildly dissipative as possible. Most useful methods belong to the Runge-Kutta class and the writer has evaluated in particular a series of multi-stage one-step implicit methods due to Nørsett (see Smith, 1976; Smith, Siemieniuch and Gladwell, 1977; for example). The best of these methods are mildly dissipative in precisely the way which is desirable for the present application. Clearly a beneficial development would be explicit methods with larger stability regions which would considerably extend the critical timesteps necessary in the Euler method. Unfortunately many such improved explicit algorithms are dissipative in character. See chapter 10 for further discussion of these methods.

11.9. ODE PACKAGES

In the above computations, no effort is made to decide on an "acceptably" accurate answer in the way that is done in many standard ordinary differential equation solving packages. Since spatial errors are so important in the present application such accuracy adjustment is of debatable importance.

11.10. CONCLUSIONS

Numerical solutions to the diffusion-convection equation have been presented for cases in which convection is dominant. The options open to the analyst have been shown to include:

space discretization
- finite-difference
- finite element (consistent)
- finite element (lumped)

time discretization
- implicit
- explicit

"tricks" { slope continuity elements
upwinding
penalty functions
dispersion corrections
higher order time schemes .

Acceptable solutions were produced for fairly coarse meshes using simple finite elements (Chapeau), consistent mass and implicit integration in time. For large nonlinear problems there is a strong motivation for using lumping and explicit integration in time. When this is done the quality of solutions deteriorates badly. Upwinding can then be an effective technique for improving the explicit solutions from lumping by introducing spurious diffusion. However the technique could be difficult to apply in complex two- and three-dimensional flows. Of the other "tricks", slope continuity elements can be beneficial as can higher order time schemes. Dispersion-correction does not seem yet to have been established on a sufficiently general basis.

12

BOUNDARY LAYER AND SIMILARITY SOLUTIONS FOR PARABOLIC EQUATIONS

12.1. INTRODUCTION

This chapter is concerned with the mathematical and numerical treatment of a class of fluid mechanics problems that may be formulated as parabolic partial differential equations, namely boundary-layer flows. The concept of a boundary layer was introduced into fluid mechanics by Prandtl in 1904 as being the region in which frictional (i.e. viscous) forces in fluid flow at high Reynolds number are significant. During the present century, boundary-layer theory has found rapidly increasing application to problems of practical importance in hydrodynamics, aerodynamics and gas dynamics, and is now being applied to both steady and unsteady problems involving laminar or turbulent flow of compressible and incompressible fluids in two or three space dimensions.

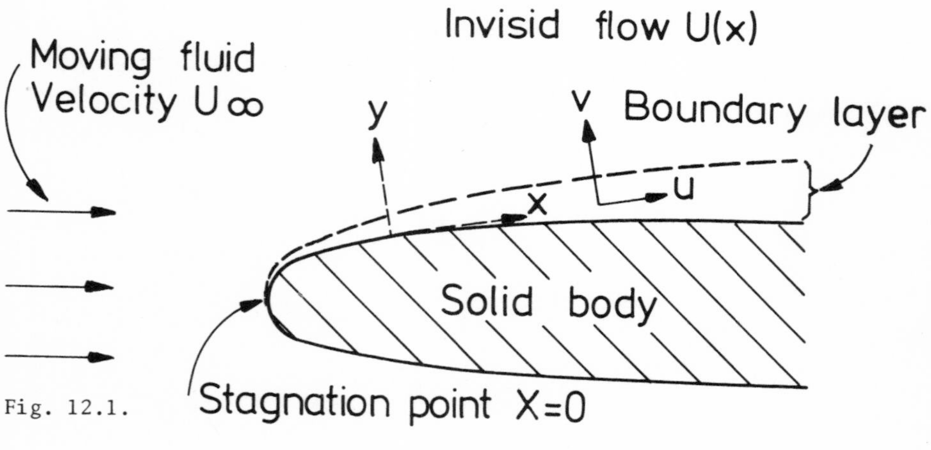

Fig. 12.1.

The mathematical concepts underlying boundary-layer theory can best be discussed briefly by considering the steady, incompressible, two-dimensional laminar flows of a fluid of low viscosity μ about a body of slender cross-section (see Fig. 12.1). Except in the vicinity of the body, the flow velocity is essentially equal to the free-stream velocity $U(x)$, and the motion may be treated as a frictionless, potential flow. Close to the surface, however, there is a sharp transition from zero velocity (no slip) at the surface to a velocity approaching U(x) a short distance δ from the surface. Thus there is a thin region, the boundary layer, in which the velocity gradient normal to the surface is very large, and in this region the frictional force

$$\tau = \mu \frac{\partial u}{\partial y}$$

may be comparable with the inertial forces.

12.2. BOUNDARY-LAYER EQUATIONS

The Navier-Stokes equations for steady, incompressible flow in two dimensions are

$$u\frac{\partial u}{\partial x} + v\frac{\partial u}{\partial y} = -\frac{1}{\rho}\frac{\partial p}{\partial x} + \nu\left(\frac{\partial^2 u}{\partial x^2} + \frac{\partial^2 u}{\partial y^2}\right) , \qquad (12.1)$$

$$u\frac{\partial v}{\partial x} + v\frac{\partial v}{\partial y} = -\frac{1}{\rho}\frac{\partial p}{\partial y} + \nu\left(\frac{\partial^2 v}{\partial x^2} + \frac{\partial^2 v}{\partial y^2}\right) , \qquad (12.2)$$

$$\frac{\partial u}{\partial x} + \frac{\partial v}{\partial y} = 0 , \qquad (12.3)$$

where $\underset{\sim}{u} = (u,v)^T$ are the velocities, p the pressure, ρ the density and $\nu = \mu/\rho$ the kinematic viscosity of the fluid. Introducing non-dimensional variables (denoted by primes) by relating all velocities to a free-stream velocity U_∞ and lengths to a characteristic length L such that $\partial u/\partial x = O(U_\infty/L)$, the equations become

$$u'\frac{\partial u'}{\partial x'} + v'\frac{\partial u'}{\partial y'} = -\frac{\partial p'}{\partial x'} + Re^{-1}\left(\frac{\partial^2 u'}{\partial x'^2} + \frac{\partial^2 u'}{\partial y'^2}\right) \qquad (12.4)$$

magnitude: $1.1 \quad \delta'.\delta'^{-1} \quad \delta'^2 \quad 1 \quad \delta'^{-2}$

$$u'\frac{\partial v'}{\partial x'} + v'\frac{\partial v'}{\partial y'} = -\frac{\partial p'}{\partial y'} + Re^{-1}\left(\frac{\partial^2 v'}{\partial x'^2} + \frac{\partial^2 v'}{\partial y'^2}\right) \qquad (12.5)$$

magnitude: $1.\delta' \quad \delta'.1 \quad \delta'^2 \quad \delta' \quad \delta'^{-1}$

$$\frac{\partial u'}{\partial x'} + \frac{\partial v'}{\partial y'} = 0 \qquad (12.6)$$

magnitude: 1 1

where $Re = U_\infty L/\nu$ is the Reynolds number. The orders of magnitude of the velocity terms have been expressed as powers of the dimensionless boundary-layer thickness $\delta' = \delta/L$, and it is seen that the viscous terms are of the same order as the inertial terms when $Re = O(\delta'^{-2})$.

For flows at large Reynolds number, therefore, we have $\delta' \ll 1$, implying a very thin boundary layer, and the Navier-Stokes equations may be simplified. In equation (12.5) $\partial p'/\partial y'$ must be $O(\delta')$, so that the pressure change across the boundary layer is $O(\delta'^2)$ and may be neglected. The pressure $p = p(x)$ is thus the pressure impressed on the boundary layer by the outer, potential flow, and is related to the free-stream velocity by

$$p(x) + \frac{1}{2}\rho U^2(x) = \text{constant}. \qquad (12.7)$$

Neglecting also the viscous term $\partial^2 u'/\partial x'^2$ in equation (12.4), we obtain the Prandtl boundary-layer equations for steady two-dimensional flow

$$u\frac{\partial u}{\partial x} + v\frac{\partial u}{\partial y} = U\frac{dU}{dx} + \nu\frac{\partial^2 u}{\partial y^2} \qquad (12.8)$$

$$\frac{\partial u}{\partial x} + \frac{\partial v}{\partial y} = 0 . \qquad (12.9)$$

The boundary conditions are simply the no-slip condition at the surface

$$u = v = 0 \quad \text{at} \quad y = 0 ,$$

and $u = U(x)$ as $y \to \infty$. In addition, an initial velocity profile $u(x_0,y)$ is required at a reference point $x = x_0$.

The above equations have been derived for flow over a flat surface; using curvilinear orthogonal coordinates, the same equations may be shown to hold (Schlichting, 1960, p.111) for flow over a curved wall $y = 0$, provided the variations in the radius of curvature R satisfy $dR/dx = O(1)$.

The simplification of the Navier-Stokes equations in the boundary layer is two-fold; firstly, there is a reduction of one in the number of equations and unknowns, and secondly the elliptic form of the Navier-Stokes equations is replaced by parabolic boundary-layer equations.

In order to satisfy the continuity equation (12.9) automatically, it is often useful to introduce the stream functions

$$u = \frac{\partial\psi}{\partial y}, \quad v = -\frac{\partial\psi}{\partial x}. \tag{12.10}$$

Substituting, (12.5) becomes a third-order equation

$$\frac{\partial\psi}{\partial y}\frac{\partial^2\psi}{\partial x\partial y} - \frac{\partial\psi}{\partial x}\frac{\partial^2\psi}{\partial y^2} = U\frac{dU}{dx} + \nu\frac{\partial^3\psi}{\partial y^3} \tag{12.11}$$

with boundary conditions $\partial\psi/\partial x = \partial\psi/\partial y = 0$ at $y = 0$ and $\partial\psi/\partial y = U(x)$ as $y \to \infty$.

12.3. DEPENDENCE ON THE REYNOLDS NUMBER

The dependence of the boundary-layer equation (12.8) on the Reynolds number arises through the inverse dependence of Re on ν. It is important to note that this dependence may be eliminated by a simple non-dimensional scaling of the transverse variables y and v, as follows

$$\begin{aligned} u' &= \frac{u}{U_\infty}, \qquad x' = \frac{x}{L}, \\ v' &= \frac{v}{U_\infty}\mathrm{Re}^{\frac{1}{2}}, \qquad y' = \frac{y}{L}\mathrm{Re}^{\frac{1}{2}}. \end{aligned} \tag{12.12}$$

Thus one solution to the boundary-layer equations using these non-dimensional variables provides the solution for all large values of the Reynolds number for which the flow is laminar. This similarity principle with respect to Reynolds number applies to all solutions of the boundary-layer equations (12.8) and (12.9).

12.4. SIMILAR BOUNDARY LAYERS

Let us now consider under what conditions, if any, the boundary-layer solutions exhibit similarity with respect to x. Solutions $u(x_1,y)$ and $u(x_2,y)$ for the velocity profiles u at x_1 and x_2 are said to be *similar* if they differ only by scale factors on u and y; that is, if

$$u(x_1,\ y/g(x_1))/U(x_1) = u(x_2,\ y/g(x_2))/U(x_2). \qquad (12.13)$$

It is readily shown (Schlichting, 1960, p.132) that such similar solutions exist if, and only if, the velocity distribution U(x) of the external potential flow is of the form

$$U(x) = U_\infty\ C\ x^m \qquad (12.14)$$

where C and m are constants.

When the external flow is of the form (12.14), the boundary-layer equations may be treated as follows. We introduce the dimensionless coordinates

$$\xi = \frac{x}{L}\ ,\ \eta = \frac{y\sqrt{Re}}{L\ g(x)} \qquad (12.15)$$

and the dimensionless stream function

$$f(\xi,\eta) = \frac{\psi(x,y)\sqrt{Re}}{LU(x)\,g(x)} \qquad (12.16)$$

and substitute in the boundary-layer equation (12.11). If the scale function g(x) on y is taken to be

$$g(x) = \left\{\frac{2}{(1+m)}\,\frac{x}{L}\,\frac{U_\infty}{U(x)}\right\}^{\frac{1}{2}} \qquad (12.17)$$

then the stream function $f = f(\eta)$ becomes independent of ξ, and satisfies the ordinary differential equation (Falkner and Skan, 1931)

$$\frac{d^3f}{d\eta^3} + f\,\frac{d^2f}{d\eta^2} + \beta\left\{1 - \left(\frac{df}{d\eta}\right)^2\right\} = 0 \qquad (12.18)$$

where $\beta = 2m/(1+m)$ is a constant, with boundary conditions

$$f = \frac{df}{d\eta} = 0 \quad \text{at} \quad \eta = 0, \qquad \frac{df}{d\eta} = 1 \quad \text{at} \quad \eta = \infty .$$

Thus, for the class of potential flow (12.14), solution of the boundary-layer equations reduces to the solution of a nonlinear, third-order, two-point boundary value problem (12.18) for $f(\eta)$. The velocity profiles $u(x,y)$, $v(x,y)$ are then given in terms of $U(x)$ and $f(\eta)$ by

$$u = U(x) \frac{df}{d\eta}$$

$$v = -\left(\frac{\nu}{\beta}\frac{dU}{dx}\right)^{\frac{1}{2}} \left\{f + (\beta-1)\eta \frac{df}{d\eta}\right\} . \qquad (12.19)$$

12.5. EXAMPLES OF SIMILAR FLOWS

(a) Flow over a flat plate, $\beta = m = 0$ (Figure 12.2a)
Potential flow $U(x)$ = constant.

(b) Accelerated flow over a wedge, half angle $\beta\pi/2$, $0 < \beta < 1$. (Figure 12.2b)
Potential flow $U(x) \propto x^m = x^{\beta/(2-\beta)}$.

(c) Retarded flows around a corner of angle $\beta\pi/2$, $\beta < 0$ (Figure 12.2c).
Potential flow $U(x) \propto x^{\beta/(2-\beta)}$.

Such boundary-layer solutions exist for $-.198838 < \beta < 0$. At $\beta \simeq -.198838$, the solution is on the point of separating at all points on the surface. Numerical solutions to the Falkner-Skan equation (12.18) have been given by Hartree (1937), and Evans (1968). The stream functions and velocity profiles for several values of β are shown in Figures 12.3 and 12.4. The solution of the nonlinear ordinary differential equation boundary value problems arising in similarity solutions is a well-developed area of research, see Gladwell (1978), and subroutines exist for solving the equations almost automatically.

12.6. THERMAL BOUNDARY-LAYER EQUATION

So far we have considered only the fluid flow field in the

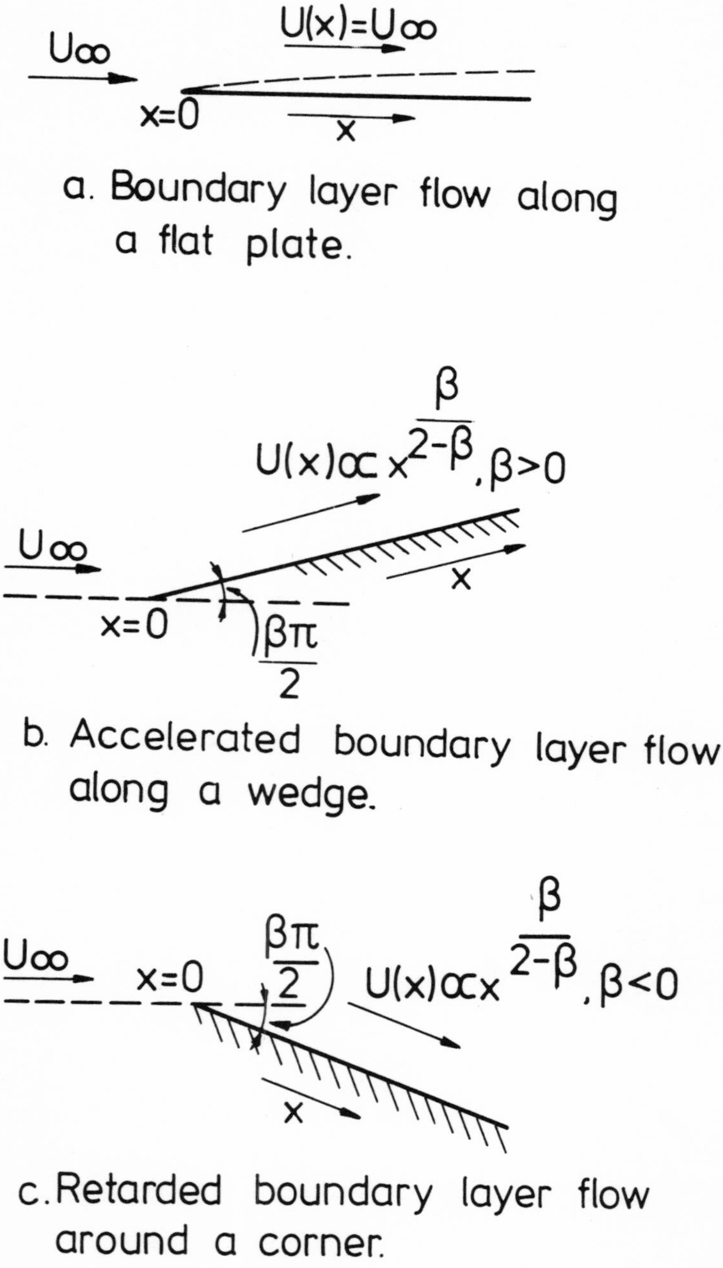
U∞
U(x)=U∞
x=0
x
a. Boundary layer flow along a flat plate.
U(x)∝x^{β/(2-β)}, β>0
U∞
x
x=0
βπ/2
b. Accelerated boundary layer flow along a wedge.
U∞
x=0
βπ/2
U(x)∝x^{β/(2-β)}, β<0
x
c. Retarded boundary layer flow around a corner.

Fig. 12.2.

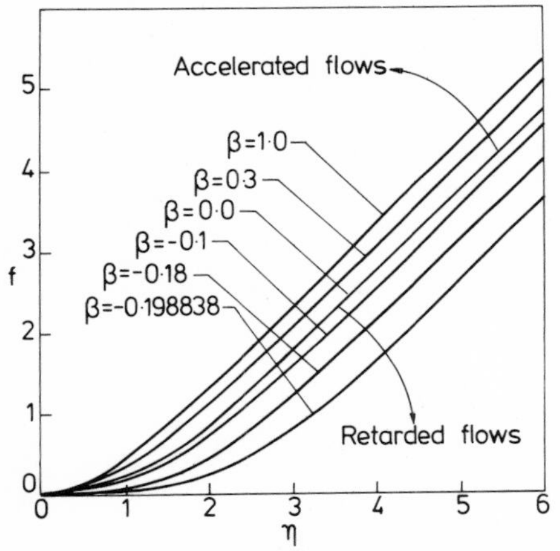

FIG.12·3. Stream function profiles for similar boundary layers.

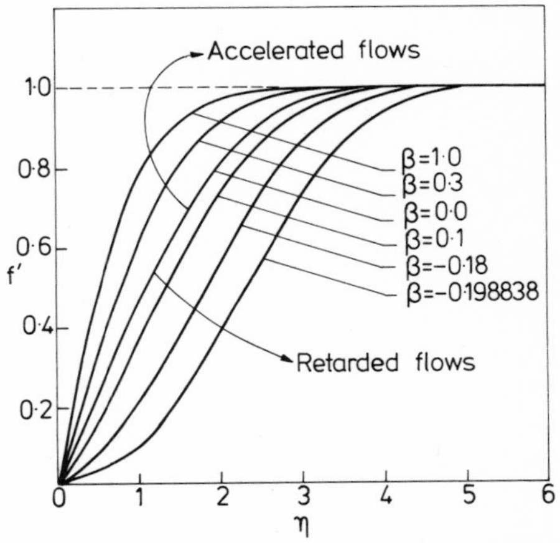

FIG.12·4 Non-dimension velocity profiles df/dη for similar boundary layers.

boundary layer near a solid surface. Many practical problems are concerned with the transfer of heat between a solid body and the surrounding fluid (by forced or natural convection), in which the bulk of the temperature transition occurs in a thin layer, the thermal boundary layer, in which flow and thermal effects interact strongly.

In its simplest form, when frictional heating and buoyancy forces are neglected, the temperature T satisfies the thermal equation

$$u \frac{\partial T}{\partial x} + v \frac{\partial T}{\partial y} = \kappa \frac{\partial^2 T}{\partial y^2} \tag{12.20}$$

where κ is the thermal diffusivity, with $T = T_0$ at $y = 0$ and $T = T_\infty$ at $y \to \infty$, and may be solved in conjunction with the flow equations (12.8) and (12.9). For constant temperatures T_0 and T_∞, the dimensionless temperature

$$\theta = \frac{T-T_0}{T_\infty - T_0}$$

may be shown to satisfy the following equation for similar boundary layers

$$\frac{d^2\theta}{d\eta^2} + \Pr f \frac{d\theta}{d\eta} = 0 \tag{12.21}$$

with $\theta = 0$ at $\eta = 0$ and $\theta = 1.0$ as $\eta \to \infty$, where Pr denotes the Prandtl number $\Pr = \nu/\kappa$. For flow past a flat plate, comparison of equations (12.8) and (12.20) shows that the velocity and temperature distributions are identical if the Prandtl number is unity; that is if

$$\frac{u}{U_\infty} = \frac{T-T_0}{T_\infty-T_0} \quad .$$

12.7. NONSIMILAR BOUNDARY LAYERS

The class of flows for which similar solutions exist is limited, and a variety of methods have been proposed to solve nonsimilar problems. We shall consider a number of methods that are based on the similarity concept.

Many of the more accurate methods have utilized a transformation of variables proposed by Gortler (1957), namely

$$\xi = \frac{1}{U_\infty L} \int_0^x U(x')\,dx', \quad \eta = y\,U(x)/(2\nu L U_\infty \xi)^{\frac{1}{2}} \quad .$$

The transformed stream function $f(\xi,\eta)$ is given by

$$f(\xi,\eta) = \psi/(2\nu L U_\infty)^{\frac{1}{2}} \quad ,$$

and the momentum equation becomes

$$\frac{\partial^3 f}{\partial\eta^3} + f\frac{\partial^2 f}{\partial\eta^2} + \beta(\xi)\left\{1 - \left(\frac{\partial f}{\partial\eta}\right)^2\right\} = 2\xi\left\{\frac{\partial f}{\partial\eta}\frac{\partial^2 f}{\partial\xi\partial\eta} - \frac{\partial^2 f}{\partial\eta^2}\frac{\partial f}{\partial\xi}\right\} \tag{12.22}$$

with boundary conditions

$$f(\xi,0) = \frac{\partial f}{\partial\eta}(\xi,0) = 0 \; ; \qquad \frac{\partial f}{\partial\eta}(\xi,\infty) = 1 \; ,$$

where

$$\beta(x) = \left(\frac{2}{U^2}\right)\frac{dU}{dx}\int_0^x U(x')\,dx' \, . \tag{12.23}$$

The "local similarity" approximation for nonsimilar boundary layers simply takes the similarity solution corresponding to the local value of $\beta(\xi)$ given by equation (12.23). Such a solution can be reasonably accurate for flows that are nearly similar (e.g. near the stagnation point $x = 0$).

Sparrow, Quack and Boerner (1970) use a local expansion of the stream function

$$f(\xi,\eta) = f(\xi_0,\eta) + (\xi - \xi_0)\frac{\partial f}{\partial \xi}(\xi_0,\eta) + \tfrac{1}{2}(\xi - \xi_0)^2 \frac{\partial^2 f}{\partial \xi^2}(\xi_0,\eta) \tag{12.24}$$

to obtain a "local nonsimilarity" solution. Substitution into equation (12.22) gives coupled third order ordinary differential equations for

$$f(\xi_0,\eta), \quad g(\eta) = \frac{\partial f}{\partial \xi}(\xi_0,\eta), \qquad h(\eta) = \frac{\partial^2 f}{\partial \xi^2}(\xi_0,\eta) \; .$$

Good results appear to be obtained both with the two-term model (f and g only) and the three-term model, except in the vicinity of flow separation. The method has also been applied to heat transfer calculations in the thermal boundary layer (Sparrow and Yu, 1971).

Many other methods of asymptotic expansion have been proposed for treating nonsimilar boundary layers, see Merk (1959) and Dewey and Gross (1967). A recent method that appears to be particularly effective for retarded flows leading to flow separation has been proposed by Kao and Elrod (1974, 1976). They use the variable β as independent coordinate in place of ξ, and expand the stream function in terms of

$$\varepsilon(\beta) = 2\xi \frac{d\beta}{d\xi} \; . \tag{12.25}$$

At the same time they introduce a 'strained' value β^* and use the expansions

$$f = f_0(\beta^*,\eta) + \varepsilon(\beta) f_1(\beta^*,\eta) + \ldots..$$

$$\beta = \beta^* + \varepsilon(\beta)\,\Delta_0(\beta^*) + \ldots.. \tag{12.26}$$

where β^* is determined by an additional condition on f_1, namely that the wall shear $d^2f_1/d\eta^2(\beta^*,0)$ be zero.

Example For a cylinder radius r in cross-flow, the potential flow is

$$U(x)/U_\infty = 2\sin(x/r)$$

where x is circumference distance measured from the stagnation point. Then $\xi = 2\{1 - \cos(x/r)\}$ and $\beta = 2\cos(x/r)/\{1 + \cos(x/r)\}$.

Figure 12.5 shows the predicted values of the wall shear stress $f''(x/r, 0)$ calculated by the methods mentioned above. The 'exact' numerical results are taken from calculations by Terrill (1960).

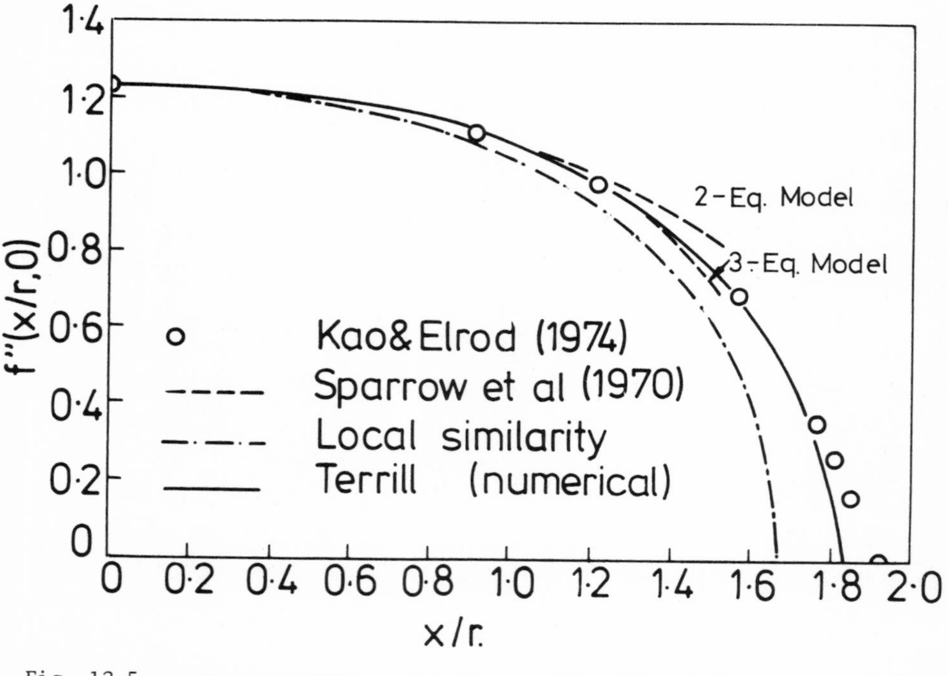

Fig. 12.5. Comparison of predicted shear stress around a cylinder in cross-flow.

12.8 NUMERICAL METHODS

In recent years a variety of numerical methods have been developed for approximating the solutions to boundary-layer problems. Two main classes of finite-difference methods have been proposed.

A procedure used by Smith and Clutter (1963), based on

the approach of Hartree and Womersley (1937), involves the replacement of the first-order stream-wise derivatives by finite differences, followed by integration of the resulting ordinary differential equations in the cross-stream direction at successive stations. The need to employ an iterative technique to match the boundary conditions at $\eta = 0$ and $\eta = \infty$ would appear to make such cross-stream integration methods unattractive, especially for problems in which the momentum and energy equations are coupled.

The second class comprises the more traditional finite-difference marching methods. Fussel and Hellums (1965) examined the solution of Gortler transformed boundary-layer equations using central-difference cross-stream replacements, and compared the solutions obtained using implicit centred and backward stream-wise differencing applied to both the nonlinear and to the linearized boundary-layer equations. They concluded that the nonlinear central-difference method was very effective, and proposed an iterative method with a five-diagonal iteration matrix. Keller and Cebeci (1971) applied a 'box' method based on centred differences in which the third order boundary-layer equation is replaced by three first order equations. An advantage of this approach is that the stream function f, velocity u (proportional to $\partial f/\partial\eta$), and shear stress ($\partial^2 f/\partial\eta^2$) are all calculated to the same order of accuracy. The use of this method will be illustrated by an example in chapter 13.

12.9. CONCLUDING REMARKS

The analysis that has been given above for two-dimensional flow is very easily extended to include axisymmetrical flow, and the boundary-layer equations for compressible flow may be treated in a similar manner to the incompressible equations. The boundary layer and similarity concepts may also be applied to turbulent flows, but the mathematical description of such flows is still in the development stage.

Although similarity solutions exist only for a limited class of problems, similarity is the point of departure for a variety of mathematical and numerical methods of solving nonsimilar flow problems. Asymptotic methods of the type described above require significantly less computation than a full numerical solution of the parabolic boundary layer equations.

13

STRONGLY NONLINEAR PARABOLIC EQUATIONS-EXAMPLES

13.1. INTRODUCTION

In this chapter we discuss some aspects of the numerical solution of problems involving nonlinear parabolic partial differential equations by treating three practical examples. In the first we examine the solution of a laminar flow boundary-layer problem of the type described in chapter 12. The second example is concerned with the solution of coupled parabolic systems arising from laminar flame theory; in this application, transient parabolic equations are often integrated in order to obtain the steady-state solutions. Finally, we discuss some solutions to a "stiff", highly nonlinear chemical kinetic problem involving five coupled parabolic equations.

13.2. EXAMPLE 1 . BOUNDARY-LAYER FLOW

In chapter 12, the boundary-layer equations for laminar incompressible flow in two dimensions were formulated and approximate methods of solution based on the concept of similarity were described. Here, we discuss a numerical technique that has been used to solve a variety of both laminar and turbulent boundary-layer equations numerically (Keller and Cebeci, 1971 and 1972), and illustrate its use in solving a nonsimilar, retarded flow problem due to Howarth (1938).

The laminar flow boundary-layer equation may be written in terms of the non-dimensional stream function $f(\xi,\eta)$ as a third-order parabolic equation (c.f. equation (12.22))

$$\frac{\partial^3 f}{\partial\eta^3} + f\,\frac{\partial^2 f}{\partial\eta^2} + \beta(\xi)\left\{1 - \left(\frac{\partial f}{\partial\eta}\right)^2\right\} = 2\xi\left\{\frac{\partial f}{\partial\eta}\,\frac{\partial^2 f}{\partial\eta\partial\xi} - \frac{\partial^2 f}{\partial\eta^2}\,\frac{\partial f}{\partial\xi}\right\} \qquad (13.1)$$

where $\beta(\xi) = [2\xi/U(\xi)]\,dU(\xi)/d\xi$, and $U(\xi)$ is the velocity of the external, potential flow. We shall solve the Howarth retarded flow problem where U is defined by

$$U(x) = 1 - x/L.$$

The boundary conditions for the problem are

$$f(\xi,0) = \frac{\partial f}{\partial \eta}(\xi,0) = 0; \quad \lim_{\eta\to\infty} \frac{\partial f}{\partial \eta}(\xi,\eta) = 1. \tag{13.2}$$

The numerical technique that we shall describe is that applied by Keller and Cebeci (1971) to laminar boundary-layer flows, and is based on a 'box' difference method for parabolic problems proposed by Keller (1971). In this method the third-order equation (13.1) is replaced by first-order equations by introducing new dependent variables g and h such that

$$\begin{aligned}
&\text{(a)} \qquad \frac{\partial f}{\partial \eta} = u\,, \\
&\text{(b)} \qquad \frac{\partial u}{\partial \eta} = v\,, \\
&\text{(c)} \qquad \frac{\partial v}{\partial \eta} = -\,fv - \beta\left\{1 - u^2\right\} + 2\xi\left\{u\,\frac{\partial u}{\partial \xi} - v\,\frac{\partial f}{\partial \xi}\right\}.
\end{aligned} \tag{13.3}$$

Although this substitution increases the number of unknowns, it ensures that the velocity profile (proportional to u) and the shear stress (proportional to v) are calculated to the same order accuracy as the stream function.

An arbitrarily spaced rectangular net of points (ξ_n,η_j) is defined on $\xi > 0$, $0 < \eta < \eta_\infty$, where

$$\xi_0 = 0, \ \xi_n = \xi_{n-1} + k_n\,,$$

and

$$\eta_0 = 0, \ \ \eta_j = \eta_{j-1} + h_j, \ \ j = 1,2,\dots,J, \ \eta_J = \eta_\infty.$$

The first order equations are then approximated by differences over each rectangle of the mesh. Equation (13.3c) uses differences centred on $(\xi_{n-\frac{1}{2}}, \eta_{j-\frac{1}{2}})$ for the first derivatives, with simple averaging to evaluate the variables at the off-grid midpoints. The equations (13.3a) and (13.3b) use differences centred on $(\xi_n, \eta_{j-\frac{1}{2}})$.

The difference equations are therefore, for $1 < j < J$,

$$\text{(a)} \qquad f_j^n - f_{j-1}^n = h_j u_{j-\frac{1}{2}}^n \equiv \tfrac{1}{2} h_j (u_{j-1}^n + u_j^n)$$

$$\text{(b)} \qquad u_j^n - u_{j-1}^n = h_j v_{j-\frac{1}{2}}^n \tag{13.4}$$

$$\text{(c)} \qquad h_j^{-1}(v_j^{n-\frac{1}{2}} - v_{j-1}^{n-\frac{1}{2}}) = -(fv)_{j-\frac{1}{2}}^{n-\frac{1}{2}} - \beta(\xi_{n-\frac{1}{2}})\{1 - (u^2)_{j-\frac{1}{2}}^{n-\frac{1}{2}}\} +$$

$$+ 2\,k_n^{-1}\,\xi_{n-\frac{1}{2}}\{u_{j-\frac{1}{2}}^{n-\frac{1}{2}}(u_{j-\frac{1}{2}}^n - u_{j-\frac{1}{2}}^{n-1}) - v_{j-\frac{1}{2}}^{n-\frac{1}{2}}(f_{j-\frac{1}{2}}^n - f_{j-\frac{1}{2}}^{n-1})\}.$$

The boundary conditions are simply

$$f_0^n = u_0^n = 0; \quad u_J^n = 1. \tag{13.5}$$

Equations (13.4) and (13.5) are treated as 3J + 3 non-linear equations in the unknowns f, u, v at each station ξ_n. Initial profiles at $\xi = 0$ may be obtained by solving the equations with the superscripts n and $n-\frac{1}{2}$ set to zero: this yields an accurate approximation to the similarity solution of the Falkner-Skan ordinary differential equation (12.18).

The nonlinear equations may be solved iteratively using Newton's method. By ordering the variables as

$$(f_0, u_0, v_0, f_1, \ldots, f_J, u_J, v_J)$$

and the equations as

$$\begin{aligned}
&1 - 2: && \text{boundary conditions at } \eta = 0,\\
&3 - 5: && (13.4c), (13.4a), (13.4b) \text{ for } j = 1,\\
& && \vdots\\
&3J, 3J+1, 3J+2: && (13.4c), (13.4a), (13.4b) \text{ for } j = J,\\
&3J+3: && \text{boundary condition at } \eta_J,
\end{aligned} \tag{13.6}$$

Keller *et al.* (1971) show that the increment $\underset{\sim}{\delta}^{(i)}$ on the ith Newton iteration requires the solution of the linear system

$$A^{(i)}\,\underset{\sim}{\delta}^{(i)} = \underset{\sim}{q}^{(i)} \quad (i = 0, 1, \ldots) \tag{13.7}$$

with the matrix $A^{(i)}$ having a block tridiagonal form

$$A = \begin{bmatrix} A_0 & C_0 & & & \\ B_1 & A_1 & C_1 & & \\ & \ddots & \ddots & \ddots & \\ & & B_{J-1} & A_{J-1} & C_{J-1} \\ & & & B_J & A_J \end{bmatrix} \tag{13.8}$$

Equation (13.7) may be solved using a block tridiagonal factorization

$$A = LU ,$$

where

$$L = \begin{bmatrix} I & & & \\ L_1 & I & & \\ & \ddots & \ddots & \\ & & L_J & I \end{bmatrix} , \quad U = \begin{bmatrix} U_0 & C_0 & & & \\ & U_1 & C_1 & & \\ & & \ddots & \ddots & \\ & & & U_{J-1} & C_{J-1} \\ & & & & U_J \end{bmatrix} \tag{13.9}$$

with

$$\left.\begin{aligned} U_0 &= A_0 , \\ L_j &= B_j \, U_{j-1}^{-1} \\ U_j &= A_j - L_j \, C_{j-1} \end{aligned}\right\} \quad j = 1, 2, \ldots, J , \tag{13.10}$$

followed by the back-substitutions

$$L\underset{\sim}{z} = \underset{\sim}{g} , \quad U\underset{\sim}{\delta} = \underset{\sim}{z} . \tag{13.11}$$

The stability of this block tridiagonal method has been studied by Varah (1972), where the number of multiplications of the block-triangularization method is compared with a band-matrix solver. In Keller *et al.* (1971), the equations are ordered with (b) and (c) in (13.3) reversed. This gives

$$A_0 = \begin{bmatrix} 1 & 0 & 0 \\ 0 & 1 & 0 \\ 0 & 1 & \tfrac{1}{2}h_1 \end{bmatrix}$$

so that U_0^{-1} is $O(h_1^{-1})$, tending to give an unstable decomposition. The ordering (13.6) is used in Keller and Cebeci (1972), where the method is applied to turbulent boundary-layer flows.

To achieve convergence to the solutions of the nonlinear equations, it may well be adequate to use a modified Newton iteration, in which A is decomposed, say, for the first iteration, and the decomposition then used in subsequent iterations

$$A^{(0)} = L^{(0)}\, U^{(0)} \,,$$

$$L^{(0)}\, \underset{\sim}{z}^{(i)} = \underset{\sim}{g}^{(i)} \,, \quad U^{(0)}\, \underset{\sim}{\delta}^{(i)} = \underset{\sim}{z}^{(i)} \quad (i = 0, 1, 2, \dots\,); \tag{13.12}$$

see chapter 18 for further methods for nonlinear equations.

The exact numerical solution of the difference equations gives a second-order accurate approximation. Furthermore, if we use equi-spaced meshes $h_j = h$, $k_n = k$, the errors in the numerical solution have expansions in even powers of the step-sizes h and k. Thus, provided the iteration errors and rounding errors arising in the numerical solution are kept sufficiently small, it may be possible to improve the accuracy of the solutions using Richardson extrapolation. For example, if solutions at $(\bar{\xi}, \bar{\eta})$ are computed using mesh sizes $(h = h_1, k)$ and $(h = h_2, k)$, the solutions may be combined, using an obvious notation, to give

$$f(\bar{\xi}, \bar{\eta};\, h_1, h_2, k) = \frac{h_1^2\, f(\bar{\xi}, \bar{\eta};\, h_2, k) - h_2^2\, f(\bar{\xi}, \bar{\eta};\, h_1, k)}{h_1^2 - h_2^2} \tag{13.13}$$

with an error $O(h_1^4 + k^2)$, and similarly for u and v. Extrapolations may also be carried out with respect to the step-size k.

Some results quoted by Keller *et al.* (1971) for solutions to Howarth's retarded flow problem are shown in Table 13.1. for the wall shear stress function $v(\xi,0)$. A uniform mesh size h is used with $\eta_\infty = 6.0$. Results obtained using relatively fine mesh sizes $h = 1/10$ and $h = 1/20$ are in good agreement with results obtained by extrapolation from solutions using very coarse grids.

13.3. EXAMPLE 2. LAMINAR FLAME PROPAGATION

This example concerns the propagation of a laminar flame through a premixed combustible gas. Initially, the time-dependent equations for laminar flames were formulated and solved in order to arrive at the steady-state solution, since considerable difficulty had been experienced in solving the two-point boundary value problem for the time-independent equations (Spalding, Stephenson and Taylor, 1971). More recently, the unsteady solutions have been used to estimate the energy required to ignite combustible gas mixtures (Bledjian, 1973; Overley, Overholser and Reddien, 1978).

Neglecting viscous effects and radiative heat transfer, and assuming for simplicity that the enthalpies are linear functions of temperature, the one-dimensional equations may be written as

$$\frac{\partial\rho}{\partial t} + \frac{1}{r^k}\frac{\partial}{\partial r}(r^k\rho v) = 0 \qquad (13.14)$$

$$\frac{\partial y_i}{\partial t} + v\frac{\partial y_i}{\partial r} = \frac{1}{\rho r^k}\frac{\partial}{\partial r}\left(r^k D_i\rho\frac{\partial y_i}{\partial r}\right) + \frac{W_i}{\rho}\quad (i = 1, 2,\ldots,n) \qquad (13.15)$$

$$C_p\frac{\partial T}{\partial t} + vC_p\frac{\partial T}{\partial r} = \frac{1}{\rho r^k}\frac{\partial}{\partial r}\left(r^k\lambda\frac{\partial T}{\partial r}\right) - \sum_{i=1}^{n}\frac{W_i h_i}{\rho} + \sum_{i=1}^{n} D_i\frac{\partial y_i}{\partial r}\frac{\partial h_i}{\partial r} \qquad (13.16)$$

where W_i denotes the chemical kinetic rate of production of component i with mass fraction y_i, D_i is a multicomponent diffusion coefficient, and h_i is the partial enthalpy

TABLE 13.1

ξ_n	x/L	Wall shear stress function $v(\xi, 0)$						
		(a) h = 1	(b) h = 1/2	(c) h = 1/3	Extrapolated from (a)(b)(c)	Extrapolated from (d)(e)	(e) h = 1/20	(d) h = 1/10
$\xi_0 = 0$	0	.506065	.478914	.473753	.469594	.469601	.469694	.469697
$\xi_{10} = .4$	.05132	.377384	.353976	.349523	.345960	.345941	.346021	.346262
$\xi_{20} = .7$	.09171	.238914	.215469	.210875	.207143	.207143	.207227	.207481
$\xi_{30} = .8$	.10557	.175931	.149222	.143751	.139256	.139240	.139343	.139651
$\xi_{40} = .86$	.11400	.129243	.096389	.089096	.082990	.082860	.083004	.083438
$\xi_{50} = .894$	.11881	.096925	.054038	.042351	.031198	.030531	.030832	.031740
No. of mesh pts. η_j		7	13	19	(39)	(182)	121	61

Computed and extrapolated values of the wall shear stress function $v(\xi,0)$ for Howarth's retarded flow problem (Keller and Cebeci, 1971)

$$h_i = C_{pi}\,(T - T_0) + h_{fi}\ ,\ h = \sum_i y_i h_i\ .$$

The convective terms may be eliminated, and the continuity equation (13.14) automatically satisfied, by introducing a stream function Ψ such that

$$\frac{\partial\Psi}{\partial r} = \rho r^k,\quad \frac{\partial\Psi}{\partial t} = -\ \rho v r^k\ . \tag{13.17}$$

Transforming to the independent variables (t,Ψ) gives

$$\frac{\partial y_i}{\partial t} = \frac{\partial}{\partial\Psi}\left(D_i\rho^2 r^{2k}\frac{\partial y_i}{\partial\Psi}\right) + \frac{W_i}{\rho}\ (i = 1,\ 2,\ \ldots,n)\ , \tag{13.18}$$

$$C_p\frac{\partial T}{\partial t} = \frac{\partial}{\partial\Psi}\left(\lambda\rho r^{2k}\frac{\partial T}{\partial\Psi}\right) + \sum_{i=1}^{n} D_i\rho^2 r^{2k}\frac{\partial y_i}{\partial\Psi}\frac{\partial h_i}{\partial\Psi} - \sum_{i=1}^{n}\frac{W_i h_i}{\rho}\ . \tag{13.19}$$

The chemical source terms W_i in equations (13.18) and (13.19) are determined by the chemical reaction scheme, and usually introduce stiff nonlinear coupling between the species variables and with the temperature. For one of the simplest flames studied, the hydrazine flame (Adams and Cook, 1960), the reaction mechanism has been taken to be

$$S_1 \to S_2 + S_2,\ \text{rate}\ r_1 \propto y_1\exp(-E_1/RT),$$
$$S_1 + S_2 \to S_2 + S_3 + S_3,\ \text{rate}\ r_2 \propto y_1y_2\exp(-E_2/RT),$$
$$S_2 + S_2 + M \to S_3 + S_3 + M\ ,\ \text{rate}\ r_3 \propto y_2^2\exp(-E_3/RT)\ ,$$

so that $W_1 = -\ r_1 - r_2$ and $W_2 = 2r_1 - 2r_3$.

The planar laminar flame equations $(k = 0)$ have been used principally for calculating steady-state solutions in which a narrow flame front separating burnt and unburnt gas travels with constant flame speed S_f (Figure 13.1 and 13.2). The initial profiles $y_i\,(0,\Psi)$ and $T(0,\Psi)$ are such that

$$y_i \to y_{ib},\ T \to T_b\ \text{as}\ \Psi \to -\infty,$$

$$y_i \to y_{iu},\ T \to T_u\ \text{as}\ \Psi \to \infty.$$

Zero flux boundary conditions are used at $\Psi = \pm\infty$.

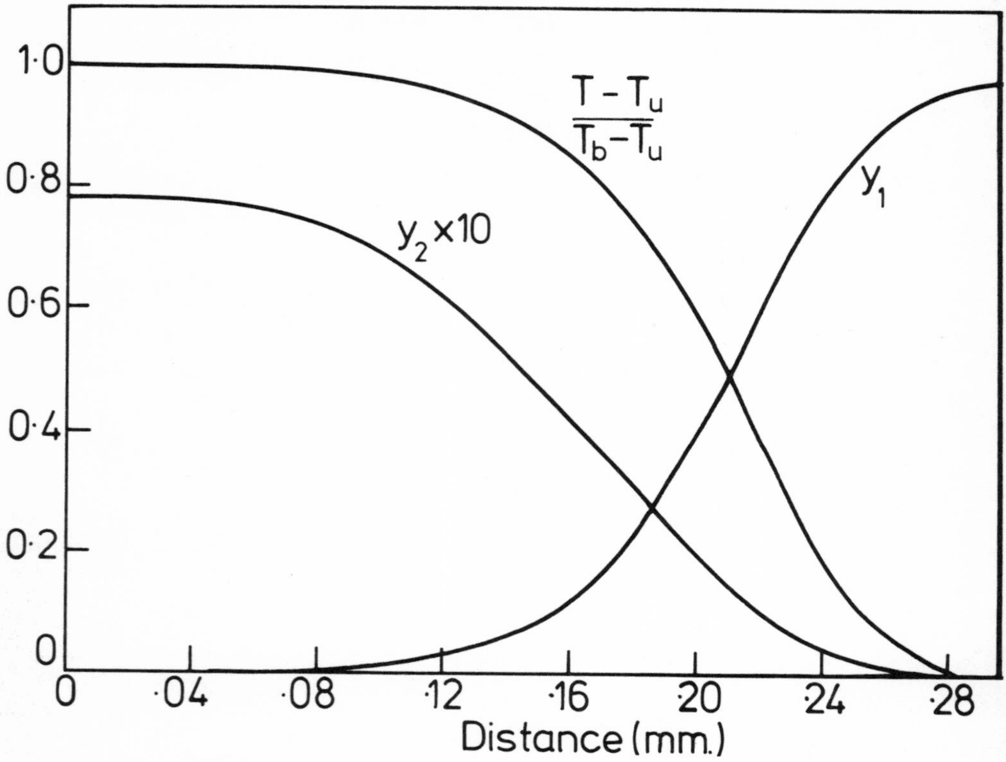

Fig. 13.1. Steady-state structure of a hydrazine decomposition flame.

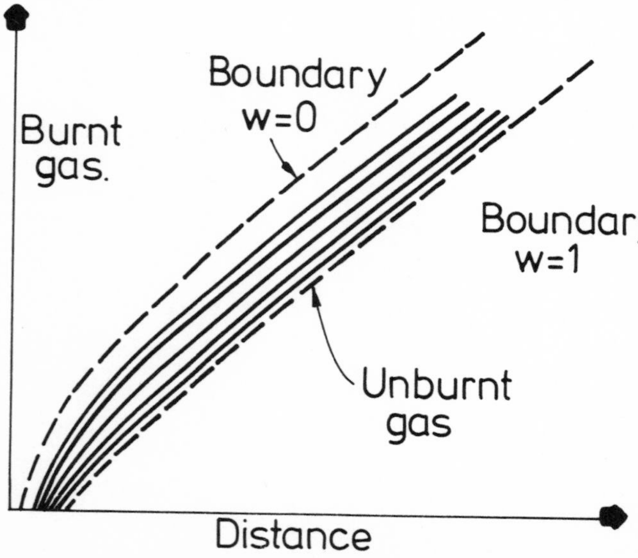

Fig. 13.2. Schematic plot of temperature isotherms during development of a steady flame.

To confine the numerical solution to the flame front region, Spalding *et al.* (1971) use a non-dimensional stream function

$$\omega = \frac{\Psi-\Psi_b}{\Psi_u-\Psi_b} \qquad (13.20)$$

where $\Psi_u(t)$ and $\Psi_b(t)$, the effective locations of the unburnt and burnt limits of the flame front region, are computed as part of the solution.

Several studies have employed the numerical method developed by Patankar and Spalding (1968) as a general technique for boundary-layer flows. The method uses a non-uniform distribution of mesh points w_j, and is based on a fully-implicit marching integration technique applied to the locally-linearized equations. The linearization neglects any coupling between the equations, so that each of the parabolic equations gives a tridiagonal set of linear equations to solve for each variable in turn. Since, in the flame equations, strong nonlinear coupling of the equations often arises through the source terms, this imposes a stability limitation on the time step that may be used. By careful ordering of the equations and the use of computed variables in the source terms, the stability restriction can to some extent be alleviated (Stephenson and Taylor, 1973).

Using this technique, solutions have been obtained for a variety of flames, ranging in complexity from a hydrazine flame (Spalding, Stephenson and Taylor, 1971), involving two species continuity equations, to a hydrogen/oxygen/nitrogen flame with eight species equations (Stephenson and Taylor, 1973).

In problems such as this, when interest lies solely in evaluating the steady-state solution as $t \to \infty$, there are strong arguments in favour of using a low-order highly-stable implicit time-integration method, since the accuracy of the steady state solution does not depend on the time steps provided stability is maintained. In the laminar flame equations, the 'stiff' nature of the source terms makes the use of a stable method even more desirable. The effort involved in solving coupled sets of linear or nonlinear equations at each step need not be very great if proper advantage is taken of the banded or block-tridiagonal form of the associated matrix, and the reduction in the number of time steps, compared with methods limited by stability, should be considerable.

The transient flame equations have also been used to calculate the energy needed to ignite the combustible gases and initiate the flame. In this case, the initial conditions are modified to describe a small hot packet of burnt gas (representing a spark) surrounded by unburnt gas. The solutions then either develop into a propagating flame, or show the spark energy dissipating without ignition. Numerical studies of the ignition energy required for planar and spherical flames have been reported by Bledjian (1973), and for spherical and prolate spheroidal flames by Overley, Overholser and Reddien (1978). The latter study uses a Crank-Nicolson predictor-corrector method with a non-uniform mesh.

13.4. EXAMPLE 3. PARALLEL-PLATE REACTOR

The final example is in many respects similar to the laminar flame model above. It concerns the burning of a combustible gas mixture contained in a reactor consisting essentially of two horizontal plates maintained at different temperatures, the hot plate ($z = 0$) being above the cold plate ($z = a$) to eliminate convective flow. The problem is part of a study of auto-ignition of fuel/air mixtures in the vicinity of hot surfaces (Bull and Quinn, 1975).

The parabolic equations may be written as

$$\rho(T)\, C_v(T)\, \frac{\partial T}{\partial t} = \beta(T)\, q\, \rho(T)\, y_2 + \frac{\partial}{\partial z}\left\{D_1(T)\, \frac{\partial T}{\partial z}\right\} \qquad (13.21)$$

$$\rho(T)\, \frac{\partial y_i}{\partial t} = R_i + \frac{\partial}{\partial z}\left\{\rho(T)\, D_i(T)\, \frac{\partial y_i}{\partial z}\right\} \quad (i = 2, 3, 4, 5) \qquad (13.22)$$

where D_1 denotes thermal conductivity, the D_i ($i = 2,3,4,5$) are diffusion coefficients and the R_i are chemical kinetic source terms. The source terms are nonlinear functions of the species concentrations y_i (Robinson, 1976) and are strongly coupled to the temperature through terms of the form

$$y_i \sim \exp(-C/T), \text{ C constant.}$$

The boundary conditions are simply

$$\begin{aligned} T(t,0) &= T_h \quad \text{(hot plate)}, \\ T(t,a) &= T_c \quad \text{(cold plate)}, \\ \frac{\partial y_i}{\partial z}(t,0) &= \frac{\partial y_i}{\partial z}(t,a) = 0, \end{aligned} \tag{13.23}$$

and the initial conditions are

$$\begin{aligned} T(0,z) &= T_h - \frac{z}{a}(T_h - T_c), \\ y_i(0,z) &= 0 \ (i = 2,3,4) \ \text{(intermediates)}, \\ y_5(0,z) &= y_5^* \quad \text{(fuel/oxygen)}. \end{aligned} \tag{13.24}$$

Experimental measurements in the parallel reactor indicated that under certain conditions a 'cool flame' occurs, giving the type of temperature/time variation shown in Figure 13.3; to obtain solutions to the model equations that matched the experiments required a considerable number of calculations with different model parameters, but reasonable agreement was eventually obtained (Figure 13.4).

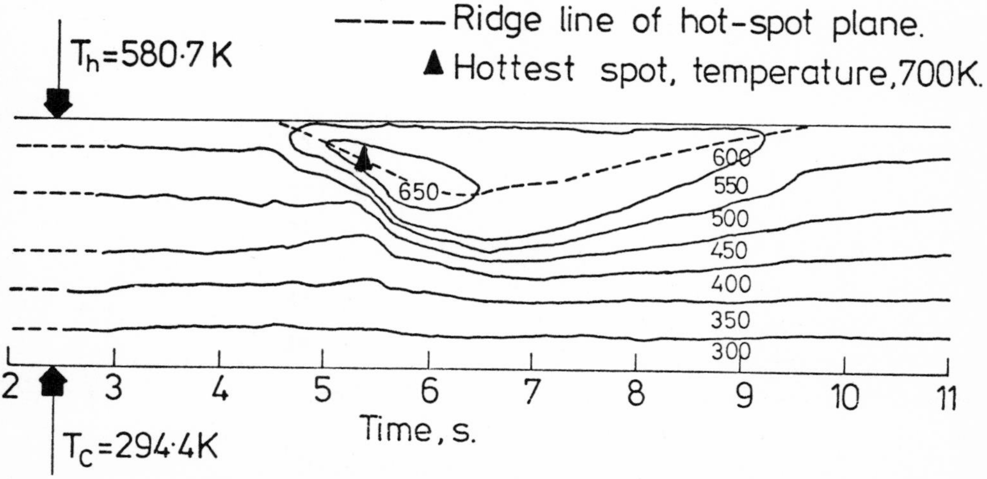

Fig. 13.3. Time evolution of the temperature of a cool flame.

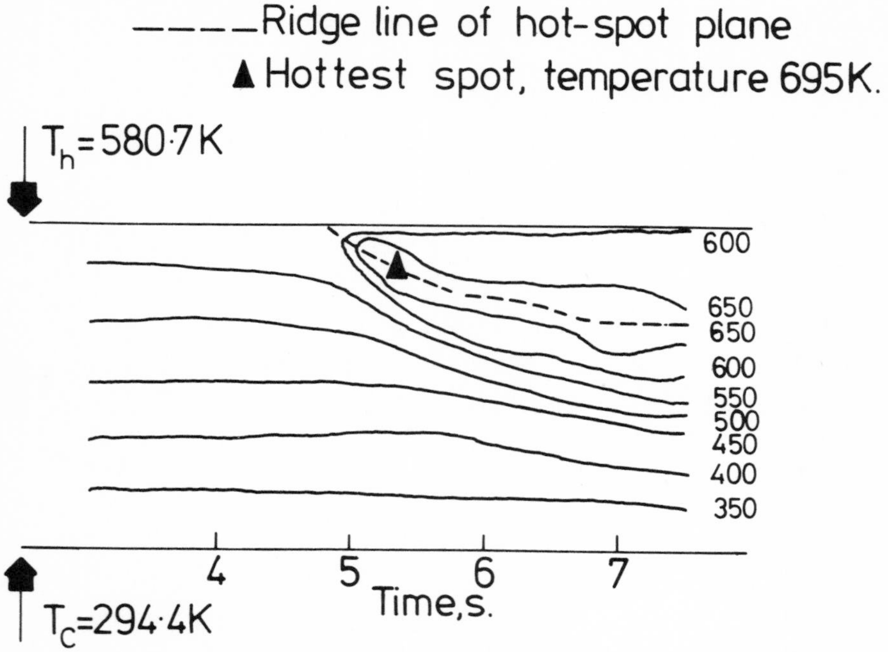

Fig. 13.4. Model prediction of the cool flame shown in Figure 3.

A 'method of lines' approach was used to obtain the numerical solutions, in which finite-difference replacements for the spatial derivatives are substituted at a prescribed set of values of the spatial variable z, and the resulting large set of ordinary differential equations are integrated using a suitable initial-value method. Recently, software has become available that seeks to simplify the task of setting up a program to solve such parabolic systems, using the method of lines. Sincovec and Madsen (1975) have developed a software interface PDEONE, suitable for a coupled system of essentially parabolic equations, that automatically evaluates the semi-discrete ordinary differential equations that result from using centred-difference approximations on a grid specified by the user; a robust variable-step, variable-order integration routine may then be used to solve the resulting ordinary differential equations, the GEARB algorithm for stiff problems developed by Hindmarsh (1973) being well-suited for use with PDEONE.

A similar software interface (Robinson, 1976) has been used in solving equations (13.21) and (13.22). The user

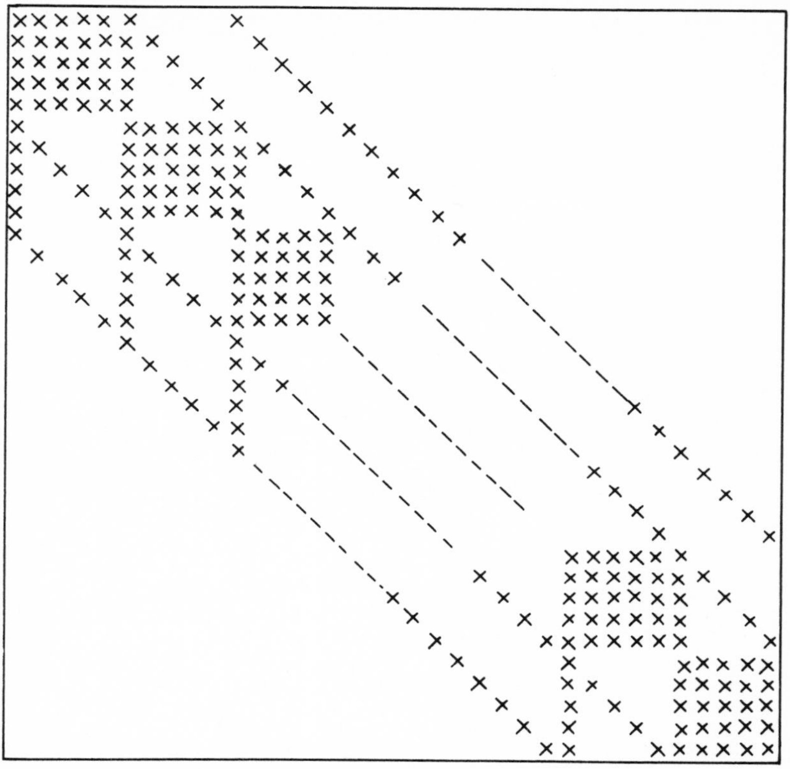

Fig. 13.5. Band matrix structure of the Jacobian matrix for the parallel plate reactor model.

specifies the form of the equations and boundary conditions, and defines the spatial grid. A system of ordinary differential equations is then generated, in which the coupling between equations is confined to a band of half-width 2N+1, where N is the number of parabolic equations (Figure 13.5). As the equations are known to be very 'stiff', a variable-step A-stable implicit method is used to integrate the differential equations.

A modified Newton iteration is performed to solve the nonlinear equations at each integration step, using a banded matrix triangular decomposition and back-substitution. In solving this problem it was found that an average of over 60 back-substitutions were carried out for each matrix decomposition. Roughly 40 matrix decompositions were needed in a typical run, many of these being necessary principally because of the wide variation in the step-size used during the integration.

PART III

FINITE-DIFFERENCE AND FINITE ELEMENT EQUATIONS

This part contains a brief description of finite-difference discretizations to linear elliptic equations and of solution techniques which are closely related to the form of the discretization. In addition, there is a discussion of solution techniques for finite element discretizations of linear elliptic equations and a chapter on solving the nonlinear equations arising in discretizations of nonlinear elliptic problems. Also included is a chapter on alternating direction methods which have applications to both elliptic and parabolic equations.

The description of finite-difference methods is deliberately rather brief as this material has been widely discussed elsewhere. The interested reader is directed to Isaacson and Keller (1966, ch.9) or Forsythe and Wasow (1960) for a discussion. The inexperienced reader may find Smith (1977) and Ames (1969) of help here.

Contents

Chapter	Title	Contributor
14	Finite-Difference Equations and Their Iterative Solution.	J.K. Reid
15	Direct Solution of Finite Element and Finite-Difference Equations.	J.K. Reid
16	Fast Direct Methods for Separable Elliptic Equations.	G. Hall
17	Alternating Direction Implicit Methods.	J. Williams
18	Solution of Nonlinear Equations.	T.L. Freeman

14

FINITE-DIFFERENCE EQUATIONS AND THEIR ITERATIVE SOLUTION

14.1. FINITE-DIFFERENCE APPROXIMATIONS

Elliptic equations may be approximated by placing a grid over the region Ω and replacing the differential operator by difference equations involving nearby grid points. For example Laplace's equation

$$\nabla^2 u \equiv \frac{\partial^2 u}{\partial x^2} + \frac{\partial^2 u}{\partial y^2} = 0 \tag{14.1}$$

may be approximated on a square grid by the 5-point finite-difference equation

$$\nabla_h^2 u \equiv [u(x+h,y) + u(x-h,y) + u(x,y+h) + u(x,y-h) - 4u(x,y)]/h^2 = 0 \tag{14.2}$$

A finite-difference operator L_h on a grid whose step sizes are proportional to the parameter h is said to be *consistent* with the differential operator L if the *local truncation error*

$$\tau_h = L_h u - Lu , \tag{14.3}$$

defined only at the grid points for which L_h is defined, tends to zero with h. It is trivial to verify this condition by using Taylor's theorem to expand $L_h u$ in terms of u and its derivatives at (x,y). For example if $L = \nabla^2$ and $L_h = \nabla_h^2$ then the local truncation error is

$$\tau_h = \frac{h^2}{12}\left(\frac{\partial^4 u}{\partial x^4} + \frac{\partial^4 u}{\partial y^4}\right) + 0(h^4) . \tag{14.4}$$

For points at or near the boundary Γ of Ω similar difference approximtions will be needed for the boundary conditions and the local truncation error must again tend to zero with h for consistency. It is, of course, sensible to choose approximations such that the truncation error at the boundary has at least as high a rate of convergence to zero as that for the differential equation itself. Sometimes this is very easy, for instance when grid points lie on a Dirichlet boundary on which the solution is known, but it can be difficult for curved boundaries particularly when the boundary condition involves normal derivatives, see Fox (1962, pp. 264-266) or Forsythe and Wasow (1960, pp. 202-204).

We may collect all the difference equations together into a single equation

$$L_h\ u_h\ = 0 \tag{14.5}$$

assumed to approximate the given differential equation

$$L\ u\ = 0. \tag{14.6}$$

In this chapter we will almost always be assuming that L and its approximation L_h are linear so that equation (14.5) takes the matrix form

$$A_h \underset{\sim}{u}_h - \underset{\sim}{b}_h = \underset{\sim}{0}\ . \tag{14.7}$$

To estimate the accuracy of a solution there are two simple approaches available. The first, *deferred approach to the limit*, depends on repeating the calculation with a changed step-size h (e.g. halved) and on knowing that the error behaves smoothly as a function of h. For example in the case of Laplace's equation (14.1) with known boundary values on the edge of a square whose side length is an integral multiple of h the error has the form

$$u_h(x,y) - u(x,y) = h^2\ e(x,y) + 0(h^4) \tag{14.8}$$

if u_h is a component of $\underset{\sim}{u}_h$, the finite-difference solution with step h. This function can be estimated as

$$h^2\ e(x,y) = \frac{4}{3}\ (u_h(x,y) - u_{h/2}(x,y))+0(h^4)\ . \tag{14.9}$$

The more accurate solution, $u_{h/2}$, may be corrected by $1/3(u_h-u_{h/2})$ and the correction taken as an error estimate.

The second simple approach to error estimation is Fox's *deferred correction* (see Fox (1962, p. 258), for example). Here the truncation errors (see equation (14.3)) are estimated directly from higher differences of the current approximate solution, that is we calculate c(x,y) in

$$L_h(u(x,y)) - D(u(x,y)) \approx c(x,y) \quad . \qquad (14.10)$$

A more accurate solution is therefore obtainable by solving the equation

$$L_h \, u_h' = c \qquad (14.11)$$

in an obvious notation, and the difference $\underset{\sim}{u}_h' - \underset{\sim}{u}_h$ may be used for an error estimate. The process may be iterated by calculating a further correction from the improved solution $\underset{\sim}{u}_h'$.

14.2. THE CLASSICAL ITERATIVE METHODS

Each finite-difference equation (14.2) involves only a few mesh values, while the total number of mesh values is likely to be very large, so the overall system is likely to be very sparse. It is very worthwhile to exploit this sparsity and the structure of the equations and iterative methods do so very effectively since they require little storage apart from the difference equations themselves and a current iterate that approximates the required finite-difference solution. Indeed it is not even necessary to store the difference equations if code can be written to generate them whenever they are wanted.

A number of iterative methods for the linear set of equations

$$A\underset{\sim}{x} = \underset{\sim}{b} \qquad (14.12)$$

are based on a splitting

$$A = M - N \qquad (14.13)$$

and the use of the iteration

$$M\underset{\sim}{x}^{(k+1)} = N\underset{\sim}{x}^{(k)} + \underset{\sim}{b} \, . \tag{14.14}$$

Of course it is necessary that the splitting be chosen so that equation (14.14) is easy to solve. In practice M is usually chosen diagonal, block diagonal, triangular or block triangular. It is very straightforward to verify that the errors $\underset{\sim}{e}^{(k)} = \underset{\sim}{x}-\underset{\sim}{x}^{(k)}$, the residuals $\underset{\sim}{r}^{(k)} = \underset{\sim}{b}-A\underset{\sim}{x}^{(k)}$ and the displacements $\underset{\sim}{\Delta}^{(k)} = \underset{\sim}{x}^{(k+1)}-\underset{\sim}{x}^{(k)}$ satisfy the equations

$$\left.\begin{aligned} \underset{\sim}{e}^{(k+1)} &= T\underset{\sim}{e}^{(k)} \\ \underset{\sim}{r}^{(k+1)} &= T\underset{\sim}{r}^{(k)} \\ \underset{\sim}{\Delta}^{(k+1)} &= T\underset{\sim}{\Delta}^{(k)} \end{aligned}\right\} \tag{14.15}$$

where T is the iteration matrix

$$T = M^{-1} N \, . \tag{14.16}$$

It follows that convergence depends on whether the largest eigenvalue of T has modulus less than unity and the rate of convergence depends on the size, λ, of this largest eigenvalue. It should be appreciated that the error $\underset{\sim}{e}^{(k)}$ may be much larger than the current displacement $\underset{\sim}{\Delta}^{(k)}$, so that convergence criteria should not be based on $\underset{\sim}{\Delta}^{(k)}$ alone. In fact if T has a single dominant eigenvalue then for large k the norms of the error, displacement and residual vectors all decrease by the factor λ at each iteration and the relation

$$\underset{\sim}{e}^{(k)} \approx \underset{\sim}{\Delta}^{(k)}/(1-\lambda) \tag{14.17}$$

follows. If λ is near to unity then the error is much larger than the displacement. The size of λ may be estimated from the ratio of successive displacement or residual norms. If no single eigenvalue dominates the rest then the norms do not decrease so regularly but we can estimate λ from the average reduction over several iterations and use (14.17) with more caution.

The simplest iteration of this kind is Jacobi's, in which

M is a block-diagonal submatrix of A. Equation (14.14) is solved as a sequence of steps

$$A_{ii}\,\underset{\sim}{x}_i^{(k+1)} = \underset{\sim}{b}_i - \sum_{j\neq i} A_{ij}\underset{\sim}{x}_j^{(k)}, \quad i = 1,2,\ldots,p. \tag{14.18}$$

The blocks A_{ii} may have order 1 (point Jacobi) or may correspond to lines of points (line Jacobi) or even planes of points. In the case of line Jacobi the matrices A_{ii} are band matrices with very small bandwidth so each equation (14.18) is very economical to solve. If storage is available it is worthwhile to store the triangular factors of A_{ii}.

It would seem sensible to use the available improved estimates $\underset{\sim}{x}_j^{(k+1)}$ in place of $\underset{\sim}{x}_j^{(k)}$ on the right of equation (14.18). This gives the Gauss-Seidel iteration

$$A_{ii}\,\underset{\sim}{x}_i^{(k+1)} = \underset{\sim}{b}_i - \sum_{j=1}^{i-1} A_{ij}\,\underset{\sim}{x}_j^{(k+1)} - \sum_{j=i+1}^{p} A_{ij}\,\underset{\sim}{x}_j^{(k)} \tag{14.19}$$

which corresponds to the choice of a block lower triangular submatrix of A for M.

Further improvement is possible by multiplying the indicated change $\underset{\sim}{x}_i^{(k+1)} - \underset{\sim}{x}_i^{(k)}$ to $\underset{\sim}{x}_i^{(k)}$ in (14.19) by an over-relaxation factor $\omega > 1$, to give a "successive over-relaxation" (SOR) iteration. It can be shown (Young, 1971) that if A is symmetric and positive definite then convergence is obtained for any ω in the range $0 < \omega < 2$. Some useful theoretical results on the effect of varying ω are available when the matrix has Young's property A. In the case of point SOR (blocks A_{ii} in (14.19) of order unity) property A essentially amounts to its having the block-tridiagonal form

$$\begin{bmatrix} A_{11} & A_{12} & & & \\ A_{21} & A_{22} & A_{23} & & \\ & \ddots & \ddots & \ddots & \\ & & \ddots & \ddots & A_{p-1p} \\ & & & A_{pp-1} & A_{pp} \end{bmatrix} \tag{14.20}$$

with the diagonal blocks A_{ii} being block-diagonal matrices whose inner block sizes correspond to the iteration blocks A_{ii} in (14.19). The form can be obtained with a five-point finite-difference operator on a rectangular grid (e.g. (14.2)) by ordering the points suitably. Checker board ordering, as illustrated in Figure 14.1, gives this form with two blocks corresponding to "black"/"white" points. Diagonal ordering, illustrated in Figure 14.2 gives it with blocks corresponding to diagonals. The theory also applies to the pagewise ordering illustrated in Figure 14.3 because the arithmetic performed is identical with that performed for diagonal ordering, though the operations are in a different order. In these cases the eigenvalues λ_i of the iteration matrix T satisfy

1	9	2	10
11	3	12	4
5	13	6	14
15	7	16	8

Figure 14.1
Checker-board ordering

1	2	4	7
3	5	8	11
6	9	12	14
10	13	15	16

Figure 14.2
Diagonal ordering

1	2	3	4
5	6	7	8
9	10	11	12
13	14	15	16

Figure 14.3
Pagewise ordering

the equations

$$(\lambda_i+\omega-1)^2 = \lambda_i\omega^2\mu_i^2 \qquad (14.21)$$

where μ_i is the corresponding eigenvalue of the Jacobi iteration matrix. In the case where the Jacobi matrix has real eigenvalues (e.g. if A is symmetric) the relation (14.21) can be used to show that the largest eigenvalue of the SOR iteration matrix is minimized by the choice

$$\omega = \omega_{opt} \equiv 2/[1 + (1-\mu^2)^{\frac{1}{2}}], \qquad \mu = \max_i |\mu_i| \ . \qquad (14.22)$$

A separate *a priori* calculation of μ may be made but alternatively it may be estimated while iterating with $\omega < \omega_{opt}$. Carré (1961) estimates the largest eigenvalue of the SOR matrix from ratios of norms of displacement vectors, uses this to estimate μ via relation (14.21) and finally uses (14.22). The procedure fails if $\omega > \omega_{opt}$ because the relation (14.21) tells us that $|\lambda_i| = \omega-1$ for all i. To reduce the likelihood of an overestimate for ω Carré replaces his estimate ω_e by $\omega_e - (2-\omega_e)/4$ but this does not always prevent it (see Hageman, 1972, for example). Reid (1966) proposed using the relationship for eigenvectors that corresponds to the relationship (14.21) for eigenvalues to estimate the dominant eigenvector of the Jacobi matrix and use its Rayleigh quotient to get an estimate for μ that is sure to be an under-estimate. A variant of this has been proposed by Hageman (1972) for σ_1-orderings (p=2 in (14.20)). He shows the relation

$$\mu^2 \geq (\underset{\sim}{s}^T A_{21} A_{11}^{-1} A_{12} \underset{\sim}{s})/(\underset{\sim}{s}^T A_{22} \underset{\sim}{s}) \qquad (14.23)$$

holds in the case $A_{21}^T = A_{12}$ where $\underset{\sim}{s}$ is an estimate of the second component of the partitioned dominant eigenvector of the SOR iteration matrix. The right-hand side of (14.23) is a good estimate for μ^2 and Hageman reports good experience. Wachspress (1966) discusses the estimation of ω in some detail.

14.3. THE METHOD OF CONJUGATE GRADIENTS

Iterations of the form (14.14) are known as "stationary" because of the fixed relationship between successive iterates. Very useful non-stationary iterations are provided by variants of the method of conjugate gradients. The classical form of this algorithm is as follows. Given $\underset{\sim}{x}^{(1)}$, set $\underset{\sim}{p}^{(1)} = \underset{\sim}{r}^{(1)} = \underset{\sim}{b} - A\underset{\sim}{x}^{(1)}$ then for k = 1,2,... calculate

$$\begin{aligned}
\alpha_k &= \underset{\sim}{r}^{(k)T}\underset{\sim}{r}^{(k)}/\underset{\sim}{p}^{(k)T} A \underset{\sim}{p}^{(k)} , \\
\underset{\sim}{x}^{(k+1)} &= \underset{\sim}{x}^{(k)} + \alpha_k \underset{\sim}{p}^{(k)} , \\
\underset{\sim}{r}^{(k+1)} &= \underset{\sim}{r}^{(k)} - \alpha_k A \underset{\sim}{p}^{(k)} , \qquad (14.24) \\
\beta_k &= \underset{\sim}{r}^{(k+1)T} \underset{\sim}{r}^{(k+1)}/\underset{\sim}{r}^{(k)T} \underset{\sim}{r}^{(k)} , \\
\underset{\sim}{p}^{(k+1)} &= \underset{\sim}{r}^{(k+1)} + \beta_k \underset{\sim}{p}_k .
\end{aligned}$$

The iteration may be used for any symmetric positive definite matrix A. The vector $\underset{\sim}{r}^{(k+1)}$ is the residual

$$\underset{\sim}{r}^{(k+1)} = \underset{\sim}{b} - A\underset{\sim}{x}^{(k+1)} \tag{14.25}$$

corresponding to $\underset{\sim}{x}^{(k+1)}$ and can be written in the form

$$\underset{\sim}{r}^{(k+1)} = P_k(A)\ \underset{\sim}{r}^{(1)} \tag{14.26}$$

where P_k is a polynomial of degree k whose value at zero is one. The power of the method lies in the fact that of all such vectors, $\underset{\sim}{r}^{(k+1)}$ minimizes the error measure

$$Q(\underset{\sim}{r}) = \underset{\sim}{r}^T A^{-1} \underset{\sim}{r}\ . \tag{14.27}$$

To understand the behaviour, suppose that A has eigensolutions $(\lambda_i, \underset{\sim}{e}^{(i)})$ and expand $\underset{\sim}{r}^{(1)}$ as

$$\underset{\sim}{r}^{(1)} = \sum_i \gamma_i\ \underset{\sim}{e}^{(i)} \tag{14.28}$$

then $\underset{\sim}{r}^{(k+1)}$ has the expansion

$$\underset{\sim}{r}^{(k+1)} = \sum_i \gamma_i\ P_k(\lambda_i)\ \underset{\sim}{e}^{(i)}\ . \tag{14.29}$$

If therefore there is a polynomial that is small on all the eigenvalues then $\underset{\sim}{r}^{(k+1)}$ will be small. Notice that there is no need for any computational eigenanalysis. Conjugate gradients automatically finds a good P_k. The method works particularly well if the eigenvalues are clustered. For example if they all lie in the range 1.99 to 2.01 then since the polynomial

$$(x-2)^2/4 \tag{14.30}$$

has value 1 at zero and maximum value .000025 in the range of the eigenvalues, two iterations will give a residual reduced by four orders of magnitude.

14.4. STONE'S STRONGLY IMPLICIT METHOD (SIP)

Stone (1968) proposed a method based on the approximate factorization

$$A \approx LU \tag{14.31}$$

of the finite-difference matrix A, where L and U are lower- and upper-triangular matrices with the same pattern of non-zeros as the corresponding parts of A. A naive way of finding such a factorization is to perform Gaussian elimination on A ignoring all fill-ins. For the 5-point operator (e.g. equation (14.2)) the rows of LU will correspond to a 7-point operator which is the original 5-point operator plus extra terms involving u(x+h,y-h) and u(x-h,y+h) if we order pagewise as in Figure 14.3. Stone proposed that instead of having errors of this form they should be of the form of multiples of

$$u(x+h,y-h) + \alpha\{u(x,y) - u(x+h,y) - u(x,y-h)\} \tag{14.32}$$

and

$$u(x-h,y+h) + \alpha\{u(x,y) - u(x-h,y) - u(x,y+h)\} \tag{14.33}$$

which is easily obtained by making perturbations within the sparsity pattern of A. In the case $\alpha=1$, the errors (14.32) and (14.33) are $O(h^2)$. Using the parameter α, with value near one, gives more flexibility. Indeed, on the basis of an intuitive argument Stone recommends using cycles of varying values. Details of the approximate factorization in the case of the 5-point operator and of suitable values for α are given in the excellent survey of finite-difference methods by Fox (1977).

Stone's actual iteration takes the form

$$\begin{aligned} L_\alpha\, U_\alpha\, \underset{\sim}{\Delta}^{(k)} &= \underset{\sim}{b} - A\, \underset{\sim}{x}^{(k)} \\ \underset{\sim}{x}^{(k+1)} &= \underset{\sim}{x}^{(k)} + \underset{\sim}{\Delta}^{(k)} \quad , \end{aligned} \tag{14.34}$$

where L_α, U_α are the approximate factors with α of (14.32), (14.33) in use.

It is tempting to look for a symmetric approximate factorization with $U = L^{\mathrm{T}}$, but Saylor (1974) shows that the coefficients of the error matrix $A-LL^{\mathrm{T}}$ can grow too fast with decreasing h for a useful algorithm to result if the $o(h^2)$ property (which can be expressed as the requirement that $(A-LL^{\mathrm{T}})\underset{\sim}{v} = \underset{\sim}{0}$ for $\underset{\sim}{v}$ corresponding to mesh values of a first order polynomial) is wanted.

The method in its original form has proved very useful and is widely used. Although originally written only for the 5-point operator, it has been generalized to other operators, for example by Jacobs (1973a, 1973b). Since the factorization is very easy to obtain, the method can be used for nonlinear problems where A depends on $\underset{\sim}{x}$.

14.5. PRE-CONDITIONED CONJUGATE GRADIENTS

An approximate symmetric factorization

$$A \approx LL^{\mathrm{T}} \tag{14.35}$$

may be used to form a "preconditioned" set of equations

$$(L^{-1}\, A\, L^{-\mathrm{T}})(L^{\mathrm{T}}\underset{\sim}{x}) = L^{-1}\underset{\sim}{b} \tag{14.36}$$

to which the method of conjugate gradients may be applied. Meijerink and van der Vorst (1977) proposed such a method, based on simply ignoring fill-ins during Choleski factorization and Kershaw (1978) reports impressive numerical results on some large practical problems. Jacobs (private communication) has recently been using the SIP factorization, with fixed α, in a similar way and his preliminary results are very encouraging.

The success of the method in all these cases is based on the matrix $L^{-1}\, A\, L^{-\mathrm{T}}$ having most of its eigenvalues near unity so that the method of conjugate gradients is able to find a low order polynomial P_k which is small on all the eigenvalues.

14.6. MULTIPLE GRID METHODS

A feature of most iterative methods is that they can deal very effectively with highly oscillatory errors but they require many iterations to eliminate smooth errors. A number of authors (including Federenko (1962), Brandt (1972, 1977), Frederickson (1975) and Nicolaides (1975)) have therefore proposed switching to a coarser grid to iterate away these smooth errors.

Typically we have a sequence of successively coarser approximations

$$A^{(k)}\underset{\sim}{x}^{(k)} = \underset{\sim}{b}^{(k)}, \qquad k = 1,2,\ldots,\ell \tag{14.37}$$

and require the solution of the finest one, with $k = 1$. An iteration begins with an approximate solution with residual $\underset{\sim}{r}^{(1)}$ and seeks a correction $\underset{\sim}{\Delta}^{(1)}$ which approximately satisfies the equation

$$A^{(1)}\underset{\sim}{\Delta}^{(1)} = \underset{\sim}{r}^{(1)} . \tag{14.38}$$

Vectors $\underset{\sim}{r}^{(k)}$, $k = 2,3,\ldots,\ell$ are formed successively from $\underset{\sim}{r}^{(k-1)}$ using interpolation where necessary but mostly by simply omitting components that appear in the fine grid but not in the coarse one and copying across those which have the same purpose in both. The sets of equations

$$A^{(k)}\underset{\sim}{\Delta}^{(k)} = \underset{\sim}{r}^{(k)}, \qquad k = \ell,\ell-1,\ldots,1 \tag{14.39}$$

are now solved successively, probably starting with a direct solution of the coarsest, $k = \ell$. From each $\underset{\sim}{\Delta}^{(k)}$ is constructed a first iterate for $\underset{\sim}{\Delta}^{(k-1)}$ by interpolation.

The overall success of the method depends on the details of the interpolations from grid to grid, the iterations used and the various convergence criteria, but the total work need be no more than proportional to the number of variables in the final grid.

15

DIRECT SOLUTION OF FINITE ELEMENT AND FINITE-DIFFERENCE EQUATIONS

15.1. INTRODUCTION

We consider the direct solution of sets of linear equations

$$A\underset{\sim}{x} = \underset{\sim}{b} \tag{15.1}$$

as they arise from finite element and finite-difference calculations. We give particular prominence to those from finite element calculations, which have the form

$$A = \sum_{k} B^{(k)}, \quad \underset{\sim}{b} = \sum_{k} \underset{\sim}{c}^{(k)} \tag{15.2}$$

where each superscript represents an element so that the summation is over all the elements. The matrix $B^{(k)}$ has non-zeros only in positions (i,j) such that the variables x_i and x_j are both associated with the kth finite element. Hence $B^{(k)}$ may be stored as a small full matrix of order the number of variables associated with the kth element. Similarly the vector $\underset{\sim}{c}^{(k)}$ may be stored as a small full vector. Of course if this condensed form of storage is used then small vectors of integers are needed to indicate which variables are associated with the columns of each small full storage matrix. Finite-difference matrices can also (somewhat artificially) be written in this form.

A finite element calculation usually involves three distinct phases:

(i) calculation of the individual matrices $B^{(k)}$ and vectors $\underset{\sim}{c}^{(k)}$ (as small full matrices and vectors);

(ii) assembly of the overall problem (15.1); and

(iii) solution of the overall problem by Gaussian elimination.

It is convenient to regard phase (i) as separate although

in practice it is often combined with phase (ii). During phase (ii) we need to be able to obtain the matrices $B^{(k)}$ and this may be done either by recovering them from storage or by generating them. Phases (ii) and (iii) may be combined and we will pay particular attention to such methods. They show significant savings if a backing store (for example disk storage) has to be used to hold the overall matrix A.

Because the problem (15.1) usually arises from the minimization of a positive function (normally energy) it usually has a matrix A which is symmetric and positive definite ($\underset{\sim}{z}^T A \underset{\sim}{z} > 0$ for any vector $\underset{\sim}{z} \neq \underset{\sim}{0}$). Except in section 15.6 we assume that A has these properties.

15.2. BAND-MATRIX TECHNIQUES

If the variables can be so ordered that those associated with any single element or finite-difference equation are never further than m positions apart then the element A_{ij} of A will always be zero if $|i-j| \geq m$, that is the matrix A will be a band matrix of semi-bandwidth m. Of course it very rarely happens that the first non-zero element in every row is exactly m positions away from the diagonal. This means that significant gains may be obtained by generalising to variable-band matrices (Jennings, 1966). The lower half of a symmetric example is shown in Figure 15.1a. The pattern arises from the triangulation of a square as shown in Figure 15.1b. Notice that the maximal semi-bandwidth of 5 is attained in only 6 of the 16 rows.

It may readily be verified that if Gaussian elimination without interchanges is applied to such a matrix then no fill-in can take place before the first non-zero in any row or column, that is the forms of band and variable-band matrices are preserved. With the matrix positive definite there is no risk of instability and we may exploit symmetry to halve the work and storage. Any zeros between the first non-zero in any row and the diagonal are stored explicitly because they soon fill in. In fact George and Liu (1975b) have remarked that if all rows after the first have a non-zero to the left of the diagonal then they all fill in totally between this non-zero and the diagonal.

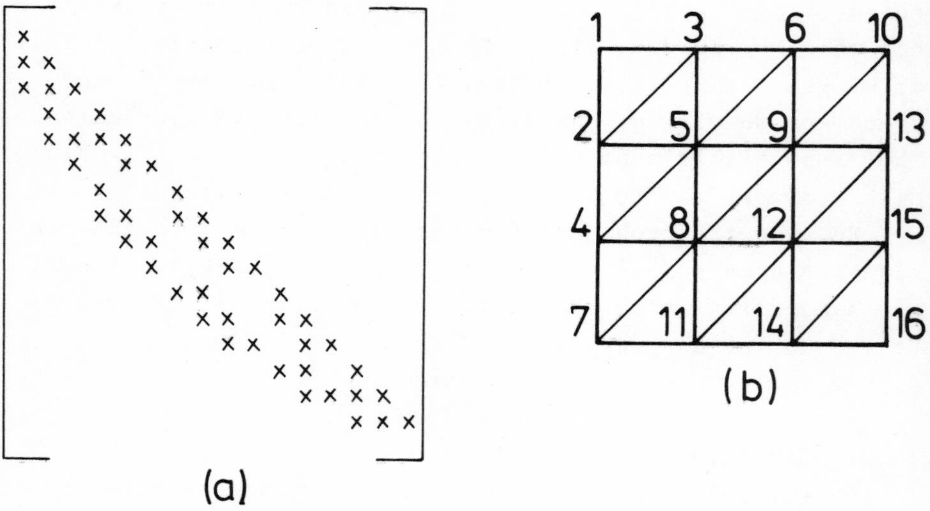

Fig. 15.1

In general we need an ordering strategy to obtain a satisfactory variable-band form. It is best to think in terms of ordering the nodes in the graph* of the matrix. Most algorithms order the nodes by placing them in "level" sets, S_i, $i = 1,2,\ldots$, the first of which consists of a single starting node. Later, all nodes which are neighbours of points in set S_i (two nodes are neighbours if they are associated with the same finite element or finite-difference equation) but are not in an earlier set are taken to comprise set S_{i+1}. If the variables are ordered to correspond to the sets $S_1,S_2,\ldots$ (or $S_k,S_{k-1},\ldots,S_1$) then it is readily seen that the resulting matrix is block tridiagonal with diagonal blocks corresponding to the sets S_i. To order within each block S_i Cuthill and McKee (1969) suggest taking first those nodes that are neighbours of the first node in S_{i-1}, then those that are neighbours of the second node in

*An edge joins node i to j in the graph of A if and only if $A_{ij} \neq 0$. The graph may be visualized by imagining nodes at or near the physical location of corresponding variables.

S_{i-1} and so on; the several other orderings that have been proposed since that time do not appear to offer any significant advantages. In any case George and Liu (1975a) point out that we can avoid storage of the fill-ins in the off-diagonal blocks by using the partitioning ideas of Rose and Bunch (1972). In terms of Choleski factorization $A = LL^T$ we have

$$A = \begin{bmatrix} A_1 & B_1 & & & \\ B_1^T & A_2 & B_2 & & \\ & B_2^T & A_3 & \cdot & \\ & \cdot & \cdot & \cdot & \\ & & \cdot & \cdot & B_{k-1} \\ & & & B_{k-1}^T & A_k \end{bmatrix} \quad L = \begin{bmatrix} L_1 & & & & \\ W_1^T & L_2 & & & \\ & W_2^T & L_3 & & \\ & \cdot & \cdot & & \\ & & \cdot & \cdot & \\ & & & W_{k-1}^T & L_k \end{bmatrix} \qquad (15.3)$$

where

$$A_1 = L_1L_1^T; \; W_i = L_i^{-1}B_i, \qquad i = 2,3,\ldots,k-1;$$

$$A_i - W_{i-1}^T W_{i-1} = L_iL_i^T, \qquad i = 2,3,\ldots,k. \qquad (15.4)$$

The Choleski factors L_i will be full and must be stored but there is no need to store W_i explicitly, as long as the sparser matrix B_i is held.

It is clear from (15.3) and (15.4) that we should be aiming for the level sets (diagonal blocks) to be small and numerous. Cuthill and McKee (1969) suggested trying several starting nodes with a small number of neighbours, see also Cuthill (1972). Gibbs, Poole and Stockmeyer (1976) suggest that each node with least neighbours in the final level set S_k should be tried as starting node. If one gives more than k level sets, it is regarded as replacing the original starting node and the new final level set is examined.

We show in Figure 15.2 a simple example of the Cuthill-McKee ordering, with the level sets indicated by dashed lines. The reader will find it instructive to examine the level

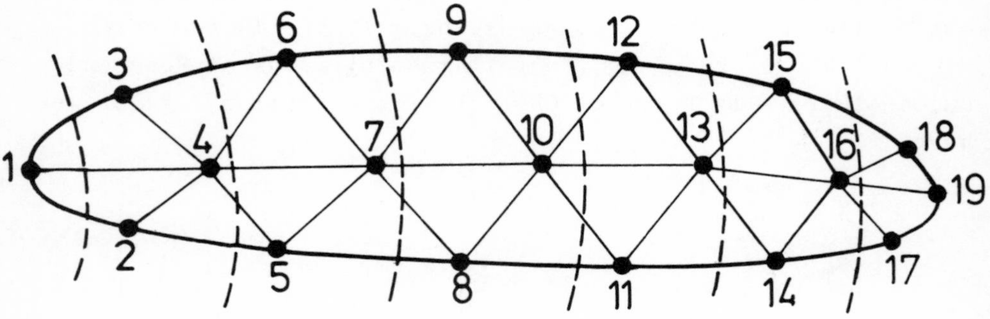

Fig. 15.2.

sets and the speed with which the Gibbs *et al*. algorithm recovers from a poor starting node.

It frequently happens that, even with a good ordering of the variables, the variable-band matrix is too large to fit into main storage and therefore has to be held on backing store (for example, on disk storage). If the greatest semi-bandwidth is m and we have main storage for $\frac{1}{2}m(m+1)$ variables and a reasonably large input-output buffer, then it is straightforward to organise the elimination without writing any elements of the factorized form out to backing store more than once. This is possible because the kth stage of elimination involves at most elements in the square submatrix of rows and columns k to k+m-1. Elements in earlier columns can have been written to backing store already and elements in later rows are not yet required. Similar remarks apply to George and Liu's variant, if m is now the order of the largest block A_i.

15.3. FRONTAL METHODS

Irons (1970) proposed a very useful technique for combining phases (ii) and (iii) of the process (as defined in section 15.1), and it has become known as his "frontal" approach. It depends on the fact that the elimination steps for A_{ij}

involve subtracting the quantities

$$A_{i\ell}^{(\ell)} A_{\ell j}^{(\ell)} / A_{\ell\ell}^{(\ell)} \tag{15.5}$$

for $\ell = 1,2,\ldots,\min(i-1,j-1)$ and the assembly operations involve adding the quantities

$$B_{ij}^{(k)} \tag{15.6}$$

for all the finite elements k that involve variables i and j. These operations can be performed in any order. All that matters is that they must all be completed before row i or column j becomes pivotal. Irons therefore eliminates each variable as soon as its row and column is fully assembled. Variables not yet eliminated that are involved in elements that have been assembled constitute the "front", which separates the region consisting of assembled elements from the rest. A full matrix can be used at each stage to hold rows and columns that correspond to the front. On assembly its order increases to accommodate any new variables. On an elimination, the pivotal row is written away to backing store and the size of the active matrix reduces by one. Notice that variables frequently leave the front in a different order to that in which they entered it, so that it may be necessary to perform a symmetric interchange on the full active matrix before an elimination. If the front never has more than m variables then we will need main storage for $\frac{1}{2}m(m+1)$ variables to hold the active matrix at its largest.

The ordering problem has now shifted from the variables to the elements. We want to order the elements to keep a small front. This is likely to be somewhat easier manually since there are likely to be fewer elements than variables. For automatic ordering the obvious analogues of the bandwidth algorithms that we have described are available.

The storage demands of the variable-band method and the frontal method are often similar. Each requires room for a symmetric "active" matrix and a buffer for input-output operations. The orders of the largest active matrices will actually be identical (and equal to the maximal semi-bandwidth) if the ordering makes the variables leave the front in the same order as they entered it (first-in, first-out). An example where they differ, to the advantage of the frontal method, is where some variables are associated with just one element so that they can be eliminated immediately after the element is assembled (this process is sometimes called "static condensation"). Similar, but less dramatic, results may be obtained when some variables are associated with just two elements.

A significant improvement to the frontal algorithm has been proposed by Speelpenning (1973). We illustrate with the solution of a problem over the propeller-shaped region shown in Figure 15.3. A good ordering from the point of view

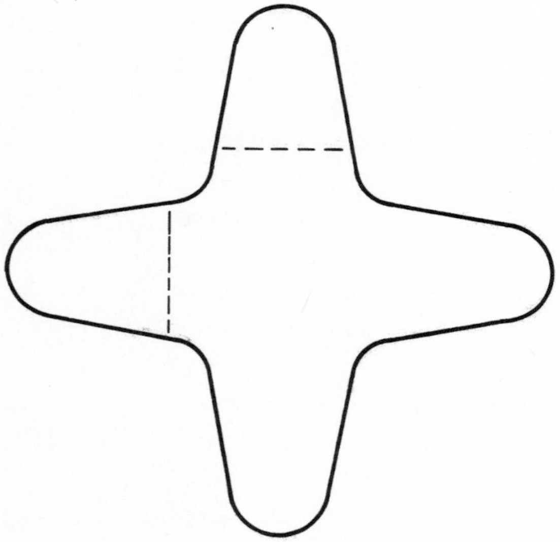

Fig. 15.3.

of arithmetic operations would be that starting from the tip of one blade and working steadily towards the hub, then doing likewise on the second blade. When elimination has reached about half-way down the second blade the front will consist of variables along the dotted lines shown, and no change will be being made to the part of the matrix that corresponds to variables near the hub. Speelpenning exploits this situation so that work is performed only on the active part of the front. He does this by regarding each operation of assembly followed by elimination of variables as creating a new finite element. Its variables are those of all the assembled elements less the eliminated variables. Normally the next element is adjacent to the last, so that the newly created super-element is immediately changed to another super-element by the operations of assembly and elimination. However when a new front is started, as in the example of Figure 15.3, the old super-element is regarded simply as an element to be assembled later and assembly and elimination is started on the new front. To save main storage the inactive super-elements may be held on backing storage with little loss of efficiency.

15.4. NESTED DISSECTION

Closely related to Speelpenning's work is the nested dissection algorithm of George (1973). The region is divided into subregions, each of which consists of the union of a set of elements which will presently be assembled into a super-element. Each subregion is further subdivided into sub-subregions, and so on to any depth of division. A simple example on a square region is shown in Figure 15.4. To solve the problem we assemble the super-elements corresponding to the smallest subregions and eliminate any variables whose row and column are fully assembled. Next

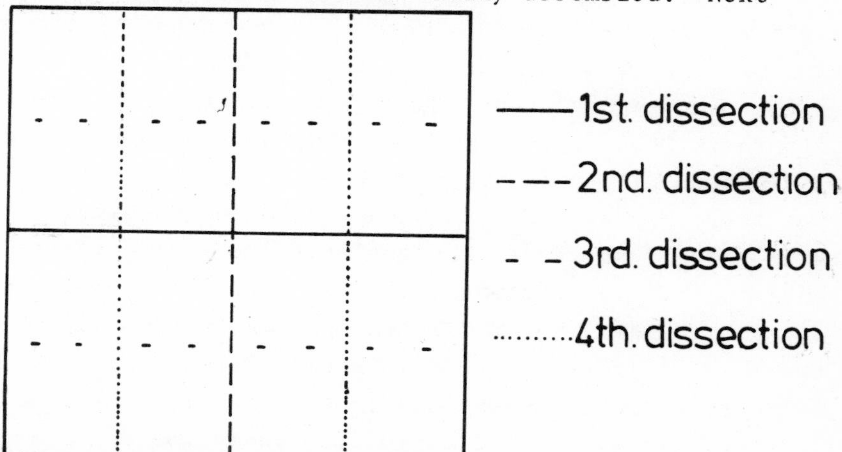

Fig. 15.4.

these super-elements are assembled into larger super-elements corresponding to the next level of subregion, and so on.

Rather advantageous operation and overall storage counts can be obtained in this way. For example George (1973) showed that if nested dissection is used to solve a problem over a square grid of q^2 square elements each having four variables at its four corners then $O(q^3)$ operations and $O(q^2 \log q)$ storage locations are needed in place of the $O(q^4)$ operations and $O(q^3)$ storage locations needed by bandwidth methods. Actual gains for practical values of q, though worthwhile, are not enormous because the multiplying factors in these operation and storage counts favour the band technique.

George (1976) and George and Liu (1976b) have proposed an algorithm for the automatic choice of a satisfactory dissection. It is based on using level sets in a similar way to the Gibbs, Poole and Stockmeyer (1976) algorithm for ordering to band form. The level sets are chosen as before and the middle set (or as much of it as is needed) is used to dissect the region. Each of the parts is treated similarly, continuing until sufficiently small sets are obtained.

15.5. GENERAL SPARSE TECHNIQUES AND HYPERMATRICES

Another possibility is to use general purpose sparse matrix software. A pivotal strategy that has proved satisfactory in a wide variety of applications is that due to Markowitz (1975). In the symmetric case it corresponds to the minimal degree algorithm and consists at each stage in choosing as pivot the diagonal non-zero with least other non-zeros in its row. This frees the user entirely from any problems associated with choosing a good ordering of the variables, but there may be considerable expense in choosing the pivots with current software, though techniques for performing this much faster are becoming available (see for example George and Liu (1976a)). For illustration we show in Table 15.1 the results of using standard sparse matrix software (Reid, 1972) to solve the equations that arise from the usual 5-point finite-difference approximation to Laplace's equation on a rectangular 30 × 31 grid. For the variable band results

TABLE 15.1

Matrix order		930
No. of non-zeros stored	Variable band Markowitz	20,245 11,328
Time	ANALYSE (Markowitz)	12,200
in	FACTOR Variable band Markowitz	1,300 600
370/165		
m-secs.	OPERATE Variable band Markowitz	160 120

we ordered the nodes by diagonals, as in Fig. 15.1b. The total storage used here by the two codes was very similar because the simpler data structure required for the variable-band code compensated for the extra non-zeros held.

An approach that has been used with some success by Argyris and his collaborators (see for example Schrem, 1971) is to group the variables and to work with the sparsity pattern of the resulting block matrices. The blocks themselves can be divided into smaller blocks, and so on. This they call the *hypermatrix* method. Demands on main storage may be kept very modest by using small blocks at the deepest block level. It seems that in practice usually only two or three levels are used.

15.6. THE FRONTAL APPROACH FOR MORE GENERAL MATRICES

If the matrix A is not symmetric and positive definite then we cannot safely choose a pivotal sequence on the basis of the sparsity pattern alone. Hood (1976) remarks that in the unsymmetric case we may use an element as pivot only if its row and column is fully assembled. He suggests continuing assembly until the front exceeds a critical size, then choosing as pivot the maximal element in the fully assembled rows and columns. This however seems unreasonably cautious; it would surely be preferable to follow each assembly with as many eliminations as possible subject to the magnitude of each pivot exceeding a suitable tolerance.

Similar considerations apply to the symmetric but nondefinite case. We accept a 1×1 pivot (on the diagonal) subject to some tolerance and otherwise use a 2×2 pivot (in two fully assembled rows) in the style of Bunch and Parlett (1971).

15.7. COMPARISON BETWEEN DIRECT AND ITERATIVE METHODS

We conclude this chapter with some remarks on the relative merits of direct and iterative methods. The significant advantage of an iterative method is that it works with the original matrix and involves no creation of additional non-zeros (fill-in). Indeed, often we do not even need to store the matrix A since it suffices to use code to generate the non-zeros when they are wanted. These methods are therefore well-suited to very large problems.

Some feeling for the relative merits may be obtained by looking at the 5-point finite-difference approximation to Laplace's equation in a square and a cube, in each case with s unknowns in each coordinate direction so that the number of unknowns are s^2 and s^3 respectively. Using direct methods the maximum semi-bandwidth is s and s^2 and the number of multiplications is approximately $1/4s^4$ and $1/7s^7$. Using successive over-relaxation with parameter ω, the number of multiplications per iteration is about $2s^2$ and $2s^3$. The maximal eigenvalues are about $(1-4\pi^2/s^2)$ at $\omega = 1$ and $(1-2\pi/s)$ at the optimum ω so that the number of iterations needed to gain a decimal is about $s^2/17$ at $\omega = 1$ and $s/2.7$ at the optimum ω. This corresponds to $s^3/1.3$ and $s^4/1.3$ multiplications per decimal in 2 and 3 dimensions respectively with the optimum ω. It can be seen from these approximate figures that direct and iterative methods are likely to be fairly competitive in two dimensions but that iterative methods are likely to be superior in three dimensions.

In a more general situation, much depends on the problem. A narrow range of eigenvalues favours iterative methods, all of which then require few iterations as was illustrated for conjugate gradients in section 14.3. Also very large numbers of variables favour iterative methods, unless the structure results in little fill-in for a direct method (for example if A is a band matrix with small bandwidth). The cost of a direct method is dependent on structure and not on conditioning.

Where several sets of equations with the same matrix are to be solved the factorization may be retained so that second and subsequent sets may be solved much more rapidly. For example the operation counts of about $1/4s^4$ and $1/7s^7$ for Laplace's equation in a square and a cube reduce to about $4/3s^3$ and $4/5s^5$. Iterative methods are able to use previously determined relaxation parameters and perhaps have a good first iterate, but this does not result in such a dramatic improvement. Direct methods are therefore superior to iterative ones more frequently for "many-off" cases than for "one-off" cases.

16

FAST DIRECT METHODS FOR SEPARABLE ELLIPTIC EQUATIONS

This chapter concerns recent work on the direct solution of the finite-difference equations for a separable elliptic equation, using techniques analogous to applying separation of variables to the differential equation. Very fast methods result for such separable problems given on rectangular regions with uniform boundary conditions along each side. The success of this work has led to extensions to problems on arbitrary regions by embedding the problem in a region for which these efficient techniques can be applied.

A vital ingredient in some approaches to separable problems has been the development of the Fast Fourier Transform which is described in the first section. In the next two sections we discuss two basic separation of variables techniques and in section 16.4 we consider problems on non-rectangular regions. In section 16.5 we briefly discuss some work on the use of these efficient methods for non-separable differential operators.

16.1. THE FAST FOURIER TRANSFORM

The original algorithm was developed by Cooley and Tukey (1965) although similar ideas exist in previous work. A full account of the development may be found in the book by Brigham (1974). We attempt here to describe the essential features of the algorithm without a detailed analysis of refinements in implementation that produce the most efficient versions. The general complex case is considered, although for the purposes of this chapter only real summations will be required. Most program libraries contain algorithms for the commonly occurring cases.

The problem may be stated as one of computing quantities $c(j)$, $j = 0,1,\ldots,N-1$, defined by

$$c(j) = \sum_{k=0}^{N-1} a(k)w^{jk} \qquad (16.1)$$

where the $a(k)$ are given (complex) constants and w is a Nth root of unity, $w^N = 1$. Direct computation using nested multiplication requires N^2 complex operations, where each operation involves one complex multiplication and one addition. This work can be substantially reduced if N is large and highly composite and first we consider the case $N = 2^n$.

Express each value of k in the form $k = 2k_1$ or $2k_1 + 1$ according as k is even or odd, so that $0 \le k_1 \le N_1 - 1$, where $N_1 = N/2 = 2^{n-1}$. Then (16.1) can be rearranged to obtain

$$c(j) = \sum_{k_1=0}^{N_1-1} a(2k_1)(w^2)^{jk_1} + \sum_{k_1=0}^{N_1-1} a(2k_1+1)(w^2)^{jk_1}w^j. \quad (16.2)$$

If j is written in the form $j = tN_1 + j_1$, where $t = 0$ or 1, we have

$$(w^2)^{jk_1} = (w^2)^{tN_1k_1}(w^2)^{j_1k_1} = (w^{2N_1})^{tk_1}(w^2)^{j_1k_1} .$$

Since $w^{2N_1} = w^N = 1$, (16.2) takes the form

$$c(j) = \sum_{k_1=0}^{N_1-1} a(2k_1)(w^2)^{j_1k_1} + w^j \sum_{k_1=0}^{N_1-1} a(2k_1+1)(w^2)^{j_1k_1}. \quad (16.3)$$

Observing that $(w^2)^{N_1} = 1$ we see that each summation in (16.3) is of the form (16.1), over half the number of terms. Since $N_1 = 2^{n-1}$ each of these sums may be split in the same way and this process can be continued until we reach sums involving one term only.

Let s_n be the number of operations required to evaluate (16.1) by this process. To compute c(j) from (16.3), $j = 0,1,\ldots,N-1$, given the values of the sums involved requires at most 2N operations, assuming the values w^j are not stored. Therefore

$$s_n \le 2s_{n-1} + 2N, \quad N = 2^n .$$

By induction on n we have

$$s_n \le 2n.2^n = 2N \log_2 N.$$

The very striking reduction in computational effort achieved by this method has led to a complete change in attitude when considering what can be achieved using Fourier methods in other computational areas as well as that considered here.

A significant reduction in cost can always be achieved if N is highly composite. By a generalization of the above reasoning it is possible to show that if $N = k_1 k_2 \ldots k_r$ then (16.1) can be evaluated in $N(k_1+k_2+\ldots+k_r)$ operations, rather than N^2. The base 2 case above leads to the simplest algorithm. A full discussion and comparison of the various possibilities may be found in Brigham (1974).

16.2. FOURIER METHODS FOR THE DISCRETE POISSON EQUATION

Consider the solution of Poisson's equation on a rectangle

$$\nabla^2 u = f, \quad 0 < x < a, \quad 0 < y < b$$

with u specified on the boundary. There are many applications where this problem has to be solved repeatedly, in which the differential operator and the region remain the same. Imposing a uniform mesh and using the standard five-point difference replacement to the differential operator leads to the discrete Poisson equation

$$B\underset{\sim}{U} = \begin{bmatrix} A & I & & & \\ I & A & I & & \\ & \cdot & \cdot & \cdot & \\ & & \cdot & \cdot & I \\ & & & I & A \end{bmatrix} \begin{bmatrix} \underset{\sim}{U}_1 \\ \underset{\sim}{U}_2 \\ \vdots \\ \underset{\sim}{U}_N \end{bmatrix} = \underset{\sim}{V} = \begin{bmatrix} \underset{\sim}{V}_1 \\ \underset{\sim}{V}_2 \\ \vdots \\ \underset{\sim}{V}_N \end{bmatrix} \tag{16.4}$$

$\underset{\sim}{U}_k = (U_{1k}, U_{2k}, \ldots, U_{Mk})^T$ are values of the mesh function on the line $y = kh$. For convenience of description we have assumed a square mesh of side $h = a/(M+1) = b/(N+1)$. In this case the $M \times M$ matrix A is of the form

$$A = \begin{bmatrix} -4 & 1 & & & \\ 1 & -4 & 1 & & \\ & \cdot & \cdot & \cdot & \\ & & \cdot & \cdot & 1 \\ & & & 1 & -4 \end{bmatrix}.$$

The eigenvalues of this matrix are

$$\lambda_j = -4 + 2\cos\frac{j\pi}{M+1}, \quad j = 1,2,\ldots,M$$

with corresponding orthonormal eigenvectors

$$\underset{\sim}{q}_j = \left(\sin\frac{j\pi}{M+1}, \sin\frac{2j\pi}{M+1}, \ldots, \sin\frac{Mj\pi}{M+1}\right)^T \times \sqrt{\frac{2}{M+1}}.$$

Using the eigenvectors as columns we can write down an orthogonal matrix Q such that

$$Q^T A Q = \text{diag}(\lambda_j) = \Lambda, \; Q^T Q = I.$$

We apply a similarity transformation using this matrix Q to each line of (16.4),

$$\underset{\sim}{U}_{k-1} + A\underset{\sim}{U}_k + \underset{\sim}{U}_{k+1} = \underset{\sim}{V}_k,$$

to obtain

$$\underset{\sim}{\bar{U}}_{k-1} + \Lambda\underset{\sim}{\bar{U}}_k + \underset{\sim}{\bar{U}}_{k+1} = \underset{\sim}{\bar{V}}_k \tag{16.5}$$

where

$$\underset{\sim}{\bar{U}}_k = Q\underset{\sim}{U}_k, \; \underset{\sim}{\bar{V}}_k = Q\underset{\sim}{V}_k.$$

Taking together the jth row of (16.5), for each value of k, leads to the separated equations

$$
\begin{aligned}
\lambda_j\bar{U}_{j1} + \bar{U}_{j2} \qquad &= \bar{V}_{j1} \\
\bar{U}_{j1} + \lambda_j\bar{U}_{j2} + \bar{U}_{j3} &= \bar{V}_{j2} \\
&\;\vdots \\
\bar{U}_{j-1,N}+\lambda_j\bar{U}_{jN} &= \bar{V}_{jN}
\end{aligned}
\tag{16.6}
$$

The problem now consists of M systems of N equations of a simple tridiagonal form. The Fourier method therefore consists of the stages:

(i) Compute $\bar{\underset{\sim}{V}}_k = Q\underset{\sim}{V}_k$, $k = 1,2,\ldots,N$,

(ii) Solve the systems (16.6), $j = 1,2,\ldots,M$,

(iii) Compute $\underset{\sim}{U}_k = Q^T\bar{\underset{\sim}{U}}_k$, $k = 1,2,\ldots,N$.

The computations in stages (i) and (iii) can be treated by special (real) cases of the FFT algorithm of section 16.1 if $M+1 = 2^n$ or is highly composite, and this is the key reason why the algorithm is so efficient.

Hockney (1965, 1970) solves the systems in stage (ii) using odd/even reduction, which we discuss in block form in the next section (this does require that N also be of the form 2^n-1). This method is then called Fourier analysis and cyclic reduction, FACR. Fischer *et al.* (1974) observe that the matrix in (16.6)

$$
\begin{bmatrix}
\lambda & 1 & & & \\
1 & \lambda & 1 & & \\
 & \ddots & \ddots & \ddots & \\
 & & \ddots & \ddots & 1 \\
 & & & 1 & \lambda
\end{bmatrix}
$$

is a Toeplitz matrix, in which the entries depend only on the difference between the row and column number. A rank one modification, in which the (1,1) element becomes $\mu = \lambda/2 - (\lambda^2/4 - 1)^{\frac{1}{2}}$, can be stably factored into the product of Toeplitz matrices,

$$\begin{bmatrix} \mu & 1 & & & \\ 1 & \lambda & 1 & & \\ & \cdot & \cdot & \cdot & \\ & & \cdot & \cdot & \cdot \\ & & & \cdot & \cdot & 1 \\ & & & & 1 & \lambda \end{bmatrix} = \begin{bmatrix} 1 & & & & \\ \mu^{-1} & 1 & & & \\ & \cdot & \cdot & & \\ & & \cdot & \cdot & \\ & & & \cdot & \cdot \\ & & & \mu^{-1} & 1 \end{bmatrix} \begin{bmatrix} \mu & 1 & & & \\ & \mu & \cdot & & \\ & & \cdot & \cdot & \\ & & & \cdot & \cdot \\ & & & & \cdot & 1 \\ & & & & & \mu \end{bmatrix}.$$

This gives an efficient method requiring very little storage. The change in the element in the upper left most position of A is compensated for by using the Sherman-Morrison formula

$$(B+\underset{\sim}{u}\,\underset{\sim}{v}^T)^{-1} = B^{-1}-B^{-1}\underset{\sim}{u}(1+\underset{\sim}{v}^T B^{-1}\underset{\sim}{u})\underset{\sim}{v}^T B^{-1}$$

where $\underset{\sim}{u} = (1,0,\ldots,0)^T$ and $\underset{\sim}{v} = (\lambda-\mu)\underset{\sim}{u} = (1/\mu)\underset{\sim}{u}$. This technique for stage (ii) is called the Fourier-Toeplitz method, see Fischer *et al.* (1974) for full details.

Periodic and Neumann boundary conditions can be handled equally simply. In the case of the nine-point replacement the block unit matrices in (16.4) become tridiagonal matrices T; however A and T are simultaneously diagonalisable so the method still applies. The operation count is basically $cMN \log_2 M$ where the constant c is small and depends on the type of boundary condition. Analysis of the methods of this and the next section is contained also in Buzbee *et al.* (1970). Using a modified FACR method, introduced in the next section, the cost for solving this problem on 128×128 mesh can be as little as three iterations of SOR (Hockney, 1970).

16.3. ODD/EVEN REDUCTION

We again consider the discrete Poisson equation (16.4) with the specific restriction $N = 2^n-1$. We also consider the more general case where the unit blocks are replaced by a matrix T for which $AT = TA$, thus including the case of the nine-point finite-difference replacement. Three successive lines of (16.4) now take the form

$$\begin{aligned} T\underset{\sim}{U}_{k-2} + A\underset{\sim}{U}_{k-1} + T\underset{\sim}{U}_{k} \qquad\qquad\qquad &= \underset{\sim}{V}_{k-1} \\ T\underset{\sim}{U}_{k-1} + A\underset{\sim}{U}_{k} + T\underset{\sim}{U}_{k+1} \qquad\quad &= \underset{\sim}{V}_{k} \\ T\underset{\sim}{U}_{k} + A\underset{\sim}{U}_{k+1} + T\underset{\sim}{U}_{k+2} &= \underset{\sim}{V}_{k+1} \quad . \end{aligned}$$

Multiplying the first and third equations by T, the second equation by $-A$ and adding leads to

$$T^2\underset{\sim}{U}_{k-2} + (2T^2 - A^2)\underset{\sim}{U}_k + T^2\underset{\sim}{U}_{k+2} = T\underset{\sim}{V}_{k-1} - A\underset{\sim}{V}_k + T\underset{\sim}{V}_{k+1}.$$

Thus we obtain a separation of the even rows leading to two linear systems

$$\begin{bmatrix} 2T^2-A^2 & T^2 & & \\ T^2 & 2T^2-A^2 & T^2 & \\ & \ddots & \ddots & \ddots \\ & & \ddots & T^2 \\ & & T^2 & 2T^2-A^2 \end{bmatrix} \begin{bmatrix} \underset{\sim}{U}_2 \\ \underset{\sim}{U}_4 \\ \vdots \\ \\ \underset{\sim}{U}_{N-1} \end{bmatrix} = \begin{bmatrix} T\underset{\sim}{V}_1 + T\underset{\sim}{V}_3 - A\underset{\sim}{V}_2 \\ T\underset{\sim}{V}_3 + T\underset{\sim}{V}_5 - A\underset{\sim}{V}_4 \\ \vdots \\ \\ T\underset{\sim}{V}_{N-2} + T\underset{\sim}{V}_N - A\underset{\sim}{V}_{N-1} \end{bmatrix}, \quad (16.7)$$

and

$$\begin{bmatrix} A & & & \\ & A & & \\ & & \ddots & \\ & & & A \end{bmatrix} \begin{bmatrix} \underset{\sim}{U}_1 \\ \underset{\sim}{U}_3 \\ \vdots \\ \underset{\sim}{U}_N \end{bmatrix} = \begin{bmatrix} \underset{\sim}{V}_1 - T\underset{\sim}{U}_2 \\ \underset{\sim}{V}_3 - T\underset{\sim}{U}_2 - T\underset{\sim}{U}_4 \\ \vdots \\ \underset{\sim}{V}_N - T\underset{\sim}{U}_{N-1} \end{bmatrix} . \qquad (16.8)$$

This enables the even lines to be solved for first, using (16.7), followed by the separated odd lines, consisting of $M \times M$ systems, from (16.8).

Since $N = 2^n-1$, the number of blocks in (16.7) has been reduced to $2^{n-1}-1$ so the odd/even reduction process can be repeated successively until we have one system of M equations for $\underset{\sim}{U}_{(N-1)/2}$. (Note that since T and A commute so do $2T^2-A^2$ and T^2.) This is the idea of the algorithm of Buneman (1969), which is analysed by Buzbee *et al.* (1970) and Hockney (1970), where Neumann and periodic boundary conditions are also considered. The block matrices arising in the successive reductions of (16.7) are kept in factored form and particular care is necessary in handling the right hand sides; for numerical stability it is necessary to avoid repeated multiplications by the matrix A. The operation count for Buneman's method, also called cyclic odd/even reduction and factorization (CORF) is of the form $cMN \log_2 N$.

An alternative to continuing the process of reduction is to to solve (16.7) by the Fourier methods of the previous section; the fact that the blocks commute implies that they have the same eigenvectors as A and T. Hockney (1970) advocates one or two cyclic reductions followed by the Fourier method, denoted by FACR(ℓ), ℓ = 1 or 2 respectively. FACR(ℓ) has a smaller constant in the operation count than Buneman's algorithm. However, Buneman's method regains some advantage due to the simplicity of the program. Overall FACR(ℓ) or Fourier-Toeplitz is usually 60%-80% faster in execution time than the Buneman algorithm. Details of these algorithms may also be found in Dorr (1970).

16.4. NON-RECTANGULAR REGIONS

The methods of the previous sections can be applied to problems where the differential operator can be written as the sum of commuting operators each of which involves differentiation with respect to one variable, see Swarztrauber (1974) for a discussion of the case of cyclic reduction. A particularly severe restriction on the methods discussed so far is that the region must be rectangular. Problems of the correct form given on a general region can be solved by fast direct methods by embedding the region in a rectangle, or an infinite strip. Such techniques have been studied by Hockney (1970), Buzee *et al.* (1971) and Proskurowski and Widlund (1976).

Consider the solution of the Dirichlet problem for

Poisson's equation, $\nabla^2 u = f$, on a quadrant of a circle, R, which is embedded on a square on which a uniform mesh is imposed.

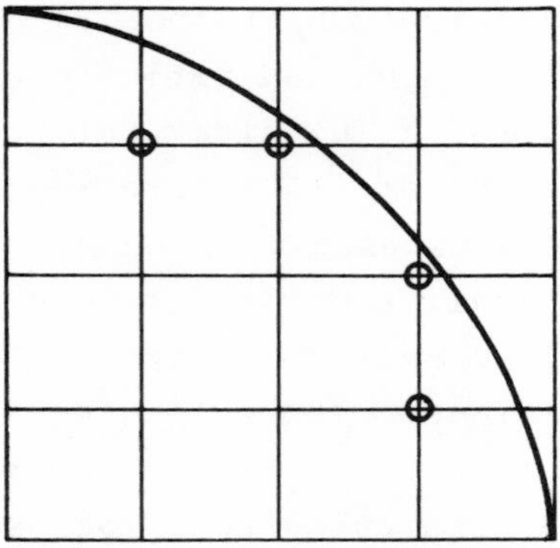

Fig. 16.1.

The equations for the quadrant involve p (say) special finite-difference equations for mesh points adjacent to the curved boundary, marked in Figure 16.1. We enlarge this system, for which the solution is required, by adding the regular (5-point) finite-difference equations for points in the square lying outside R. It is necessary to extend the definition of the function f to the full interior of the square mesh and to extend the boundary conditions to the full boundary of the square. This can be done in any convenient way and has little effect on the performance of the algorithm. We now have a system of, say, n equations which we write in the form

$$A\underset{\sim}{U} = \begin{bmatrix} A_1 \\ \\ A_2 \end{bmatrix} \underset{\sim}{U} = \begin{bmatrix} \underset{\sim}{V}_1 \\ \\ \underset{\sim}{V}_2 \end{bmatrix} = \underset{\sim}{V} \,. \tag{16.9}$$

Here A_1 is $p \times n$ and the first p equations represent the special finite-difference equations associated with the special marked points. This ordering of the equations is merely for convenience of description and it is important to note that this ordering could not be used in practice without destroying the structure of the equations arising in the algorithm, which is necessary for using the fast direct solvers.

If we replace these special equations by the regular equations appropriate to the square region, ignoring the curved boundary, we obtain the system

$$B\bar{\underset{\sim}{U}} = \begin{bmatrix} B_1 \\ \\ A_2 \end{bmatrix} \underset{\sim}{U} = \begin{bmatrix} \bar{\underset{\sim}{V}}_1 \\ \\ \bar{\underset{\sim}{V}}_2 \end{bmatrix} = \bar{\underset{\sim}{V}} \quad , \qquad (16.10)$$

which can be solved by fast direct techniques.

B is a rank p modification of A and we may write

$$A = B + \begin{bmatrix} A_1 - B_1 \\ \\ O \end{bmatrix} = B + FG$$

where $F = [{}^{W}_{O}]$ and $G = W^{-1}(A_1 - B_1)$. W is an arbitrary non-singular p × p matrix, which may be chosen as I_p, and F, G^T are n × p matrices. The Woodbury formula gives

$$A^{-1} = B^{-1}\{I - F(I + GB^{-1}F)^{-1}GB^{-1}\} \ . \qquad (16.11)$$

Using the above definitions of F and G we can show that

$$(I + GB^{-1}F)^{-1} = C^{-1}W$$

where $C = A_1B^{-1}[{}^{W}_{O}]$ is called the capacitance matrix in Buzbee *et al.* (1971), whose treatment we are following. Further manipulation of (16.11) leads to

$$A^{-1}\underset{\sim}{V} = B^{-1}\{\bar{\underset{\sim}{V}} + [{}^{W}_{O}]C^{-1}(\underset{\sim}{V}_1 - A_1\bar{\underset{\sim}{U}})\} \qquad (16.12)$$

Hence the desired (extended) solution $\underset{\sim}{U} = A^{-1}\underset{\sim}{V}$ can be obtained by the following algorithm:

(i) Compute and factor C, the capacitance matrix,

(ii) Solve $B\bar{\underset{\sim}{U}} = \bar{\underset{\sim}{V}}$ by a fast direct method,

(iii) Solve $C\underset{\sim}{Y} = \underset{\sim}{V}_1 - A_1\underset{\sim}{\overline{U}}$,

(iv) Solve $B\underset{\sim}{U} = \underset{\sim}{\overline{V}} + [{W \atop O}]\underset{\sim}{Y}$ by a fast direct method.

In many applications we solve with the same differential operator on the same region many times. In such cases the computation and factorization of p × p matrix C is required only once and the factors can be stored on backing store. This is usually the most expensive stage of the algorithm. Since p is typically $O(\sqrt{n})$, stage (iii) involves forward and back substitutions on a relatively small problem.

The references mentioned at the beginning of this section also deal with the modified algorithm for the case B singular, arising in Neumann problems.

Proskurowski and Widlund (1976) embed the region in an infinite strip parallel to the x-axis with periodic boundary conditions. In this way they obtain, after Fourier separation of variables, an advantageous form of tridiagonal matrix which can be factored immediately into (almost) Toeplitz matrices (this avoids the need to use the Sherman-Morrison formula mentioned in section 16.2). They also give an algorithm for reducing the cost of obtaining the capacitance matrix, see also Buzbee and Dorr (1974) for a discussion of this point.

Buzbee *et al.* (1971) also consider an algorithm appropriate for thin L-shaped regions for which embedding in a rectangle would lead to a great increase in the size of the problem.

In Pereyra *et al.* (1977) an algorithm is described for solving the Dirichlet problem for general bounded regions in R^n, using the capacitance matrix method of Proskurowski and Widlund (1976), and employing deferred correction to the solutions of the finite-difference equations.

16.5. NON-SEPARABLE EQUATIONS

The success of the direct methods discussed in this chapter has led to their use in iterative methods for solving more general problems. Concus and Golub (1973) consider the Dirichlet problem for elliptic equations using a shifted form of the iterative procedure

$$\nabla^2 u_{n+1} = \nabla^2 u_n + t(Lu_n - f)$$

for solving the non-separable problem Lu = f; this form of iterative procedure was advocated by D'Jakanov (1969). They describe the method as extremely efficient, if some initial scaling is carried out and Chebyshev acceleration is employed.

Nonlinear problems on irregular regions are treated in Bartels and Daniel (1974) by methods involving some of the ideas of this chapter.

17

ALTERNATING DIRECTION IMPLICIT METHODS

17.1. INTRODUCTION

Here we are concerned with a certain treatment of finite-difference equations arising from discretizations of elliptic and parabolic partial differential equations. We introduce the techniques by first treating elliptic equations and outline various results for typical model problems. The case of time-dependent (parabolic) equations is then discussed and here the idea of alternating direction implicit (ADI) methods is extended to more general so-called "splitting methods". Finally we briefly indicate some recent developments in which the idea of ADI methods is applied in the context of Galerkin methods for parabolic equations.

The first ADI method was given by Peaceman and Rachford (1955) and subsequently extended by Douglas and Rachford (1956); interestingly these methods for solving elliptic difference equations were derived as by-products of numerical methods for solving parabolic equations. The idea of alternating directions arises naturally from a consideration of the classical block (or line) Jacobi method (see Isaacson and Keller (1966, ch.9) for a clear exposition of the classical iterative methods applied to Poisson's equation on a rectangle). It is the possible improved convergence rate of the ADI method over the successive overrelaxation technique which gives the method its importance; this is guaranteed for certain model problems. For an excellent full treatment and analysis of model problems for the ADI method the reader is referred to Varga (1962, chs. 7, 8). For further developments and particular applications to elliptic equations arising in reactor physics, see Wachspress (1966); a further treatment which contains references and developments up to 1971 can be found in the book by Young (1971, ch. 17). For an up-to-date outline of ADI methods and more general splitting methods as applied to time-dependent problems see Gourlay (1977).

17.2. ELLIPTIC EQUATIONS, THE MODEL PROBLEM

Consider Poisson's equation on the unit square Ω, $0 < x, y < 1$, with boundary Γ,

$$-\nabla^2 u = f(x,y), \qquad (x,y) \in \Omega, \tag{17.1}$$

$$u(x,y) = g(x,y), \qquad (x,y) \in \Gamma. \tag{17.2}$$

Proceeding in the standard fashion, let the closed square be approximated by the uniform mesh $(x_j, y_k) \equiv (j\Delta x, k\Delta y)$, $j = 0,1,\ldots,N+1$, $k = 0,1,\ldots,N+1$, with $\Delta x = \Delta y = 1/(N+1)$ and at the interior mesh points let (17.1) be approximated by the five-point formula

$$\{-U_{j+1\,k} + 2U_{j\,k} - U_{j-1\,k}\} + \{-U_{j\,k+1} + 2U_{j\,k} - U_{j\,k-1}\} = \Delta x^2 f_{jk},$$
$$j,k = 1,2,\ldots,N, \tag{17.3}$$

and (17.2) by

$$U_{j\,k} = g_{jk}, \begin{cases} j = 0,\ N+1,\ k = 1,2,\ldots,N \\ k = 0,\ N+1,\ j = 1,2,\ldots,N. \end{cases} \tag{17.4}$$

Given the data $f_{jk} = f(x_j, y_k)$, $g_{jk} = g(x_j, y_k)$ we now seek the solution of these N^2 finite-difference equations (17.3), where $U_{jk} \approx u(x_j, y_k)$. It can be shown, Isaacson and Keller (1966) ch. 9) that (17.3) has a unique solution for any f, g and x; also as $\Delta x \to 0$ the finite-difference approximations converge uniformly to the solution of (17.1).

Let the unknowns be ordered along horizontal mesh lines and let $\underset{\sim}{U}_k$ denote the unknowns along the kth mesh line,

$$\underset{\sim}{U}_k = (U_{1k}, U_{2k}, \ldots, U_{Nk})^T, \quad k = 1,2,\ldots,N.$$

Then with the aid of (17.4), (17.3) can be written in the familiar block tridiagonal form

$$\begin{bmatrix} D & -I & & & \\ -I_N & D & -I_N & & \\ \cdot & \cdot & \cdot & & \\ & \cdot & \cdot & \cdot & \\ & & \cdot & \cdot & \cdot \\ & & -I_N & D & -I_N \\ & & & -I_N & D \end{bmatrix} \begin{bmatrix} \underset{\sim}{U}_1 \\ \underset{\sim}{U}_2 \\ \cdot \\ \cdot \\ \cdot \\ \underset{\sim}{U}_{N-1} \\ \underset{\sim}{U}_N \end{bmatrix} = \Delta x^2 \begin{bmatrix} \underset{\sim}{F}_1 \\ \underset{\sim}{F}_2 \\ \cdot \\ \cdot \\ \cdot \\ \underset{\sim}{F}_{N-1} \\ \underset{\sim}{F}_N \end{bmatrix} \qquad (17.5)$$

where the right-hand side is obtained from the data, I_N is the $N \times N$ unit matrix and $D = 4I_N - (L_N + L_N^T)$ is tridiagonal with

$$L_N = \begin{bmatrix} 0 & & & & \\ 1 & 0 & & & \\ & 1 & 0 & & \\ & & \cdot & \cdot & \\ & & & \cdot & \cdot \\ & & & \cdot & \cdot \\ & & & 1 & 0 \end{bmatrix} .$$

Writing (17.5) as

$$A\underset{\sim}{U} = \Delta x^2 \underset{\sim}{F}, \quad \underset{\sim}{U} = (\underset{\sim}{U}_1, \ldots, \underset{\sim}{U}_N)^T , \qquad (17.6)$$

it is clear from (17.3) that A can be written as the sum of two matrices $A = H+V$ where H and V arise from the representation of the respective bracketed terms. Thus the component of the vector $H\underset{\sim}{U}$ corresponding to the mesh point (x_j, y_k) is given by

$$-U_{j+1\,k} + 2U_{j\,k} - U_{j-1\,k},$$

which is of course a finite-difference approximation to the one-dimensional operator $-u_{xx}$ along different horizontal lines. Similarly the components of $V\underset{\sim}{U}$ correspond to the discretization of $-u_{yy}$ along different vertical lines. Hence when viewed as an approximation to the operator $-\nabla^2$, $A = H+V$ is the natural "splitting" of A into its one-dimensional components. This idea of splitting an operator into its "simpler" parts is of fundamental importance in this chapter. Defining

$$L = \begin{bmatrix} L_N & & & \\ & L_N & & \\ & & \ddots & \\ & & & L_N \end{bmatrix} \qquad B = \begin{bmatrix} 0 & & & & \\ I_N & 0 & \cdot & & \\ & \cdot & & \cdot & \\ & & \cdot & & \cdot \\ & & & I_N & 0 \end{bmatrix}$$

then $H = 2I - (L + L^T)$, $V = 2I - (B + B^T)$, and we now wish to exploit this splitting of A to devise various iterative schemes for the solution of (17.6).

17.3. THE ADI METHOD

From (17.6)

$$(H + V)\underset{\sim}{U} = \Delta x^2 \underset{\sim}{F}$$

or

$$(H + rI)\underset{\sim}{U} = (rI - V)\underset{\sim}{U} + \Delta x^2 \underset{\sim}{F}$$

where r is any positive scalar, which suggests the simple block iterative method

$$(H + rI)\underset{\sim}{U}^{(\nu+1)} = (rI - V)\underset{\sim}{U}^{(\nu)} + \Delta x^2 \underset{\sim}{F}, \quad \nu = 0,1,2,\ldots, \qquad (17.7)$$

with $\underset{\sim}{U}^{(0)}$ given. Now $H + rI$ is block diagonal where each block is given by $(2+r)I_N - (L_N + L_N^T)$, hence to obtain the new approximation $\underset{\sim}{U}_k^{(\nu+1)}$ on the kth mesh line we only need solve a system of N linear equations with a symmetric diagonally dominant tridiagonal matrix. This can be done directly in an efficient and stable manner. Scheme (17.7) corresponds to sweeping through the mesh improving the approximations one horizontal line at a time; when r = 2 the scheme is the block Jacobi method which is known to be convergent. Similarly we have the scheme $(r > 0)$

$$(V + rI)\underset{\sim}{U}^{(\nu+1)} = (rI - H)\underset{\sim}{U}^{(\nu)} + \Delta x^2\underset{\sim}{F},\ \nu = 0,1,2,\ldots, \qquad (17.8)$$

with $U^{(0)}$ given, but now $V + rI$ is block tridiagonal, producing left hand sides

$$\begin{aligned} &(2 + r)\underset{\sim}{U}_1^{(\nu+1)} - \underset{\sim}{U}_2^{(\nu+1)}, \\ -&\underset{\sim}{U}_{k-1}^{(\nu+1)} + (2 + r)\underset{\sim}{U}_k^{(\nu+1)} - \underset{\sim}{U}_{k+1}^{(\nu+1)}, \quad k = 2,\ldots,N-1, \qquad (17.9) \\ &\qquad -\underset{\sim}{U}_{N-1}^{(\nu+1)} + (2 + r)\underset{\sim}{U}_N^{(\nu+1)}. \end{aligned}$$

The equations (17.8) can be simplified as follows. Writing down the first component of each of (17.9) yields, in matrix form, the left hand side

$$\{(2 + r)I_N - (L_N + L_N^T)\}\hat{\underset{\sim}{U}}_1^{(\nu+1)},$$

where $\hat{\underset{\sim}{U}}_1 = (U_{11}, U_{12},\ldots,U_{1N})^T$ the unknowns along the first vertical mesh line. Similarly the kth component of each of (17.9) yields equations with left hand side

$$\{(2 + r)I_N - (L_N + L_N^T)\}\hat{\underset{\sim}{U}}_k^{(\nu+1)}.$$

Hence reordering in this way produces, as before, N systems each with a tridiagonal matrix. Iteration (17.8) is therefore easily implemented and improves the approximations one vertical line at a time; again the choice $r = 2$ gives the convergent block Jacobi scheme for this ordering.

Before we describe the ADI scheme we note that it is possible to extend the above processes to more general differential equations. Adding the term $\sigma(x,y)u$, $\sigma > 0$ on Ω, to the left hand side of (17.1) leads to the term $\Delta x^2\sigma_{jk}\,U_{jk}$ being added to the finite-difference equations (17.3) which in turn implies the splitting $A = H + V + \Sigma$ for the resulting system where Σ is a positive diagonal matrix. More generally, we can treat the self-adjoint equation

$$-(P(x,y)u_x)_x - (P(x,y)u_y)_y + \sigma(x,y)u(x,y) = f(x,y), (x,y) \in \Omega, \tag{17.10}$$

where Ω is an open bounded connected region with boundary conditions

$$\alpha(x,y)u + \beta(x,y) \frac{\partial u}{\partial n} = \gamma(x,y), \ (x,y) \in \Gamma, \tag{17.11}$$

where $\alpha(x,y) \geq 0$, $\beta(x,y) \geq 0$, $\alpha + \beta > 0$ on Γ, and P and σ are positive and continuous in $\Omega + \Gamma$. Further there is no need to insist on a uniform mesh in each coordinate direction. Hence in summary (17.10) and (17.11) lead to finite difference equations $A\underset{\sim}{U} = \underset{\sim}{k}$ where $A = H + V + \Sigma$ is block tridiagonal, where we are assuming, as before, that the mesh points have been ordered along horizontal mesh lines and that the five-point approximation is used, see Varga (1962, section 6.3) for full details. For example for the quadrant in Figure 17.1 with $\beta \neq 0$ in (17.11), the diagonal blocks are of size 4×4, 3×3, 2×2 and 1×1 respectively. In general H and V are symmetric and positive definite and H is

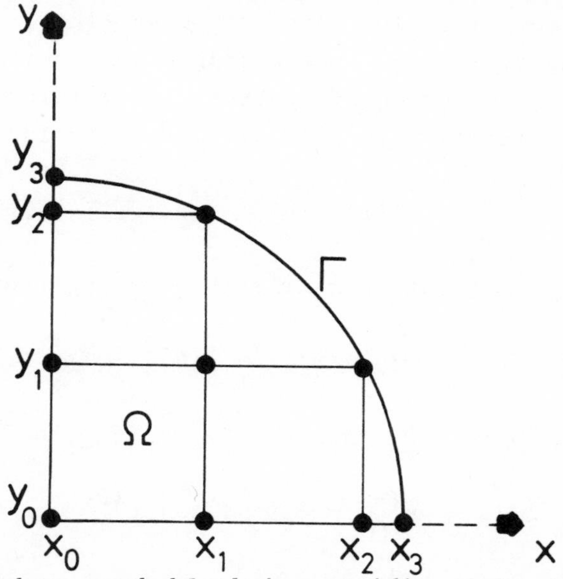

Fig. 17.1.

block diagonal where each block is a tridiagonal symmetric and positive definite matrix, and similarly for the permuted form of V. Henceforth we shall suppose that A is of order

$n \times n$.

The ADI iteration combines the features of the schemes (17.7) and (17.8) by writing $(H + V + \Sigma)\underset{\sim}{U} = \underset{\sim}{k}$ as a pair of equations

$$\begin{aligned}(H + \tfrac{1}{2}\Sigma + rI)\underset{\sim}{U} &= (rI - V - \tfrac{1}{2}\Sigma)\underset{\sim}{U} + \underset{\sim}{k} \;, \\ (V + \tfrac{1}{2}\Sigma + rI)\underset{\sim}{U} &= (rI - H - \tfrac{1}{2}\Sigma)\underset{\sim}{U} + \underset{\sim}{k}\end{aligned} \tag{17.12}$$

for any constant $r > 0$. With $H_1 = H + \frac{1}{2}\Sigma$, $V_1 = V + \frac{1}{2}\Sigma$, the first ADI method (Peaceman and Rachford, 1955) is given by

$$\begin{aligned}(H_1 + r_{\nu+1}I)\underset{\sim}{U}^{(\nu+\frac{1}{2})} &= (r_{\nu+1}I - V_1)\underset{\sim}{U}^{(\nu)} + \underset{\sim}{k} \\ (V_1 + r_{\nu+1}I)\underset{\sim}{U}^{(\nu+1)} &= (r_{\nu+1}I - H_1)\underset{\sim}{U}^{(\nu+\frac{1}{2})} + \underset{\sim}{k}, \quad \nu \geq 0,\end{aligned} \tag{17.13}$$

where $\underset{\sim}{U}^{(0)}$ is a starting approximation and the r_ν are positive constants (called acceleration parameters) whose values are chosen to maximise the rate of convergence. From the previous discussion it is clear that the first stage of (17.13) corresponds to iterating along horizontal mesh lines and the second stage to iterating along vertical mesh lines - hence the term alternating directions. Both stages can easily be carried out by the direct solution of systems with tridiagonal diagonally dominant matrices (with $r_\nu > 0$). From (17.13) we can write

$$\underset{\sim}{U}^{(\nu+1)} = T(r_{\nu+1})\underset{\sim}{U}^{(\nu)} + \underset{\sim}{g}(r_{\nu+1}) \;, \quad \nu \geq 0,$$

where the ADI iteration matrix is given by (setting $r_{\nu+1} = r$)

$$T(r) = (V_1 + rI)^{-1}(rI - H_1)(H_1 + rI)^{-1}(rI - V_1) \; . \tag{17.14}$$

With $\underset{\sim}{e}^{(\nu)} = \underset{\sim}{U}^{(\nu)} - \underset{\sim}{U}$, then $\underset{\sim}{e}^{(\nu+1)} = T(r_{\nu+1})\underset{\sim}{e}^{(\nu)}$ and hence

$$\underset{\sim}{e}^{(\nu)} = \left(\prod_{j=1}^{\nu} T(r_j)\right)\underset{\sim}{e}^{(0)} \;, \quad \nu \geq 1 \;, \tag{17.15}$$

where $\underset{\sim}{e}^{(0)}$ is the error in the starting approximation. The success of this ADI scheme lies in being able to vary the acceleration parameter r_j from one iteration to the next;

success is determined by the behaviour of $\|\Pi_{j=1}^{\nu} T(r_j)\|$. Convergence is easily shown for the simple case of constant parameters $r_j = r > 0$, $\nu \ge 1$. Let $\bar{T}(r) = (V_1 + rI)T(r)(V_1 + rI)^{-1}$, then by similarity $T(r)$ and $\bar{T}(r)$ have the same eigenvalues. If now $\rho(T_r)$ denotes the spectral radius of T_r, then from (17.14), using elementary norm properties we have

$$\rho(T(r)) = \rho(\bar{T}(r)) \le \|\bar{T}(r)\| \le$$

$$\le \|(rI - H_1)(H_1 + rI)^{-1}\| \|(rI - V_1)(V_1 + rI)^{-1}\|$$

But H_1 and V_1 are symmetric and positive definite, and so employing the spectral matrix norm (2-norm) yields

$$\|(rI - H_1)(H_1 + rI)^{-1}\|_2 = \max_{1 \le j \le n} \left| \frac{r - \lambda_j}{r + \lambda_j} \right| < 1 ,$$

where λ_j, $1 \le j \le n$ are the (positive) eigenvalues of H_1. A similar argument applies to the norm involving V_1, hence $\rho(T(r)) < 1$ for all $r > 0$, and therefore the iteration (17.13) is convergent for all $\underset{\sim}{U}^{(0)}$.

For the model problem (17.1) and (17.2) with a uniform mesh it is possible to determine the optimum parameter r_b such that

$$\rho(T(r_b)) \le \rho(T(r)) \quad \text{for all } r > 0 .$$

It turns out that $\rho(T(r_b)) = \rho(\mathcal{L}_{\omega_b})$, the optimised spectral radius for the point SOR iteration; hence the two schemes have an identical asymptotic rate of convergence - this is for all Δx. The ADI iteration does however involve far more computation, and for its efficient implementation it is essential to vary the acceleration parameters r_ν.

17.4. ADI - VARIABLE ACCELERATION PARAMETERS

For the model problem it can be shown that $H_1 V_1 = V_1 H_1$, that is H_1 and V_1 commute; equivalently H_1 and V_1 have a common set of (orthonormal) eigenvectors. This is the key property in determining the best values of r_ν. For the general application of (17.13) ($r_\nu > 0$) we assume first of all that the positive definite matrices H_1 and V_1 commute and we let $(\lambda i, \underset{\sim}{\alpha}_i)_{i=1}^n$, $(\mu_i, \underset{\sim}{\alpha}_i)_{i=1}^n$ denote their respective eigensystems ($\{\underset{\sim}{\alpha} i\}^n$, orthonormal). For ν iterations of (17.13), relation (17.14) yields, for $1 \le i \le n$,

$$\left(\prod_{j=1}^{\nu} T(r_j)\right)\underset{\sim}{\alpha}_i = \left\{\prod_{j=1}^{\nu} \left(\frac{r_j-\lambda_i}{r_j+\lambda_i}\right)\left(\frac{r_j-\mu_i}{r_j+\mu_i}\right)\right\}\underset{\sim}{\alpha}_i \; .$$

It follows that $\Pi_{j=1}^{\nu} T(r_j)$ is symmetric and therefore

$$\left\| \prod_{j=1}^{\nu} T(r_j)\right\|_2 = \rho\left(\prod_1^{\nu} T(r_j)\right) = \max_{1\le i\le n} \prod_{j=1}^{\nu} \left|\frac{r_j-\lambda_i}{r_j+\lambda_i}\right|\left|\frac{r_j-\mu_i}{r_j+\mu_i}\right| < 1 \; , \tag{17.16}$$

establishing the convergence of the iteration. Generally λ_i and μ_i, $1 \le i \le n$ are unknown, but if bounds α, β are available,

$$0 < \alpha \le \lambda_i, \; \mu_i \le \beta, \quad 1 \le i \le n \; ,$$

the problem of minimising the spectral radius in (17.16) can be tackled in practice (more realistically estimates for α, β could be obtained by variants of the power method for obtaining the dominant eigenvalue of a matrix). Clearly,

$$\max_{1\le i\le n} \prod_{j=1}^{\nu} \left|\frac{r_j-\lambda_i}{r_j+\lambda_i}\right|\left|\frac{r_j-\mu_i}{r_j+\mu_i}\right| \le \max_i \prod_{j=1}^{\nu} \left|\frac{r_j-\lambda_i}{r_j+\lambda_i}\right| \max_i \prod_{j=1}^{\nu} \left|\frac{r_j-\mu_i}{r_j+\mu_i}\right|$$

$$\le \left\{ \max_{\alpha\le x\le\beta} \prod_{j=1}^{\nu} \left|\frac{r_j-x}{r_j+x}\right|\right\}^2 ,$$

so that

$$\rho\left(\prod_1^{\nu} T(r_j)\right) \le \max_{\alpha\le x\le\beta} |R_\nu(x)|^2$$

where $R_\nu(x)$ is the rational function $R_\nu(x) \equiv \Pi_{j=1}^{\nu}(r_j-x)/(r_j+x)$. Hence given ν the problem is now one of minimising the uniform norm of the rational function R ; this problem has a unique solution which is characterised by the standard Chebyshev equioscillation property (see Varga, 1962, p. 223). The optimum parameters $\overline{r}_j$, $1 \le j \le \nu$, are distinct and satisfy

$\alpha < \bar{r}_j < \beta$, $1 \leq j \leq \nu$, but for general ν it is not possible to give explicit formulas for them in terms of α, β; they could however be determined using techniques for nonlinear Chebyshev approximation, see Meinardus (1967). Varga (1962), however, determined formulas in the case $\nu = 2^k$ which appear very attractive computationally. Based on the average rate of convergence for ν iterations Varga presents a simplified analysis and obtains approximations for the parameters as well as the optimum value of ν; ν is determined from $(\sqrt{2} - 1)^{-2\nu} = \beta/\alpha$, from which the acceleration parameters are given by

$$\hat{r}_j = \alpha(\beta/\alpha)^{(2j-1)/2\nu}, \quad 1 \leq j \leq \nu. \tag{17.17}$$

With these parameters the resulting ADI iteration for the model problem is significantly faster (for large N) than the point SOR iteration. In fact this is so for *any fixed value* of $\nu > 1$ using (17.17).

The important problem now is to decide how restrictive is the condition that H_1 and V_1 commute. In fact it is very restrictive (Birkhoff & Varga 1959) and in practice requires that the region Ω be rectangular (with sides parallel to the coordinate axes) and that the differential equation be of the form,

$$-(f_1(x)u_x)_x - (f_2(y)u_y)_y + \sigma u = s(x,y) \ ,$$

where f_1, $f_2 > 0$ on $\Omega + \Gamma$, $\sigma \equiv$ positive constant. Recall however, that the basic convergence of the ADI iteration for fixed acceleration parameter $r > 0$ was proved without the condition $H_1V_1 = V_1H_1$ so that in practice we can expect success in far more general situations than this specialised result would suggest.

17.5. VARIANTS OF THE ADI SCHEME, THREE DIMENSIONAL PROBLEMS

Many variants of the basic Peaceman-Rachford scheme have been proposed. For example, modifying the second stage of (17.13),

$$\begin{aligned}(H_1 + r_{\nu+1} I)\underset{\sim}{U}^{(\nu+\frac{1}{2})} &= (r_{\nu+1}I - V_1)\underset{\sim}{U}^{(\nu)} + \underset{\sim}{k} \\ (V_1 + r_{\nu+1}I)\underset{\sim}{U}^{(\nu+1)} &= (V_1 - (1 - w)r_{\nu+1}I)\underset{\sim}{U}^{(\nu)} + (2-w)r_{\nu+1}\underset{\sim}{U}^{(\nu+\frac{1}{2})}\end{aligned} \tag{17.18}$$

where w is a parameter. For w = 0, we have the Peaceman-Rachford scheme (17.13) (after using the first stage to express $H_1\underset{\sim}{U}^{(\nu+\frac{1}{2})}$ in terms of $V_1\underset{\sim}{U}^{(\nu)}$ and thereby economising on the evaluation of the right hand side vectors); for w = 1 the scheme is due to Douglas and Rachford (1956). For H_1, V_1 symmetric and positive definite and for a fixed acceleration parameter $r > 0$, the resulting generalised ADI scheme is convergent for any $0 \le w < 2$.

An interesting development due to Wachspress and Habelter (1960) (see Wachspress, 1966) was the introduction of a positive diagonal matrix F to replace the identity matrices in (17.18). The idea being to try to choose F so that the associated H_1 and V_1 matrices more nearly commute and so conform to the previous theory for estimating the acceleration parameters.

An important feature of the Douglas-Rachford scheme, w = 1, is that it generalises to equations with three space variables. For example consider the Dirichlet problem on the region Ω

$$u_{xx} + u_{yy} + u_{xx} = 0, \quad (x, y, z) \in \Omega ,$$

where u(x,y,z) is specified on the boundary Γ. Using the standard seven-point formula yields the finite-difference equation

$$(X + Y + Z)\underset{\sim}{U} = \underset{\sim}{k}$$

where X, Y, and Z are symmetric positive definite, and which after appropriate reordering are block diagonal (each diagonal block being tridiagonal). The Douglas scheme (1962) is given by

$$(X + r_{\nu+1}I)\underset{\sim}{U}^{(\nu+1/3)} = (r_{\nu+1}I - X - 2Y - 2Z)\underset{\sim}{U}^{(\nu)} + 2\underset{\sim}{k}$$

$$(Y + r_{\nu+1}I)\underset{\sim}{U}^{(\nu+2/3)} = Y\underset{\sim}{U}^{(\nu)} + r_{\nu+1}\underset{\sim}{U}^{(\nu+1/3)}$$

$$(Z + r_{\nu+1}I)\underset{\sim}{U}^{(\nu+1)} = Z\underset{\sim}{U}^{(\nu)} + r_{\nu+1}\underset{\sim}{U}^{(\nu+2/3)}$$

and corresponds to sweeping through the mesh parallel to the three coordinate axes in turn; each stage consists of solving tridiagonal systems. When X, Y and Z commute the scheme is convergent for any fixed iteration parameter $r > 0$. Little seems to be known about the choice of optimum acceleration parameters, but see Douglas (1962) for a discussion.

17.6. PRACTICAL APPLICATION OF THE ADI METHOD FOR ELLIPTIC PROBLEMS

It is extremely difficult to make strong statements concerning the relative efficiency of the ADI schemes for general elliptic problems. The Peaceman-Rachford scheme has been successfully used a good deal in industrial problems (particularly in the U.S.A.); Price and Varga (1962) report on numerical experiments which broadly indicate that for large problems (small mesh spacing) (even when the commutative theory does not hold) this ADI scheme is superior to the SOR schemes. The work and experience of Wachspress suggests that there are two fundamental factors which effect the efficiency of ADI methods, namely the deviation from model problem conditions and the size of the problem. Compared to SOR, ADI is best suited to large problems, and the greater the departure from model conditions the greater this break-even size becomes.

It is clear that more research is required to realise the full potential of this important technique.

17.7. PARABOLIC PROBLEMS

For time-dependent problems the primary concern is with the stability of the approximation processes coupled with an efficient computation each time step. Consider the model problem,

$$u_t = u_{xx} + u_{yy} \quad (x,y) \in \Omega,\ t > 0 \qquad (17.19)$$

where Ω is a finite region in the plane. Let, for $t > 0$, conditions be specified on the boundary of Ω and an initial condition be given,

$$u(x,y,0) = g(x,y) \quad (x,y) \in \Omega .$$

Assume that Ω is approximated using a uniform mesh $\Delta x = \Delta y$ and let $\lambda = \Delta t/\Delta x^2$ where Δt is the time increment. In the standard notation, $U_{jk}^n \approx u(x_j,y_k,t_n)$, we shall employ the five-point formula to approximate the right hand side of (17.19).

The explicit scheme

$$\frac{U_{jk}^{n+1}-U_{jk}^n}{\Delta t} = \frac{1}{\Delta x^2}(\delta_x^2 U_{jk}^n + \delta_y^2 U_{jk}^n)$$

is subject to the severe stability restriction $0 < \lambda < 1/4$, however, the fully implicit scheme

$$\frac{U_{jk}^{n+1}-U_{jk}^n}{\Delta t} = \frac{1}{\Delta x^2}(\delta_x^2 U_{jk}^{n+1} + \delta_y^2 U_{jk}^{n+1}) \tag{17.20}$$

is *unconditionally* stable. The form of the linear equations in (17.20) is familiar from the treatment of elliptic equations and can be written

$$\left(\frac{1}{\lambda} I + A\right)\underset{\sim}{U}^{n+1} = \frac{1}{\lambda}\underset{\sim}{U}^n, \quad n \geq 0, \quad \underset{\sim}{U}^0 = \underset{\sim}{g} . \tag{17.21}$$

Here, for simplicity, we have assumed zero boundary conditions, otherwise the right hand side would include a time dependent vector $\underset{\sim}{s}_n$. Hence *at each time step* we are involved with the solution of a system of linear equations, and for large problems the scheme (17.20) is often rejected on the basis of computational cost. Recalling the structure of A via the splitting $A = H + V$, the schemes (written directly in matrix form)

$$\left(\frac{1}{\lambda} I + H\right)\underset{\sim}{U}^{n+1} = \left(\frac{1}{\lambda} I - V\right)\underset{\sim}{U}^n , \quad n \geq 0,$$

$$\left(\frac{1}{\lambda} I + V\right) \underset{\sim}{U}^{n+1} = \left(\frac{1}{\lambda} I - H\right) \underset{\sim}{U}^{n}, \quad n \geq 0,$$

are each computationally feasible (involving solution of sets of tridiagonal systems, see (17.7), (17.8)) and conditionally stable. If they are used *alternately* then the overall scheme is unconditionally stable and is clearly an ADI method where $r = 1/\lambda$ has the role of a constant acceleration parameter. Douglas and Rachford (1956) proposed the scheme (see (17.18) with $w = 1$)

$$\left(\frac{1}{\lambda} I + H\right) \hat{\underset{\sim}{U}}^{n+1} = \left(\frac{1}{\lambda} I - V\right) \underset{\sim}{U}^{n},$$
$$\left(\frac{1}{\lambda} I + V\right) \underset{\sim}{U}^{n+1} = \frac{1}{\lambda} \hat{\underset{\sim}{U}}^{n+1} + V\underset{\sim}{U}^{n} \qquad (17.22)$$

where $\hat{\underset{\sim}{U}}^{n+1}$ is regarded as an "intermediate" quantity, this method is both unconditionally stable and computationally feasible - involving solving along horizontal lines for $\hat{\underset{\sim}{U}}^{n+1}$ and then along vertical lines for the final approximation $\underset{\sim}{U}^{n+1}$. (The analysis of stability for (17.22) is equivalent to that of convergence for the iteration (17.18) with $w = 1$ and constant acceleration parameter $r = 1/\lambda$).

An observation by Douglas and Rachford was that eliminating $\hat{\underset{\sim}{U}}^{n+1}$ from (17.22) yields

$$\left(\frac{1}{\lambda} I + H + V\right) \underset{\sim}{U}^{n+1} = \frac{1}{\lambda} \underset{\sim}{U}^{n} - \lambda HV(\underset{\sim}{U}^{n+1} - \underset{\sim}{U}^{n}), \quad n \geq 0.$$

Hence, (17.22) is equivalent to a perturbed form of the fully implicit scheme (17.21). The importance of this observation is that by perturbing in a suitable way an implicit scheme (which has the required accuracy and stability properties) we have obtained a scheme (17.22), with similar properties, but which is far simpler computationally. This idea forms the basis for constructing many splitting schemes. For example Douglas and Gunn (1964) extend the basic idea of alternating directions in order to realize in practice schemes of the form

$$(I + A)\underset{\sim}{U}^{n+1} = B\underset{\sim}{U}^{n} \qquad (17.23)$$

where $A = \Sigma_{i=1}^{q} A_i$ and $\{A_i\}_{i=1}^{q}$ are easily "inverted" (for example by solving sets of tridiagonal equations as in (17.22)). They use

$$(I + A_1)\underset{\sim}{U}_{(1)}^{n+1} = B\underset{\sim}{U}^n - \sum_{j=2}^{q} A_j \underset{\sim}{U}^n$$

$$(I + A_i)\underset{\sim}{U}_{(i)}^{n+1} = \underset{\sim}{U}_{(i-1)}^{n+1} + A_i \underset{\sim}{U}^n, \quad i = 2,3,\ldots,q, \ \underset{\sim}{U}^{n+1} = \underset{\sim}{U}_{(q)}^{n+1}$$

and show that it is equivalent to (17.23) with a perturbed right hand side.

At this stage it should be noted that we have assumed zero boundary values on Γ; more generally boundary conditions must be incorporated into the above processes. This raises the problem of how to generate boundary values for "intermediate" quantities, for example $\hat{\underset{\sim}{U}}^{n+1}$ in (17.22). Using the same boundary values as $\underset{\sim}{U}^{n+1}$ can lead to the introduction of errors (the true form of the equivalent perturbed scheme is not being implemented). The problem can be dealt with exactly only for rectangular regions and the reader is referred to Mitchell (1969) and Gourlay (1977) for further discussion and references. Nevertheless, ADI methods have been successful for general regions, particularly in two space dimensions.

Finally in this section we briefly mention the approach known as the hopscotch method which is a further attempt to reduce the implicitness of difference schemes whilst maintaining their order of accuracy and stability. To illustrate, consider the rather general scheme for (17.19) given by

$$U_{jk}^{n+1} - U_{jk}^{n} = \lambda\{\theta(\delta_x^2+\delta_y^2)U_{jk}^{n+1} + (1-\theta)(\delta_x^2+\delta_y^2)U_{jk}^{n}\}$$

where θ is a scalar parameter. When $\theta = 0/1$ we have the explicit/implicit schemes given at the beginning of this section; $\theta = \frac{1}{2}$ corresponds to the Crank-Nicolson scheme. Hopscotch generalises the role of θ by making θ a function of space and time, so that here we write θ as θ_{jk}^{n}. Naturally conditions must be placed on the values θ_{jk}^{n} to preserve accuracy and stability and, in practice, these values are taken to be zero-one variables. A simple example, known as odd-even hopscotch, is given by Gourlay (1970):

$$U^{n+1}_{jk} - U^n_{jk} = \lambda\{\theta^{n+1}_{jk}\ (\delta_x^2+\delta_y^2)U^{n+1}_{jk} + \theta^n_{jk}(\delta_x^2+\delta_y^2)U^n_{jk}\}$$

$$\theta^n_{jk} = \begin{cases} 1 \text{ if } n+j+k \text{ is an even integer,} \\ 0 \text{ otherwise.} \end{cases}$$

Another choice, known as line hopscotch, is given by

$$\theta^n_{jk} = \begin{cases} 1 \quad \text{if } n+j \text{ is even} \\ 0 \quad \text{otherwise;} \end{cases}$$

according to Gourlay (1977), it is generalisations of this form of the technique that are the most powerful. Both these choices exhibit unconditional stability.

17.8. ALTERNATING DIRECTION GALERKIN METHODS

So far we have been essentially concerned with the finite-difference approach to solving differential equations. In recent years variational methods for solving these problems have been developed extensively and in particular offer greater accuracy (at least in theory) over standard finite-difference methods. Here we wish to briefly show how the ADI technique has been used in the implementation of Galerkin methods for time-dependent problems. Again, the basic idea is to replace the equations arising from a multidimensional problem by a sequence of "simpler" systems, similar in form to those arising from one-dimensional problems. A fundamental paper in this area which deals with the application of this idea (on rectangular domains) to parabolic, hyperbolic and elliptic problems is by Douglas and Dupont (1971) For a more recent survey of the general area, see Fairweather (1978) and the associated references.

We shall give a brief introduction using the notation of ch.3. So, suppose we wish to solve the equation $Lu = f$, $u \in \mathcal{H}$, where $\mathcal{H}$ is some inner product space. Then let $\mathcal{H}_N \subset \mathcal{H}$ be a finite dimensional subspace with basis $\varphi_1, \varphi_2, \ldots, \varphi_N$. We now seek an approximation $U = \Sigma_{i=1}^N \alpha_i\varphi_i$ to u where the unknown coefficients are determined by demanding that the resulting residual is "small" in the sense that Lu - f is orthogonal to $\mathcal{H}_N$. Therefore $(Lu,\varphi_i) = (f,\varphi_i)$, $i = 1,2,\ldots,N$ and the solution of these equations yields the Galerkin approximation to u with respect to the subspace $\mathcal{H}_N$. For time-dependent problems the idea is applied at a series of time steps $t_n = n\Delta t$, so that the coefficients $\{\alpha_i\}$ will be functions of time.

For the model problem on the unit square Ω

$$u_t = \nabla^2 u\ , \quad t > 0,$$

$$u(x,y,t) = 0, \quad (x,y) \in \Gamma, \quad u(x,y,0) = u_0(x,y), (x,y) \in \Omega,$$

let $(u,v) = \int_\Omega uv\ dx\ dy$ and let $\mathcal{H} = \mathcal{H}_1^0$ consist of those functions vanishing on Γ and which have smooth derivatives in Ω. The Galerkin approximation satisfies

$$(U_t,\varphi_i) + a[U,\varphi_i] = 0, \quad i = 1,2,\ldots,N, \quad t > 0\ ,$$

with

$$(U,\varphi_i) = (u_0,\varphi_i), \quad t = 0\ ,$$

and

$$a[U,\varphi_i] \equiv (\nabla U,\nabla\varphi_i)\ .$$

Writing $U(x,y,t) = \Sigma_{i=1}^N\ \alpha_i(t)\varphi_i(x,y)$ yields a system of ordinary differential equations of the form,

$$C\ \frac{d\underset{\sim}{\alpha}}{dt} + A\underset{\sim}{\alpha} = \underset{\sim}{0}, \quad \underset{\sim}{\alpha}(0) \text{ specified}, \quad \underset{\sim}{\alpha} = (\alpha_1,\alpha_2,\ldots,\alpha_N)^T.$$

We discretize the Galerkin equations using a difference approximation, at time $t_{n+\frac{1}{2}}$,

$$\left(\frac{U^{n+1}-U^n}{\Delta t}\ ,\ \varphi_i\right) + (\nabla U^n,\nabla\varphi_i) = 0, \ i = 1,2,\ldots,N.$$

To ensure their use for unrestricted Δt, we stabilise these equations by adding in a suitable term

$$\left(\frac{U^{n+1}-U^n}{\Delta t}\ ,\ \varphi_i\right) + a[(U^{n+1}+U^n),\varphi_i] = 0 \quad i = 1,2,\ldots,N. \quad (17.24)$$

The vital step now is to introduce a tensor product basis for $\mathcal{H}_N$, that is $\varphi_k = \psi_i(x)\xi_j(y)$, $k = j+(i-1)m$, $i,j = 1,2,\ldots,m$,

$N = m^2$, so that $U^n(x,y) = \Sigma_{ij}\ \eta_{ij}^n\ \psi_i(x)\xi_j(y)$. To enable a convenient matrix representation of (17.24) we require the notion of a matrix tensor product; let $A = (a_{ij})$ and B be two $m \times m$ matrices then

$$A \otimes B = \begin{bmatrix} a_{11}B & a_{12}B & \cdots & a_{1m}B \\ a_{21}B & a_{22}B & \cdots & a_{2m}B \\ \vdots & \vdots & & \vdots \\ a_{m1}B & a_{m2}B & \cdots & a_{mm}B \end{bmatrix} .$$

Then with $(f,g)_x = \int_0^1 fg dx$, $(f,g)_y = \int_0^1 fg dy$, define the $m \times m$ matrices,

$$C_x = ((\psi_i,\psi_j)_x) , \qquad C_y = ((\xi_i,\xi_j)_y) ,$$

$$A_x = ((\psi_i',\psi_j')_x) , \qquad A_y = ((\xi_i',\xi_j')_y) ,$$

which are symmetric and positive definite. With the unknown coefficients ordered as follows

$$\underset{\sim}{\eta} = (\eta_{11},\eta_{12},\ldots,\eta_{1m};\eta_{21},\eta_{22},\ldots;\ \ldots,\eta_{mm})^T$$

the equations (17.24) can be written

$$[C_x \otimes C_y + \tfrac{1}{2}\Delta t(A_x \otimes C_y + C_x \otimes A_y)]\underset{\sim}{\eta}^{n+1}$$

$$= [C_x \otimes C_y - \tfrac{1}{2}\Delta t(A_x \otimes C_y + C_x \otimes A_y)]\underset{\sim}{\eta}^n,\ n \geq 0 .$$

Finally, in order to apply the ADI idea it is necessary to be able to factor the left hand side, this can be done by perturbing the left hand side by a quantity of order Δt^2, namely

$$-\tfrac{1}{4}\ \Delta t^2\ A_x \otimes A_y\ (\underset{\sim}{\eta}^{n+1} - \underset{\sim}{\eta}^n) .$$

Hence the alternating direction modification of (17.24) is

$$\begin{aligned}(C_x \otimes I + \tfrac{\Delta t}{2} A_x \otimes I)(I \otimes C_y + \tfrac{\Delta t}{2} I \otimes A_y)\, \underset{\sim}{\eta}^{n+1} \\ = (C_x \otimes I - \tfrac{\Delta t}{2} A_x \otimes I)(I \otimes C_y - \tfrac{\Delta t}{2} I \otimes A_y)\underset{\sim}{\eta}^{n} \\ = \underset{\sim}{g}^{n} \quad \text{(say)}\end{aligned} \tag{17.25}$$

for $n \geq 0$. This is implemented as

$$(C_x \otimes I + \tfrac{\Delta t}{2} A_x \otimes I)\gamma^n = \underset{\sim}{g}^{n}$$

$$(I \otimes C_y + \tfrac{\Delta t}{2} I \otimes A_y)\, \underset{\sim}{\eta}^{n+1} = \underset{\sim}{\gamma}^{n}$$

that is, as two sets of one-dimensional problems; the left hand side matrices will, by a suitable choice of basis functions, have a particularly simple block structure, based on symmetric positive definite matrices.

This treatment can be extended to deal with nonlinear equations (on rectangular regions) of the form,

$$u_t = \nabla\, (a(x,y,t,u)\nabla u) + f(x,y,t,u,u_x,u_y).$$

An important feature of this technique is that the resulting system (17.25) has a constant left hand side, the work being in updating the right hand sides $\underset{\sim}{g}^{n}$; see Douglas and Dupont (1971) for an extensive analysis of this problem.

18

SOLUTION OF NONLINEAR EQUATIONS

18.1. INTRODUCTION

In this chapter we consider the problem of finding a solution $\underset{\sim}{x}^* = (x_1^*, x_2^*, \ldots, x_n^*)^T$ of the system of n nonlinear equations

$$\begin{aligned} f_1(x_1, x_2, \ldots, x_n) &= 0 , \\ f_2(x_1, x_2, \ldots, x_n) &= 0 , \\ &\vdots \\ f_n(x_1, x_2, \ldots, x_n) &= 0 , \end{aligned} \tag{18.1}$$

which can be written more concisely as

$$\underset{\sim}{f}(\underset{\sim}{x}) = \underset{\sim}{0}, \tag{18.2}$$

where $\underset{\sim}{x}$ and $\underset{\sim}{f}(\underset{\sim}{x}) \in \mathbb{R}^n$. See Wait (1979) for a general discussion of this problem.

We shall not be concerned with questions such as the existence and uniqueness of the solution $\underset{\sim}{x}^*$ of (18.2). We shall instead assume the existence of $\underset{\sim}{x}^*$ and also that some user-supplied approximation $\underset{\sim}{x}^{(1)}$ to $\underset{\sim}{x}^*$ is available. In general it is impossible to solve a system of equations of the form (18.2) directly, and some iterative method of solution is necessary. This iterative method will have the following stages:

(a) Initialisation: user-supplied approximation $\underset{\sim}{x}^{(1)}$;

(b) Iteration: $\underset{\sim}{x}^{(i+1)} = \underset{\sim}{\phi}(\underset{\sim}{x}^{(i)})$, $i = 1,2,\ldots$;

(c) Termination: convergence criterion for (b).

This chapter is concerned primarily with stage (b) in the case where the system (18.2) arises during the numerical solution of elliptic and parabolic partial differential equations.

Such a system of nonlinear equations arises, for example, in the finite element method solution, via a minimum variational principle (or via a Galerkin approximation), of the elliptic equation

$$\nabla^2 u = e^u , \tag{18.3}$$

defined on Ω, the unit square $0 < x,y < 1$, with boundary conditions $u = 0$ on the boundary Γ. Using triangular elements as in Figure 18.1 with linear basis functions, it can be seen that the Galerkin method gives the system of n (the number of internal nodes) nonlinear equations

$$A\underset{\sim}{U} = \underset{\sim}{F}(\underset{\sim}{U}), \tag{18.4}$$

where the $n \times n$ matrix A has the block form

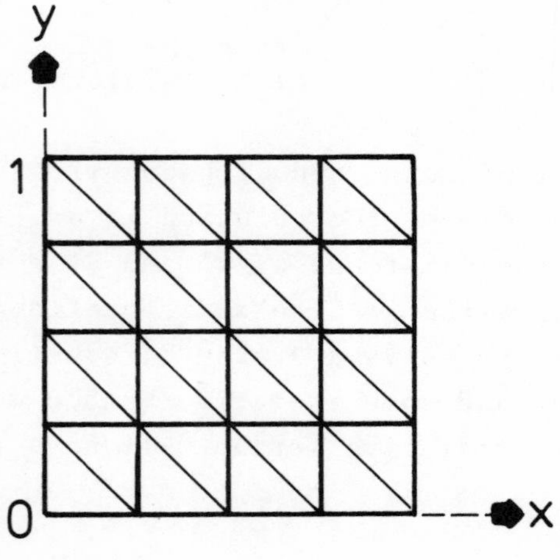

Fig. 18.1.

$$A = \begin{pmatrix} B & -I & & & \\ -I & B & \cdot & & \\ & \cdot & \cdot & \cdot & \\ & & \cdot & \cdot & -I \\ & & & -I & B \end{pmatrix},$$

with

$$B = \begin{pmatrix} 4 & -1 & & & \\ -1 & 4 & \cdot & & \\ & \cdot & \cdot & \cdot & \\ & & \cdot & \cdot & -1 \\ & & & -1 & 4 \end{pmatrix}$$

and

$$F_i(\underset{\sim}{U}) = -\iint_\Omega \varphi_i(x,y)\exp\{\sum_{j=1}^{n} U_j\varphi_j(x,y)\}dxdy \ , \quad (18.5)$$

where $\varphi_i(x,y)$ is the basis function corresponding to the ith internal node. $\underset{\sim}{F}(\underset{\sim}{U})$ can be evaluated analytically, although we do not quote the resulting rather cumbersome expression. A similar system of nonlinear equations results from a five-point finite-difference replacement.

The system (18.4) can, of course, be written in the form (18.2) as

$$\underset{\sim}{f}(\underset{\sim}{U}) = A\underset{\sim}{U} - \underset{\sim}{F}(\underset{\sim}{U}) = \underset{\sim}{0} \ . \quad (18.6)$$

It should be noted that

(i) $\underset{\sim}{f}(\underset{\sim}{U})$ consists of a linear and a nonlinear part, as might be expected from a discretization of (18.3),

(ii) the matrix A is symmetric, positive definite, banded and sparse,

(iii) each function $F_i(\underset{\sim}{U})$ of (18.5) depends only on a small number (determined by the index i) of the variables U_j.

These properties are typical of the nonlinear system resulting from a finite-difference or finite element discretization of a nonlinear elliptic partial differential equation; we shall consider how each property can be used to advantage in the numerical solution of (18.6). It should be noted however that global and regional methods lead to smaller nonlinear systems with less structure.

18.2. FUNCTIONAL ITERATION

The method of functional iteration is based on rewriting (18.2) as

$$\underset{\sim}{x} = \underset{\sim}{g}(\underset{\sim}{x}) \,, \tag{18.7}$$

where $\underset{\sim}{g}(\underset{\sim}{x}) \in \mathbb{R}^n$ is a nonlinear function. The method then uses the iteration

$$\underset{\sim}{x}^{(i+1)} = \underset{\sim}{g}(\underset{\sim}{x}^{(i)}), \quad i = 1,2,... \tag{18.8}$$

The convergence properties of (18.8) are well-known; for example Ortega and Rheinboldt (1970, p. 385) states that if $\underset{\sim}{g} : D \subset \mathbb{R}^n \to \mathbb{R}^n$ maps a closed set $D_0 \subset D$ into itself, and

$$\| \underset{\sim}{g}(\underset{\sim}{x}) - \underset{\sim}{g}(\underset{\sim}{y}) \| \le \alpha \| \underset{\sim}{x} - \underset{\sim}{y} \| \,,$$

for all $\underset{\sim}{x}, \underset{\sim}{y} \in D_0$ and for some $\alpha < 1$, then for any $\underset{\sim}{x}^{(1)} \in D_0$, the iteration (18.8) converges in the norm $\|.\|$ to the unique fixed point $\underset{\sim}{x}^*$ of $\underset{\sim}{g}(\underset{\sim}{x})$ in D_0.

Now (18.6) can be rewritten in the form (18.7) as

$$\underset{\sim}{U} = A^{-1}\underset{\sim}{F}(\underset{\sim}{U}) \,.$$

The method of functional iteration for solving (18.6) then consists of the iteration

$$\underset{\sim}{U}^{(i+1)} = A^{-1}\underset{\sim}{F}(\underset{\sim}{U}^{(i)}) \,, \quad i = 1,2,...,$$

or (in practical terms)

$$A\underset{\sim}{U}^{(i+1)} = \underset{\sim}{F}(\underset{\sim}{U}^{(i)}) \,, \quad i = 1,2,... \tag{18.9}$$

We see that the iteration (18.9) requires simply the solution of a sequence of $n \times n$ linear systems with a constant matrix A and a varying right hand side vector $\underset{\sim}{F}(\underset{\sim}{U}^{(i)})$, $i = 1,2,...$.

This very simple form results from using property (i) of the system (18.6). We can also use property (ii) to efficiently factorize A (as long as n is not too large) as

$$A = L\, D\, L^{T} ,$$

where L is a unit diagonal, lower triangular matrix, and D is a diagonal matrix. Then each iteration of (18.9) consists of the evaluation of $\underset{\sim}{F}(\underset{\sim}{U}^{(i)})$ and forward and backward substitutions to give $\underset{\sim}{U}^{(i+1)}$. If however the matrix A is very large, then it may be necessary to resort to iterative techniques to solve the linear system (18.9).

In Riley (1977) the iteration (18.9) was used to solve the nonlinear system (18.4); a starting approximation

$$\underset{\sim}{U}^{(1)} = \underset{\sim}{0}$$

and a convergence criterion

$$\| \underset{\sim}{U}^{(i+1)} - \underset{\sim}{U}^{(i)} \|_{\infty} \leq 10^{-7}$$

were used in conjunction with this iteration. For the limited range of values of n considered, it was found that the number of iterations required to achieve convergence was always 6 though some variation with n is to be expected in general.

This example serves to illustrate how useful functional iteration can be in the solution of nonlinear elliptic partial differential equations. However it should be remembered that

(a) the iteration may fail to converge (see, for example, the finite element method solution of (18.3) via a maximum variational principle (Riley, 1977)); indeed the conditions for convergence given earlier are rather restrictive.

(b) the iteration will, in general, only have a linear rate of convergence.

We therefore consider in the next section more sophisticated methods of solution which use additional information about the function $\underset{\sim}{f}(\underset{\sim}{x})$.

18.3. NEWTON'S METHOD

Newton's method for solving (18.2) makes use of the iteration

$$\underset{\sim}{x}^{(i+1)} = \underset{\sim}{x}^{(i)} + \underset{\sim}{p}^{(i)} , \quad i = 1,2,..., \tag{18.10}$$

where

$$\underset{\sim}{p}^{(i)} = -J(\underset{\sim}{x}^{(i)})^{-1}\underset{\sim}{f}(\underset{\sim}{x}^{(i)}) , \tag{18.11}$$

and $J(\underset{\sim}{x})$ is the $n \times n$ Jacobian matrix of $\underset{\sim}{f}(\underset{\sim}{x})$ with components

$$J_{ij}(\underset{\sim}{x}) = \frac{\partial f_i}{\partial x_k}(\underset{\sim}{x}) .$$

We shall use the notation $\underset{\sim}{f}^{(i)}$, $J^{(i)}$ to denote $\underset{\sim}{f}(\underset{\sim}{x}^{(i)})$, $J(\underset{\sim}{x}^{(i)})$.

There are many theorems concerning the convergence and its rate for Newton's method, perhaps the best known being the Newton-Kantorovich theorem (Ortega and Rheinboldt, 1970, p.421). This theorem gives conditions on $\underset{\sim}{f}(\underset{\sim}{x})$, $J(\underset{\sim}{x})$ and $\underset{\sim}{x}^{(1)}$ which guarantee the convergence of the iteration (18.10); it also shows that under suitable continuity conditions the method will generally have a quadratic rate of convergence.

Newton's method can be further improved by the introduction of a damping factor $\alpha^{(i)}$ in (18.10) as follows

$$\underset{\sim}{x}^{(i+1)} = \underset{\sim}{x}^{(i)} + \alpha^{(i)}\,\underset{\sim}{p}^{(i)} \tag{18.12}$$

where

$$\underset{\sim}{p}^{(i)} = -J^{(i)^{-1}}\,\underset{\sim}{f}^{(i)} , \tag{18.13}$$

and $\alpha^{(i)}$ is chosen such that

$$\| \underset{\sim}{f}^{(i+1)} \| < \varepsilon \| \underset{\sim}{f}^{(i)} \| \ , \qquad (18.14)$$

with $0 < \varepsilon \leq 1$.

There are many ways of choosing $\alpha^{(i)}$; one simple choice, which was suggested by Armijo (1966) in the context of minimisation, is

$$\alpha^{(i)} = 2^{-k}, \quad k = 0,1,2,\ldots,$$

where k is the smallest integer such that

$$\| \underset{\sim}{f}(\underset{\sim}{x}^{(i)} + 2^{-k} \underset{\sim}{p}^{(i)}) \| \leq \varepsilon \| \underset{\sim}{f}(\underset{\sim}{x}^{(i)}) \| \ .$$

Other more sophisticated ways of choosing $\alpha^{(i)}$ are possible, see for example Gill and Murray (1974), although it should be noted that these more sophisticated techniques may require the norm in (18.14) to be continuously differentiable. It should be noted that as long as $\alpha^{(i)} = 1$ is ultimately always chosen then the quadratic rate of convergence of Newton's method is maintained.

The search direction $\underset{\sim}{p}^{(i)}$ of (18.13) should not be found by inverting the matrix $J^{(i)}$. Provided the matrix $J^{(i)}$ is not too large then Gaussian elimination with partial pivoting could be used to factorise $J^{(i)}$ as

$$PJ^{(i)} = LU \ ,$$

where P is a permutation matrix, L is a unit diagonal, lower triangular matrix and U is an upper triangular matrix, and forward and backward substitutions used to find $\underset{\sim}{p}^{(i)}$.

We note that each iteration of Newton's method requires the evaluation and factorization of the $n \times n$ Jacobian matrix $J^{(i)}$. Hence, in general, Newton's method requires more operations per iteration than the functional iteration method of section 18.2, although in compensation it does have a quadratic rate of convergence which may be particularly useful if very accurate results are required.

However, in the discretization of a nonlinear partial

differential equation it frequently happens that the Jacobian matrix is a band matrix, and its evaluation and factorization are not as expensive as they might at first appear.

With this in mind we return to the nonlinear system (18.6). The Jacobian matrix of $\underset{\sim}{f}(\underset{\sim}{U})$ has elements

$$J_{ij}(\underset{\sim}{U}) = A_{ij} - \frac{\partial F_i}{\partial U_j}(\underset{\sim}{U}) \ .$$

In view of property (iii), the matrix with elements $\partial F_i/\partial U_j(\underset{\sim}{U})$ is a sparse band matrix. Indeed this fact, together with property (ii), leads to the conclusion that the Jacobian matrix is sparse and banded.

Variants of Newton's method follow from the observation that the linear system (18.11), which must be solved on each iteration of Newton's method, has much in common with the system of linear equations which arises in the discretisation of a linear partial differential equation. This suggests that rather than using a direct method to solve (18.11), one of the iterative methods, described for example in Varga (1962), might be used instead. This then leads to the composite methods which we consider in the next section.

18.4. NEWTON-SOR AND SOR-NEWTON METHODS

The Newton-SOR method uses the iteration (18.10) and (18.11) with the SOR method used to solve the linear system (18.11) for $\underset{\sim}{p}^{(i)}$.

The SOR method for solving the linear system

$$A\underset{\sim}{x} = \underset{\sim}{b} \tag{18.15}$$

consists of the iteration

$$\underset{\sim}{x}^{(k+1)} = \underset{\sim}{x}^{(k)} - \omega(D-\omega L)^{-1}(A\underset{\sim}{x}^{(k)}-\underset{\sim}{b}),\ k = 0,1,2,\ldots, \tag{18.16}$$

where $A = D - L - U$, and D, L and U are diagonal, strictly lower triangular, and strictly upper triangular matrices respectively, ω is a relaxation parameter, and $(D - \omega L)^{-1}$ is assumed to exist (see chapter 14). It can easily be shown that (18.16) can be rewritten as

$$\underset{\sim}{x}^{(k+1)} = \underset{\sim}{x}^{(0)} - \omega(H^k + H^{k-1} + \ldots + I)(D - \omega L)^{-1}(A\underset{\sim}{x}^{(0)} - \underset{\sim}{b}),\ k = 0,1,\ldots, \tag{18.17}$$

where

$$H = (D-\omega L)^{-1}((1-\omega)D + \omega U) .$$

We will now incorporate the iteration (18.17) into the Newton iteration (18.10), (18.11). If, on the ith iteration of Newton's method, we let $\underset{\sim}{x}^{(i)}$ be the starting approximation for the SOR method and we use k_i SOR iterations, then the Newton - SOR method is given by

$$\underset{\sim}{x}^{(i+1)} = \underset{\sim}{x}^{(i)} - \omega_i(H_i^{k_i-1} + H_i^{k_i-2} + \ldots + I)(D_i - \omega_i L_i)^{-1}\underset{\sim}{f}^{(i)},\ i = 1,2,\ldots, \tag{18.18}$$

where $J^{(i)} = D_i - L_i - U_i$, $H_i = (D_i - \omega_i L_i)^{-1}((1-\omega_i)D_i + \omega_i U_i)$, and D_i, L_i, U_i are diagonal, strictly lower triangular and strictly upper triangular matrices, respectively, and D_i is assumed to be nonsingular. (18.18) provides a notationally convenient way of describing the Newton-SOR method, although it is not the way to implement the method in practice.

The number, k_i, of SOR iterations used on each Newton iteration, may vary with i if the SOR method is iterated to convergence. On the other hand k_i could be fixed *a priori* For example, $k_i = 1$, $i = 1,2,\ldots$, leads to the one-step Newton - SOR method, which has the possible advantage of only requiring the evaluation of the lower triangular part of the Jacobian matrix. A more general k-step Newton - SOR method results from setting $k_i = k$, $i = 1,2,\ldots$, in (18.18).

The method just described uses the SOR method in its traditional role of solving a system of linear equations. We can however adapt the ideas of SOR to the solution of a system of nonlinear equations. We observe that on each iteration (18.16) of SOR, we solve the jth equation of (18.15) for x_j, with the remaining (n-1) variables fixed, and then we set

$$x_j^{(k+1)} = x_j^{(k)} + \omega_k(x_j - x_j^{(k)}) \ , \qquad (18.19)$$

for j = 1,2,...,n.

Extending this idea, the kth iteration of a nonlinear SOR method for solving (18.1) is given by solving

$$f_j(x_1^{(k+1)},\ldots,x_{j-1}^{(k+1)},\ x_j,x_{j+1}^{(k)},\ldots,x_n^{(k)}) = 0 \qquad (18.20)$$

for x_j and defining $x_j^{(k+1)}$ by (18.19), for j = 1,2,...,n. We see that each iteration of the given nonlinear SOR method requires the solution of n one-dimensional nonlinear equations (18.20), and that the method will only have meaning if each of these equations has a unique solution in the domain under consideration. In general the solution of (18.20) can only be obtained by using an iterative method, such as, for example, the Newton method. Thus the rôles of Newton's method and SOR have been reversed in this, the SOR - Newton method. For simplicity we shall only consider the one-step SOR - Newton method in which one Newton step, with starting value $x_j^{(k)}$ is used to give an approximation $\hat{x}_j$ to the solution x_j of (18.19). The method is given by

$$x_j^{(k+1)} = x_j^{(k)} - \omega_k \frac{f_j(\underset{\sim}{x}^{(k,j)})}{\partial f_j/\partial x_j(\underset{\sim}{x}^{(k,j)})} \ , \quad j = 1,2,\ldots,n,\ k = 1,2$$

where

$$\underset{\sim}{x}^{(k,j)} = (x_1^{(k+1)},\ldots,x_{j-1}^{(k+1)},\ x_j^{(k)},\ldots,x_n^{(k)})^T \ .$$

We note that each iteration requires the evaluation of just n partial derivatives.

The two methods described in this section show how iterative methods for nonlinear and linear equations can be combined to form composite methods for nonlinear systems. There are of course many other composite methods combining a nonlinear iterative method such as a Newton, secant or Steffensen method with a linear iterative method such as a Jacobi, SOR or ADI method. These other possibilities are considered, for example, in Ortega and Rheinboldt (1970, section 7.4) and Rheinboldt (1974, sections 3.5. 3.6).

Considerations of the conditions which guarantee convergence of the above composite methods and their rates of convergence are given in Ortega and Rheinboldt (1970, section 10.3). Based on these rates of convergence, Ortega and Rheinboldt remark that the one-step SOR - Newton method is likely to be the most efficient of the SOR methods if derivatives are relatively easy to evaluate. On the other hand if derivatives are difficult to evaluate a multistep Newton - SOR method might be more efficient.

18.5. QUASI-NEWTON METHODS

The methods of the previous two sections have been developed under the assumption that the elements of the Jacobian matrix can be evaluated. However for some nonlinear systems of equations it is either impossible or at least computationally very expensive to evaluate the Jacobian matrix. In such a case we could use a functional iteration method, which does not use the Jacobian matrix at all. On the other hand we could use Newton's method with the Jacobian matrix $J^{(i)}$ replaced by some approximation $B^{(i)}$. One way of forming $B^{(i)}$ is to use finite-difference approximations to the non-zero elements of $J^{(i)}$, see Curtis, Powell and Reid (1974) and Curtis and Reid (1974). The use of finite-differences can reduce the rate of convergence (Ortega and Rheinboldt, 1974, section 11.2) but in practice this often does not matter.

Another way is to form a new approximation $B^{(i+1)}$ to the Jacobian matrix by making a low rank change to the current approximation $B^{(i)}$ using information available on the ith iteration. This is, of course, the philosophy of quasi-Newton methods. A general quasi-Newton method has the form:

$$\underset{\sim}{p}^{(i)} = -B^{(i)^{-1}} \underset{\sim}{f}^{(i)} , \qquad (18.21)$$

$$\underset{\sim}{x}^{(i+1)} = \underset{\sim}{x}^{(i)} + \alpha^{(i)} \underset{\sim}{p}^{(i)} , \qquad (18.22)$$

$$B^{(i+1)} = B^{(i)} + \text{low rank matrix}, \qquad (18.23)$$

for $i = 1,2,\ldots$, where $\underset{\sim}{x}^{(1)}$ is a user supplied approximation to $\underset{\sim}{x}^*$, and $B^{(1)}$ is a user supplied approximation to the Jacobian matrix $J^{(1)}$. The solution of (18.21) and the choice of $\alpha^{(i)}$ in (18.22) have been considered in earlier sections We here concentrate on the problem of updating the approximate matrix $B^{(i)}$.

A Taylor expansion for $\underset{\sim}{f}(\underset{\sim}{x}^{(i+1)})$ about $\underset{\sim}{x}^{(i)}$ gives

$$\underset{\sim}{f}(\underset{\sim}{x}^{(i+1)}) = \underset{\sim}{f}(\underset{\sim}{x}^{(i)}) + J(\underset{\sim}{x}^{(i)})(\underset{\sim}{x}^{(i+1)}-\underset{\sim}{x}^{(i)})+\ldots$$

If we let

$$\underset{\sim}{\gamma}^{(i)} = \underset{\sim}{f}^{(i+1)} - \underset{\sim}{f}^{(i)} ,$$

$$\underset{\sim}{\delta}^{(i)} = \underset{\sim}{x}^{(i+1)} - \underset{\sim}{x}^{(i)} ,$$

and truncate the above series, we obtain

$$\underset{\sim}{\gamma}^{(i)} \approx J^{(i)}\underset{\sim}{\delta}^{(i)} . \tag{18.24}$$

It would seem reasonable to require $B^{(i)}$, the approximation to $J^{(i)}$, to satisfy (18.24) as an equality. However it is not possible to achieve this, and instead we require

$$B^{(i+1)}\underset{\sim}{\delta}^{(i)} = \underset{\sim}{\gamma}^{(i)} , \tag{18.25}$$

this equation being known as the quasi-Newton equation. If we now let

$$B^{(i+1)} = B^{(i)} + \underset{\sim}{a}\,\underset{\sim}{b}^T ,$$

where $\underset{\sim}{a},\underset{\sim}{b}$ are n-vectors, and require that $B^{(i+1)}$ satisfies (18.25), then we obtain the updating formula

$$B^{(i+1)} = B^{(i)} - \frac{(B^{(i)}\underset{\sim}{\delta}^{(i)}-\underset{\sim}{\gamma}^{(i)})\underset{\sim}{q}^{(i)^T}}{\underset{\sim}{q}^{(i)^T}\underset{\sim}{\delta}^{(i)}} \tag{18.26}$$

where the vector $q^{(i)}$ is arbitrary except for the obvious restriction $q^{(i)T}\delta^{(i)} \neq 0$. Perhaps the best choice is $q^{(i)} = \delta^{(i)}$, which was first suggested by Broyden (1965). The convergence properties of the resulting method, usually known as "Broyden's good method" are considered, for example, in Gay (1977).

The quasi-Newton methods just described have, as far as nonlinear partial differential equations are concerned, the unfortunate feature that they do not preserve sparsity, since the rank one change is in general a full matrix. However, a quasi-Newton method suitable for sparse nonlinear systems is considered by Schubert (1970) and Broyden (1971). This method considers a general sparse Jacobian matrix without reference to the structure associated with nonlinear partial differential equations. Perhaps the most likely future development is a quasi-Newton method which does take account of this structure.

18.6. CONTINUATION

In this section we consider briefly a device for improving the chances of convergence of the iterative methods discussed so far. The need for such a device is seen when one realises that all the methods of the earlier sections are only guaranteed to converge if the starting approximation $x^{(1)}$ is sufficiently close to the solution x^*.

Continuation, or Davidenko's method, converts the problem (18.2) into a sequence of problems by introducing an extra parameter θ. Hopefully at each stage the solutions of the preceeding problems in the sequence can be used to provide a good starting approximation for the next problem and ultimately convergence to the solution of the original problem will be achieved.

An auxiliary function $g(x,\theta)$ is constructed so that

$$g(x,1) = f(x) \tag{18.27}$$

and for which the solution of the equation

$$g(x,0) = 0$$

is known. Two simple examples are given by

$$\underset{\sim}{g}(\underset{\sim}{x},\theta) = \underset{\sim}{f}(\underset{\sim}{x}) - (1-\theta)\underset{\sim}{f}(\underset{\sim}{x}_0) ,$$

and

$$\underset{\sim}{g}(\underset{\sim}{x},\theta) = (1-\theta)(\underset{\sim}{x}-\underset{\sim}{x}_0) + \theta\underset{\sim}{f}(\underset{\sim}{x}) .$$

The solution of $\underset{\sim}{g}(\underset{\sim}{x},0) = \underset{\sim}{0}$ is hopefully a good initial estimate of the solution of

$$\underset{\sim}{g}(\underset{\sim}{x},\theta_1) = \underset{\sim}{0} \tag{18.28}$$

say. This intermediate equation may be solved using any of the methods described earlier, although, because a good initial estimate of the solution is available, a method with fast ultimate convergence might be best. The solution $\underset{\sim}{x}^*(\theta_1)$ of (18.28) is then used as the initial estimate for the solution of

$$\underset{\sim}{g}(\underset{\sim}{x},\theta_2) = \underset{\sim}{0}$$

and so on. In this way

$$\underset{\sim}{g}(\underset{\sim}{x},\theta) = \underset{\sim}{0}$$

is solved for a sequence of θ values

$$0 = \theta_0 < \theta_1 < \theta_2 < \ldots < \theta_N = 1$$

to give the solution $\underset{\sim}{x}^*(\theta_N)$ of (18.2). This approach can be improved by using extrapolation to give initial estimates of the solutions of the intermediate problems.

The difficulty with the continuation method lies in knowing how to control the parameter θ. However, often in partial differential equations which arise in practice there is a physical parameter which can be used as the continuation parameter θ and which is easy to control from physical reasoning. For example, in fluid flow problems a solution of the problem is often known for certain values of the Reynolds number and this can be used as the continuation parameter. On the other hand, when solving parabolic problems we can view the time-step as a natural continuation parameter.

This concludes our very brief consideration of continuation methods. More detail can be found in Abbott (1977), Rheinboldt (1974, section 7.2) and Wasserstrom (1973).

PART IV

FREE AND MOVING BOUNDARY PROBLEMS

This part contains a discussion of methods for a problem of great practical importance, namely the solution of elliptic and parabolic equations where the problem includes the determination of part of the boundary of the domain of the differential equation. In general, the discussion is of the principles involved and the reader is referred to earlier part of this book for the computational treatment of the elliptic and parabolic problems which arise.

Contents

Chapter	Title	Contributor
19	Free Boundary Problems in Elliptic Equations.	I. Gladwell
20	The Stefan Problem: Moving Boundaries in Parabolic Equations.	L. Fox

FREE BOUNDARY PROBLEMS IN ELLIPTIC EQUATIONS

19.1. INTRODUCTION

We consider the solution of Laplace's equation on a closed region Ω where some part Γ_1 of the boundary Γ of Ω is to be determined as part of the problem. In this section we introduce a free boundary problem that has been considered at length in the literature. In later sections we consider a number of techniques for the solution of this type of problem in some detail.

Consider the case illustrated in Fig. 19.1. This is a model for flow between two reservoirs separated by a dam of

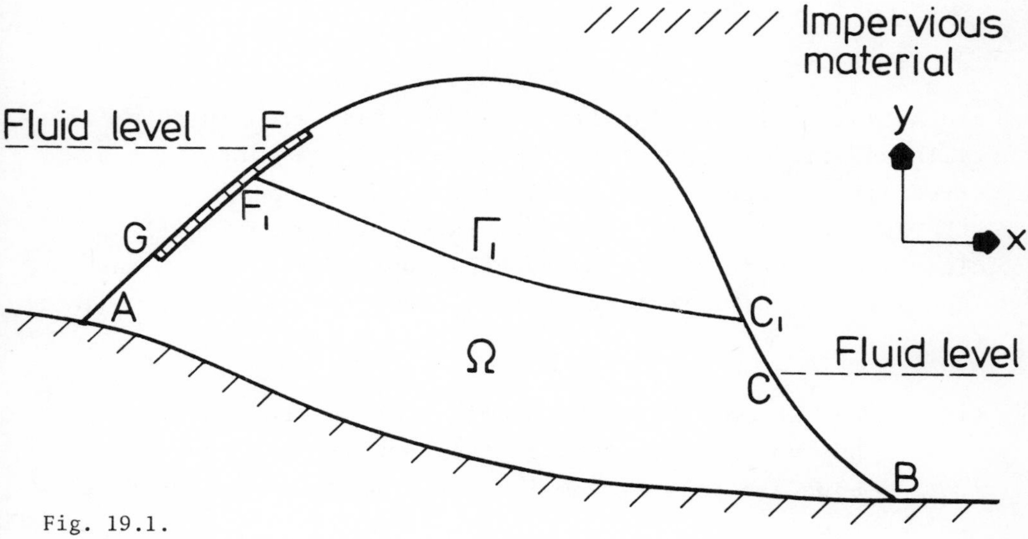

Fig. 19.1.

pervious material, where part of the dam face and the base are considered impervious. That is AB and GF are impervious boundaries but the material of the dam Ω is pervious. The free boundary F_1C_1 (denoted by Γ_1) is to be determined. Note that there is seepage on C_1C. We may define the velocity potential, assuming two-dimensional flow, by

$$u(x,y) = y + p(x,y)/\gamma \qquad (19.1)$$

where p is the pressure and γ is the specific weight of the fluid and then Darcy's law states that the velocity

$$\underset{\sim}{w} = -k\,\underset{\sim}{\nabla}\underset{\sim}{u} \qquad (19.2)$$

where, when the fluid is incompressible, k is a constant which can assume the value unity, without loss of generality. Since the fluid is incompressible we have

$$\text{div}\,\underset{\sim}{w} = \nabla^2 u = 0 \qquad (19.3)$$

(Laplace's equation) in Ω. Hence u has a harmonic conjugate v (determined up to an additive constant) such that the Cauchy-Riemann equations

$$u_x - v_y = u_y + v_x = 0$$

are satisfied in Ω. The lines u = constant are equipotentials and are orthogonal to the lines v = constant which are streamlines tangent to the velocity $\underset{\sim}{w}$.

The boundary conditions associated with equation (19.3) clearly depend on the boundary behaviour of the fluid. We identify four different types of boundary condition as follows.

(i) On the impervious boundaries, AB and FG, the fluid has no velocity through the boundary, hence the boundary is a streamline, that is the boundary condition is

$$v = \text{constant} \qquad (19.4)$$

where the constant is unknown. Alternatively we have the boundary condition

$$\underset{\sim}{w}^T \underset{\sim}{n} = \frac{\partial u}{\partial n} = 0 \qquad (19.5)$$

where $\underset{\sim}{n}$ is the unit outward normal to the boundary.

(ii) At the boundaries of the reservoirs with the dam, AG and BC, the boundary conditions are simply determined as

$$u = \text{constant} \tag{19.6}$$

under the usual assumptions that the velocity of the fluid in the reservoir is negligible and that the potential is the same at all points in the reservoir. By setting the atmospheric pressure to zero, the constant in (19.6) becomes the elevation of the surface of the reservoir.

(iii) On the seepage boundary CC_1, the fluid is in a region free of fluid and soil, and hence the pressure involved is atmospheric pressure, and, from (19.1), the boundary condition becomes

$$u = y \; . \tag{19.7}$$

(iv) On the free boundary Γ_1, the atmospheric pressure also applies, hence there is a boundary condition

$$u = y \; . \tag{19.8}$$

Also, Γ_1 is a streamline and hence

$$v = \text{constant} \tag{19.9}$$

where the constant is unknown (but can be given any value leaving the constant in (19.4) to be determined), or alternatively

$$\frac{\partial u}{\partial n} = 0 \; . \tag{19.10}$$

The position of the points F_1 and C_1 is unknown, but physically it is clear that the streamline F_1C_1 is "smooth" and monotonically decreasing. The flow across any line joining two streamlines $v = c_1$ and $v = c_2$ is always $c_1 - c_2$. Hence the flow across AG equals the flow across BC. See Baiocchi *et al.* (1974) for a more detailed discussion and Baiocchi *et al.* (1973) for a full analysis.

To summarise, we see that we can formulate the free boundary problem in at least two ways as follows.

Formulation 1

Find Γ_1, Ω, u, v, q such that

(i) the curve Γ_1, $y = \phi(x)$, is "smooth" and decreasing, F_1 is between G and F, and C_1 is above C;

(ii) Ω is the part of the dam below Γ_1;

(iii) u,v are smooth in Ω such that $u_x - v_y = u_y + v_x = 0$ in Ω; and

(iv) the boundary conditions

$$u = y_1 \text{ on AG}, \quad u = y_2 \text{ on BC}$$
$$u = y \text{ on } \Gamma_1, \quad u = y \text{ on } C_1C,$$
$$v = 0 \text{ on } \Gamma_1, \quad v = q \text{ on AB}$$

are satisfied where y_1 and y_2 are the elevation of the horizontal free surfaces of the upper and lower reservoirs respectively.

Note that the equations in (iii) are sometimes replaced by

$$\nabla^2 u = \nabla^2 v = 0. \tag{19.11}$$

Formulation 2

Find Γ_1, Ω, u such that

(i) is as in formulation 1;

(ii) is as in formulation 1;

(iii) u is "smooth" in Ω and satisfies $\nabla^2 u = 0$; and

(iv) the boundary conditions on v in formulation 1 are replaced by

$$\frac{\partial u}{\partial n} = 0 \quad \text{on} \quad \Gamma_1 \text{ and } AB$$

and the other boundary conditions of formulation 1 are to be satisfied.

In much of what follows we shall illustrate the principles involved in our various numerical techniques by considering the model problem illustrated in Figure 19.2 where the notation

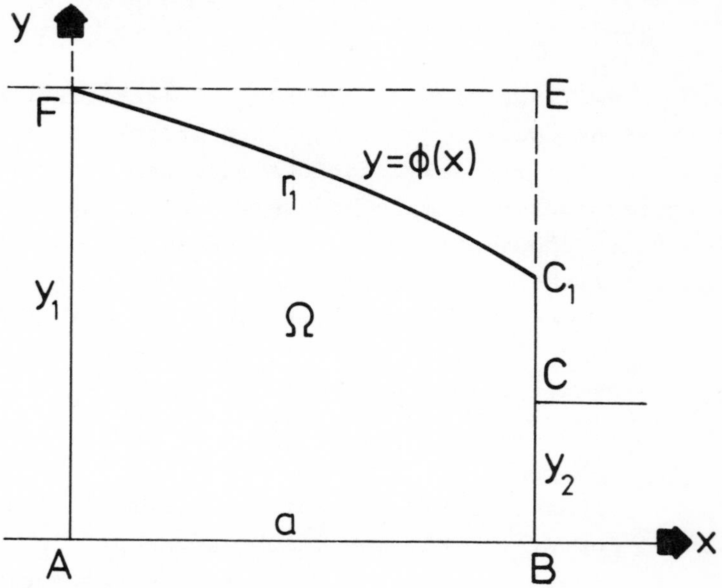

Fig. 19.2.

is identical to that in Figure 19.1. Here there is no impervious boundary FG and hence F and F_1 coincide. This model problem has been considered by many authors; indeed maybe too many as some techniques proposed for solving it do not generalise easily to the more difficult and realistic problem given in Figure 19.1.

In the following sections, we discuss a variety of techniques for solving the free boundary problem. These fall, essentially, into two classes. The first class comprises techniques which treat the problem more or less as it stands, whereas the second class comprises techniques which first transform the problem into another and then solve the new problem. The first class of techniques can usually be generalised quite simply to more difficult problems though they may not be equally successful when applied there. The second class of techniques may not have convenient generalisations to more difficult problems, or, if they do, the transformed problem may be much more difficult to solve than the original one.

We give references for the techniques described where appropriate, but more complete bibliographies may be found in Baiocchi *et al.* (1974), Furzeland (1977b) (both of which contain some discussion of techniques for the solution of free boundary problems), and Cryer (1977).

19.2. DIRECT SOLUTION OF THE FREE BOUNDARY PROBLEM

In previous chapters on elliptic problems we have seen how to solve differential equations on fixed regions Ω with Dirichlet and/or Neumann boundary conditions using finite-difference and finite element methods. To use these techniques for the model free boundary problem in Figure 19.2 (using formulation 2 above) we may proceed as follows:

(i) Choose a test curve Γ_1': $y = \eta(x)$ approximating the free boundary Γ_1 so that Ω is approximated by Ω';

(ii) use a finite-difference or finite element method to compute a solution in Ω' to the problem in formulation 2 using just one of the boundary conditions (19.8) and (19.10), that is using either $u = y$ or $\partial u/\partial n = 0$ as the boundary condition on Γ_1';

(iii) use the computed solution from (ii) to adjust the position of Γ_1' to a new curve Γ'': $y = \psi(x)$ so that on Γ_1'' the other boundary condition of (19.8) and (19.10) (that is the one not used so far) is approximately satisfied;

(iv) check if η and ψ agree to the required accuracy and stop if they do with Γ_1'' as the final approximation to Γ_1, otherwise set $\eta = \psi$ (that is $\Gamma_1' = \Gamma_1''$) and return to step (i).

We now consider some of the details of the implementation of this algorithm, but it is worth remarking first that it is very difficult, if not impossible, to prove that an algorithm of this type will converge. Indeed, it is not immediately clear that the problem as posed in section 19.1 has a solution, though, of course, one might accept numerical convergence of the algorithm above as a constructive proof of existence.

When implementing the algorithm above we must answer a number of questions. These include:

(a) How do we choose an initial approximation Γ_1' to the boundary Γ_1?

(b) Do we use finite differences or finite elements, and for what order of accuracy should we aim?

(c) When using finite differences or finite elements do we use a fixed mesh and calculate the position of Γ_1' with respect to this mesh at each iteration, or do we fix Γ_1' in the mesh and let it vary as we vary the mesh from iteration to iteration?

(d) Which of equations (19.8) and (19.10) should we use in steps (ii) and (iii) respectively of the algorithm?

(e) How do we calculate the position of the new boundary Γ_1'', $y = \psi(x)$, using the boundary conditions?

(f) Do we take any account of the singularity at the point C_1 (and F_1 in Figure 19.1) when computing the solution?

Of course, the answers to these questions are interrelated. Our initial approximation to Γ_1 can only be made from physical intuition. However this intuition may be aided by solving an associated time-dependent moving boundary (Stefan) problem initially and using the steady-state solution of this problem.

We return to this technique in the next section.

For the simple model problem there seems to be little to choose between finite differences and finite elements, though for three-dimensional problems and irregularly shaped regions finite elements should be preferred, as usual. In the literature both methods have been widely used, see Cryer (1970b) and Mogel and Street (1974) for a description of finite-difference calculations and Neumann and Witherspoon (1970), Larock and Taylor (1976) and Bathe and Khoshgoftaar (1978) for corresponding finite element calculations. The question of the accuracy of the finite-difference or finite element approximation has not been treated carefully by some authors, but others have analysed the problem thoroughly. Aitchison (1972) has shown that the solution of the model problem has a logarithmic singularity at C_1 (and there will also be a singularity at F_1 in the case of the problem in Figure 19.1). Indeed, with C_1 as the origin the equation for the *free boundary* is

$$y = \frac{x}{\pi} \ln(-x) + \sum_{k=1}^{\infty} B_k x^k \tag{19.12}$$

for some constants B_k. This singularity in the solution (and possibly also the singularity in (19.12)) should be built into the solution in one of the usual ways, see chapter 3. Otherwise, only the lowest order finite-difference or finite element techniques will be worthwhile, presuming the solution in the neighbourhood of C_1 is of interest. Mogel and Street (1974) discuss the treatment of the singularity in their rather different problem in the context of a finite-difference calculation using refined and irregular meshes.

We will now concentrate on finite element treatment of the model free boundary problem. We consider a fixed-mesh finite element solution of the model problem. At some stage in the algorithm we may have the situation illustrated in Figure 19.3. Here we are using quadrilateral elements with those elements below the line labelled L being fixed and rectangular shaped, and those above L running from L up to the free boundary which is here approximated by a piecewise linear approximation Γ_1'. At each iteration in the algorithm

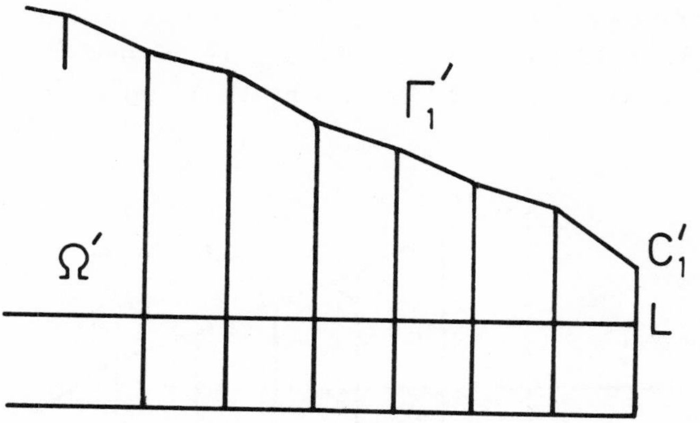

Fig. 19.3.

we adjust only the nodes on the free surface leaving the remainder of the nodes fixed. We may use any type of approximation we choose in the elements but we should bear in mind that the singularity at C_1' should be treated specially in the element adjacent to C_1' (the approximation to C_1). This type of fixed mesh approximation is easy to program but it has some disadvantages. First, the position of C_1 must be known quite accurately for L to be fixed (we could choose L = C safely though this would probably lead to the loss of accuracy). If in the iteration we find that C_1' should be below L then we must start again; in particular, great care should be taken not to implement the algorithm in such a way that C_1' is forced to lie above L at all iterations. Artificial preservation of monotonicity in the slope of the free boundary can lead to similar computational difficulties. The other major disadvantage is that the elements above L may become distended and this can lead to a great loss in accuracy. However, there is much to be said for this approach if the free boundary is likely to be relatively flat and a good initial approximation is available, as very little recalculation is required from one iteration to the next. Note that it is not necessary to preserve the vertial inclination of the mesh lines above L as we proceed from L to the free surface. One may instead choose to give the elements equal surface area, for example, or use some other criterion to make the errors in each element approximately equal, see Neumann and Witherspoon (1970). Also, additional fixed horizontal mesh lines could be inserted above L to reduce the problems with distended elements, see Figure 19.4. Such mesh lines must

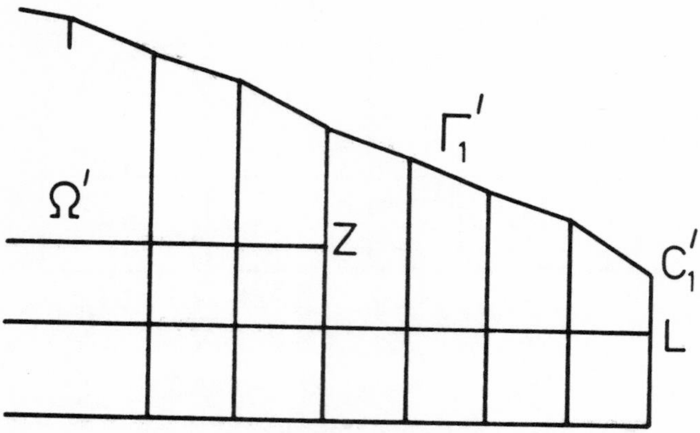

Fig. 19.4.

not meet the free boundary and nodes such as Z must be treated by special techniques such as interpolation. The requirement that the horizontal mesh lines may not meet the free boundary can lead to complicated programming.

We may instead use a variable mesh with fixed vertical mesh lines but nodes on the mesh lines chosen in some suitable automatic way which enables their position to be varied simply with the position of the nodes on the free boundary. In Figure 19.5, we illustrate a case where just two elements are used in the vertical direction and where the nodes are

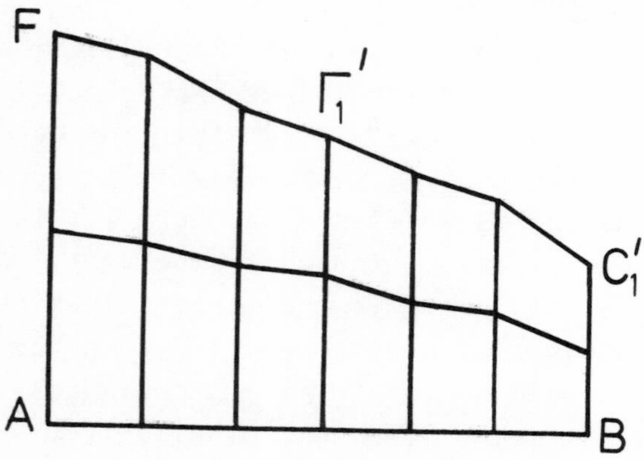

Fig. 19.5.

chosen to divide the vertical mesh lines into two equal parts. Of course, in practice, many more elements may be used but the principle of equal subdivisions would remain the same. In this case, as we vary the position of the free boundary nodes, the other nodes all vary proportionally. This element choice leads to easy programming with none of the difficulties of the fixed mesh described above. However, there could be problems of inaccuracy if BC_1 is very short relative to AF or if the singularity at C_1 is not treated specially. The cost of running a program employing this method can be expensive as the full set of finite element equations must be set up at each iteration.

An alternative finite element technique discussed by Bathe and Khoshgoftaar (1978) puts the problem of representing the free boundary in with the (nonlinear equation solving) problem of determining its position, at least in part. They set up a rectangular finite element mesh on the rectangle ABEF = R of Figure 19.2 and set up their finite element equations by minimising

$$J(U) = \iint_R k(U_x^2+U_y^2)\,dxdy + \text{boundary terms} \qquad (19.13)$$

where U is the finite element solution and $k = k(x,y)$ has value unity for $(x,y) \in \Omega'$ (the current approximation to Ω) and zero for $(x,y) \in R - \Omega'$, and where the "boundary terms" in (19.13) include the effect of the choice of one of (19.8) or (19.10). The elements intersected by the current free boundary Γ_1' must be treated carefully when forming the equations. The position of the free boundary is then adjusted as before, but the elements intersected by the free boundary may change. The major difficulty with this approach lies in preventing the approximate free boundary Γ_1' from oscillating between positions intersecting different elements from iteration to iteration. This difficulty could easily occur due to inaccuracy in the finite element approximation.

The choice of which boundary condition (19.8) or (19.10) to satisfy when solving the differential equation at step (ii) of the iteration and which to use when determining Γ_1'' at step (iii) must be problem dependent. In the references given in this chapter some choice is made, and in most cases the reasons for the choice are discussed; there seems to be no definite rule, each choice being more convenient in some cases than others; see Fox and Sankar (1973) for a brief discussion. In either case, the boundary condition can be approximated using the computed solution for the current Γ_1' either by interpolation (for (19.8)) or differencing (for (19.10)) when using finite differences, or by evaluating the finite element solution and/or its derivative. By evaluating the solution at a number of choices of Γ_1' it is possible to use inverse interpolation or some similar process to determine Γ_1''. Fox and Sankar (1973) make it clear that it is unwise to treat the mesh points which determine Γ_1'' individually; that is, they advise against improving just one mesh point on Γ_1' at a time by solving a simple inverse interpolation problem for a new position for this mesh point with all the other mesh points fixed. They indicate that this approach can lead to oscillations in the positions of the mesh points on Γ_1' from one complete sweep of the points to the next. Instead, they advise that all the mesh points should be adjusted simultaneously and they recommend a n-dimensional secant algorithm for varying the n mesh points defining Γ_1'. This algorithm requires solving fixed boundary problems for n+1 different choices of Γ_1' initially but then solving one new fixed boundary problem per iteration. Of course, other nonlinear equation system solvers could be used; see chapter 18.

Meyer (1977a, 1978) has proposed a Riccati transformation (invariant imbedding) approach to solving the free boundary problem. So that he can solve a variety of problems in irregular regions he works in polar coordinates. The model problem of Figure 19.2 can be treated without this complication and cartesian coordinates will serve to illustrate the method.

Divide AB in Figure 19.2 into n+1 equal parts of length $h = AB/(n+1) = a/(n+1)$, and draw vertical mesh lines. Discretize equation (19.3) in the x-direction using these mesh lines and apply the boundary conditions on AF and BC_1 to give a semi-discretized system of ordinary differential equations

$$A\underset{\sim}{u}'' = B\underset{\sim}{u} + \underset{\sim}{c}, \quad (\underset{\sim}{u}'' = d^2\underset{\sim}{u}/dy^2) \;, \tag{19.14}$$

where A and B are $n \times n$ matrices, $\underset{\sim}{c}$ is a vector and $\underset{\sim}{u} = [u_1, u_2, \ldots, u_n]^T$ with $u_i \simeq u(ih)$, see chapter 10 for a discussion of semi-discretizations. If we use the simple central-difference approximation for $\partial^2 u/\partial x^2$ then A is the identity matrix.

$$B = \frac{1}{h^2}\begin{bmatrix} 2 & -1 & & & \\ -1 & 2 & -1 & & \\ \cdot & \cdot & \cdot & & \\ & \cdot & \cdot & \cdot & \\ & & \cdot & \cdot & \cdot \\ & & -1 & 2 & -1 \\ & & & -1 & 2 \end{bmatrix}, \tag{19.15}$$

and $\underset{\sim}{c}$ is determined by the boundary values on AF and BC_1. Using a similar discretization for the boundary conditions on AB and Γ_1 we obtain, with the system (19.14), the boundary conditions

$$y_i'(0) = 0,\; u_i(y_i) = y_i, u_i'(y_i) = f_i(\underset{\sim}{u}, \underset{\sim}{y}),\; i = 1,2,\ldots,n, \tag{19.16}$$

where $\underset{\sim}{f}$ is a function which represents the discretization of (19.10) in the x-direction, and y_j is the position where the jth mesh line cuts the free boundary. The equations (19.14) and (19.16) constitute a system of second order two-point boundary value problems with the position of one end-point to be determined (the values $\underset{\sim}{u}$ may be continued linearly outside Ω to evaluate $\underset{\sim}{f}$ when necessary). One well-established technique for solving two-point boundary value problems is by a shooting method but this system of ordinary differential equations is inherently unstable when solved by conventional shooting methods; see Hall and Watt (1975) for a discussion of shooting methods. We must use a stabilized shooting technique if we are to proceed this way.

Alternatively we may attempt to determine the positions

y_i one component at a time using a "Gauss-Seidel" iteration where we solve only one equation of (19.14) at a time using the corresponding boundary conditions in (19.16) and with the other components of $\underset{\sim}{u}$ and $\underset{\sim}{y}$ take their latest computed values. We may then use a Riccati transformation to stabilize the shooting method. With $v_i = u_i'$, we set

$$v_i(y) = R_i(y)u_i(y) \tag{19.17}$$

when solving for y_i. By differentiating (19.17) and substituting in the ith equation of (19.14), we obtain a Riccati equation

$$R_i'(y) = 1 + c_1R_i(y) + c_2R_i^2(y), \quad R_i(0) = 0 , \tag{19.18}$$

for R_i, where c_1 and c_2 are constants. Meyer (1977a, 1978a) suggests integrating this equation using the trapezoidal rule and solving (19.17) and the last two equations of (19.16) for y_i by eliminating $u_i(y_i)$ and $u_i'(y_i) = v_i(y_i)$ and then solving the single transcendental equation arising from the last of (19.16). By this technique we can determine Γ_1' but the position of C_1 is not determined as it is a boundary point. To obtain C_1 we could fit the expression (19.12) to the free boundary Γ_1' computed by the Riccati method.

19.3. TRANSFORMATION METHODS

Probably the simplest useful transformation of the model in Figure 19.2 is to a time-dependent one-phase Stefan problem. We replace the elliptic equation (19.3) by the parabolic equation

$$u_t = \nabla^2 u \tag{19.19}$$

and boundary condition (19.10) by

$$\frac{\partial u}{\partial n} = s(t) \tag{19.20}$$

where $s(t) \to 0$ as $t \to \infty$ and where $s(0)$ is such that we can choose a physically meaningful profile for Γ_1' at $t = 0$. Of course, we must choose s(t) so that this Stefan problem has the solution of the model problem as its steady state. We then solve the Stefan problem which essentially involves solving a free boundary problem at each time step. See chapter 20 for a discussion of techniques for solving Stefan problems. We are, in effect, providing a natural continuation parameter (time) for solving a nonlinear system (the free boundary problem); see chapter 18 for a brief discussion of continuation methods.

We next briefly discuss two transformation techniques that have been widely used for special classes of problems. However, they are not generally applicable either because, as in the first case, the transformation used breaks down mathematically for some problems, or, for the second case, the equations which are to be solved after the transformation may be too complicated.

The first transformation is conformal and from Ω into the "hodograph plane"; see Bear (1972) for details. Using the mapping

$$\tau = -u_x(x,y), \quad \eta = -u_y(x,y) , \qquad (19.21)$$

the domain Ω in Figure 19.2 is mapped into the semi-infinite strip $\tilde{\Omega}$ in the $z = \tau + i\eta$ plane in Figure 19.6. The straight

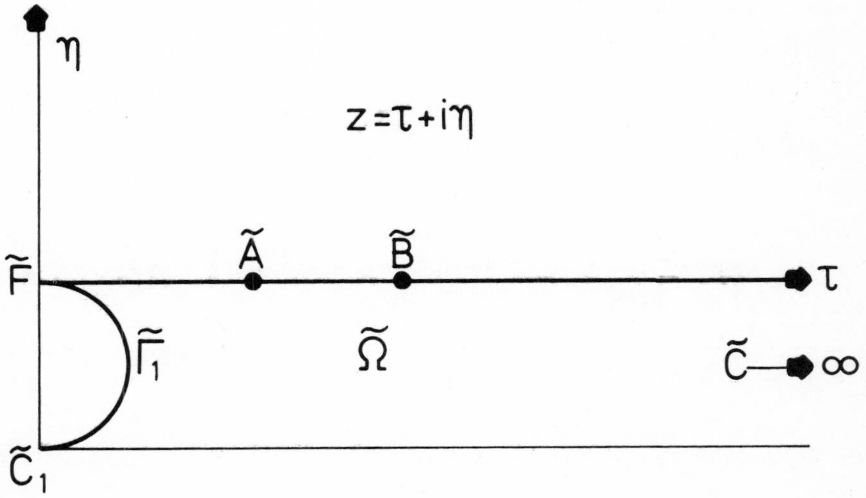

Fig. 19.6.

lines bounding $\tilde{\Omega}$ are $\eta = 0$ and $\eta = -1$ and the free surface Γ_1 is mapped into $\tilde{\Gamma}_1$, a semicircle with equation $\tau^2 + \eta^2 + \eta = 0$ $\tau \geqslant 0$. In this new formulation the problem is to determine the position of $\tilde{A}$ and $\tilde{B}$ consistent with the transformed boundary conditions. If $\tilde{A}$ and $\tilde{B}$ can be found then the solution of the original problem is obtained using the inverse of the mapping (19.21). This approach does not seem to have been particularly successful in numerical computations.

A second technique is to invert the model problem so that x and y are the dependent variables and (in terms of formulation 1 of the problem) u and v are the independent variables. In the new variables, the free surface of the model problem is fixed and a pair of nonlinear first-order partial differential equations must be solved for x and y. Three-dimensional potential flow problems have been solved by this method; see Jeppson (1972) which should be consulted for details of the transformations involved. Note that this technique strongly resembles the "isotherm migration" method of chapter 20.

Finally, we consider a promising and relatively new method based on transforming the free boundary problem to a problem in elliptic variational inequalities which can be solved, approximately, using a finite element method with quadratic programming. Consider the model problem with the notation of formulation 1 of section 19.1. Define the function $g(x,y)$ as follows:

$$\begin{aligned} &g = 0 \quad \text{on} \quad FE, \; g = 0 \quad \text{on} \quad EC, \\ &g = (y_1-y)^2/2 \quad \text{on} \quad AF, \quad g = (y_2-y)^2/2 \quad \text{on} \quad BC, \\ &g = y_2^2/2 - (y_1^2-y_2^2)(a-x)/2a \quad \text{on} \quad AB \; . \end{aligned} \tag{19.22}$$

Let R be the rectangle ABEF with boundary S, and let $\mathcal{H}_1(R)$ be the set of functions ϕ such that ϕ, ϕ_x, ϕ_y are square-integrable over R, then with

$$K = \{\phi \,|\, \phi \in \mathcal{H}_1(R), \quad \phi = g \quad \text{on} \quad S, \; \phi \geq 0\} \; , \tag{19.23}$$

we define a function $\psi \in K$ by

$$J(\psi) \leq J(\phi), \text{ for all } \phi \in K, \tag{19.24}$$

where

$$J(\phi) = \iint_R (\phi_x^2 + \phi_y^2 + \phi)\,dxdy \ . \tag{19.25}$$

The solution of the model problem is then given by

$$\Omega = \{(x,y) | (x,y) \in R \ , \quad \psi(x,y) > 0\}$$

$$u = y - \psi_y(x,y), \quad v = -\psi_x(x,y) \ , \tag{19.26}$$

$$q = (y_1^2 - y_2^2)/2 \ .$$

Note that on $R-\Omega$, $\psi \equiv 0$. See Aitchison (1977) for a discussion of this result and Baiocchi *et al*. (1974) for a general development. In chapter 20, a derivation is given of a similar result for Stefan problems. Alternative formulations for this problem are given by Baiocchi *et al*. (1974) who use variational inequalities to prove existence and uniqueness of solutions of the original problem, and to develop other properties such as monoticity of Γ.

We may approximate the solution, ψ, of the variational problem using finite elements. We choose a linear finite-dimensional finite element subspace $T_h \subset \mathcal{H}_1(R)$ such that the functions $\phi_h \in T_h$ interpolate to the function g at the finite element nodes on the boundary S; see Aitchison (1977) and Baiocchi *et al*. (1974) for more details of particular choices of T_h. We may write

$$\phi_h = \phi_0(x,y) + \sum_{i=1}^{n} a_i \phi_i(x,y) \tag{19.27}$$

where $\{\phi_i\}_{i=0}^n$ is a basis for T_h with ϕ_0 interpolating g at the boundary nodes and ϕ_i, $i = 1,2,\ldots,n$, taking zero values there. We impose the additional restriction on the functions ϕ_i, $i = 1,2,\ldots,n$, that ϕ_i takes the value 1 at the ith node (for some ordering of the nodes) and the value zero at all other nodes, and that ϕ_0 takes the value zero at all nodes in the interior of R. With this restriction, the requirement that $\phi_h \geq 0$ can be written

$$\underset{\sim}{a} \geq \underset{\sim}{0} \tag{19.28}$$

where $\underset{\sim}{a} = [a_1, a_2, \ldots, a_n]^T$. When we minimize J over $\phi_h \in T_h$ with the restriction $\phi_h \geq 0$ to obtain ψ_h, we must solve the problem

$$\min_{\underset{\sim}{a}} \underset{\sim}{a}^T A \underset{\sim}{a} + \underset{\sim}{b}^T \underset{\sim}{a} + \underset{\sim}{c} \tag{19.29}$$

subject to (19.28). Here A is a symmetric positive definite matrix with

$$A_{ij} = \iint_D (\underset{\sim}{\nabla}\underset{\sim}{\phi}_i)^T (\underset{\sim}{\nabla}\underset{\sim}{\phi}_j)\, dxdy \tag{19.30}$$

and $\underset{\sim}{b}$ and $\underset{\sim}{c}$ are vectors depending on ϕ_0. The problem (19.28), (19.29) is quadratic programming problem which can be solved by a variety of techniques. An SOR method due to Cryer (1971) and discussed in chapter 20 has long been a recommended technique. A recent report by O'Leary (1978) proposes a method based on conjugate gradients and compares a number of other methods for this problem.

When the solution of problem (19.28), (19.29) has been computed, we can determine the position of the approximate free boundary Γ_1' by studying the vector $\underset{\sim}{a}$. Clearly the nodes corresponding to non-zero components of $\underset{\sim}{a}$ lie inside Ω' and the others lie outside. Inverse interpolation can be used to determine the position of Γ_1' more precisely from ψ_h. Because ψ_h is zero everywhere on BE, we cannot determine C_1 this way. However, Aitchison (1977) demonstrates how to determine C_1 by fitting the free boundary (19.12) to information obtained from ψ_h.

Error bounds for finite element solutions obtained via variational inequalities have been obtained, in an abstract setting, by Falk (1974) and Kikuchi (1977) among others.

Finally, we remark that Baiocchi *et al.* (1974) discuss the variational inequalities corresponding to the more general problem illustrated in Figure 19.1. They show that for this case but with the dam sides vertical and the base horizontal, the variational functional remains the same but the boundary conditions are more complicated, containing q as a true unknown, and on FG there is a natural boundary condition. They also consider the general case in some detail.

20

THE STEFAN PROBLEM: MOVING BOUNDARIES IN PARABOLIC EQUATIONS

20.1. INTRODUCTION

We shall consider a typical Stefan problem in the context of heat conduction, though there are other scientific disciplines in which similar problems arise and some account of them appears in the proceedings of a recent conference edited by Ockendon and Hodgkins (1975). The fundamental feature of such problems is the presence of an interface, an internal boundary, which moves with time and has to be determined in the course of the computation. At such an interface there will be some sort of discontinuity which complicates the solving process. In heat conduction, for example, the interface might separate a material in its solid and liquid states, and since freezing or melting generates or absorbs latent heat without a change of temperature there will be a discontinuity in the rate of heat flow on the interface.

A typical problem in one space dimension is represented by Figure 20.1, in which region I is occupied by a solid, region

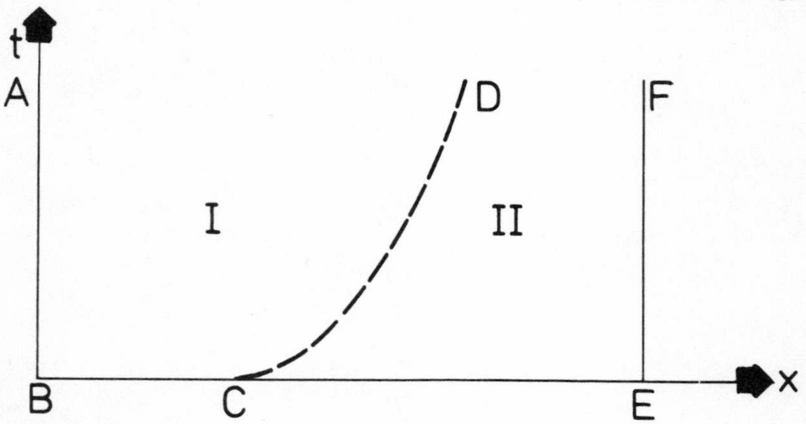

Fig. 20.1.

II by the corresponding liquid, and we have to determine the temperature in each region and the position of the interface $x = s(t)$ denoted by the line CD.

The differential equations are given by

$$\rho(u_1)\sigma(u_1)\frac{\partial u_1}{\partial t} = \frac{\partial}{\partial x}\left\{k(u_1)\frac{\partial u_1}{\partial x}\right\}, \quad 0 < x < s(t),$$
$$\rho(u_2)\sigma(u_2)\frac{\partial u_2}{\partial t} = \frac{\partial}{\partial x}\left\{k(u_2)\frac{\partial u_2}{\partial x}\right\}, \quad s(t) < x < 1, \tag{20.1}$$

where AB is $x = 0$ and EF is $x = 1$. Here u represents the temperature, and the thermal quantities ρ (density), σ (specific heat) and k (conductivity) are known functions of u. We assume that $u_1 < u_2$, except at the interface $x = s(t)$, where

$$u_1 = u_2 = u_m,$$
$$k(u_2)\frac{\partial u_2}{\partial x} - k(u_1)\frac{\partial u_1}{\partial x} = -\rho L \frac{dx}{dt}. \tag{20.2}$$

Here u_m is the (constant) melting temperature, L is the latent heat per unit mass, and ρ is the (constant) density at $u = u_m$. In fact, in theory equations (20.1) imply that the density is everywhere constant since otherwise there is some flow of the substance which is neglected in (20.1). We shall, however, assume that this flow is negligible and allow at least some variation in the density.

The double condition (20.2) at the interface is of course needed in order to compute its position. Other boundary conditions on AB($x = 0$) and EF($x = 1$), and initial conditions on BC and CE ($t = 0$), see Figure 20.1, will also be specified and may be any of those commonly encountered in problems of heat conduction. The problem outlined is a "two-phase" problem since both u_1 and u_2 have to be determined, and the corresponding one-phase problem occurs for example when $u_2 = u_m$ throughout the whole of region II.

In Ockendon and Hodgkins (1975) accounts were given by Crank and by Fox of various numerical methods of solution which had been published up to that time. Here we try to bring this discussion up to date by describing some new methods and some developments of old methods. A prior knowledge of the 1975 papers would be useful for the reader, but the material of this chapter is self-contained and does not need the early work for a full understanding. Where the latter is necessary the old material is included here in some form or other.

There are four main classes of methods, and each is given separate attention. The "front-tracking" methods effectively solve the relevant equations directly by some numerical method and compute the moving boundary in the process. Sections 20.2 and 20.3 treat this type of technique respectively by a finite-difference method of lines and by two kinds of finite element method. In the second class some coordinate transformations are used, involving new dependent or independent variables or both, which transform the moving boundary into a fixed boundary in the new formulation. These are discussed in sections 20.4 and 20.5, first with a change of independent variables and then with a change of both variables in the so-called "Isotherm migration method". In the third class we introduce the idea of a weak solution and reformulate the problem in its "enthalphy" form, the solution of which can proceed without explicit treatment of the moving boundary conditions and which finds the position of the moving boundary as an *a posteriori* adjunct to the computation. This is treated in section 20.6. A similar idea is used in section 20.7, in which the formulation and computation is performed in terms of certain "variational inequalities" and a corresponding quadratic programming problem. Section 20.8 gives a tentative evaluation of the respective methods.

In all cases comments are made about the extension of methods to more than one space dimension and to nonlinear situations. The Stefan condition, of course, means that every Stefan problem is nonlinear, but in the sequel the phrase "nonlinear" will refer to any nonlinearities in the differential equations and/or the boundary conditions.

20.2. FRONT-TRACKING METHODS. (i) METHOD OF LINES

The main difficulty of methods which try to compute explicitly the position of the moving boundary is the need for special formulae to cope with terms like ds/dt and $\partial u/\partial x$ in the interface equation (20.2). This is particularly complicated with implicit finite-difference equations since the position of the boundary is not yet known at the new time. Both Fox and Crank, in Ockendon and Hodgkins (1975), gave brief accounts of methods for dealing with this problem in one space dimension and in the one-phase case. Fox also mentioned the "method of lines" exploited by Meyer (1970) for this purpose, and more recently Meyer (1973, 1976, 1977b) has applied it successfully to two-phase problems. At least for linear equations and in one space dimension this appears to be one of the better methods, and here we mention its main properties and results.

Meyer (1976) considers the general linear equations of type (20.1) given by

$$\frac{\partial}{\partial x}\left\{k_1(x,t)\frac{\partial u_1}{\partial x}\right\} - c_1(x,t)\frac{\partial u_1}{\partial t} = F_1(x,t),\ b_1 < x < s(t),$$
$$\frac{\partial}{\partial x}\left\{k_2(x,t)\frac{\partial u_2}{\partial x}\right\} - c_2(x,t)\frac{\partial u_2}{\partial t} = F_2(x,t),\ s(t) < x < b_2\ , \quad (20.3)$$

with interface conditions, at $x = s(t)$, of type (20.2) given by

$$u_1 = \mu_1(t),\ u_2 = \mu_2(t),$$
$$k_1\frac{\partial u_1}{\partial x} - k_2\frac{\partial u_2}{\partial x} + \lambda\frac{ds}{dt} = \mu_3(t)\ , \quad (20.4)$$

with boundary conditions

$$u_1(b_1,t) = \beta_1(t)\frac{\partial u_1}{\partial x}(b_1,t) + \alpha_1(t)\ ,$$
$$\frac{\partial u_2}{\partial x}(b_2,t) = \beta_2(t)\,u_2(b_2,t) + \alpha_2(t)\ , \quad (20.5)$$

for $t > 0$, and with initial conditions of any suitable kind.

The method of lines uses a time-discretization followed by the solution at the new time of a two-point boundary value problem with an extra condition to determine the position of the interface. The fully-implicit method replaces (20.3) by the approximation

$$\frac{\partial}{\partial x}\left[k_i\frac{\partial U_i^n}{\partial x}\right] - c_i\frac{(U_i^n - V_i^{n-1})}{\Delta t} = F_i,\ i = 1,2, \quad (20.6)$$

where V_i^{n-1} is U_i^{n-1} or some convenient interpolation or extrapolation thereof near the interface. If the moving boundary is as shown in Figure 20.2, for example, Meyer proposes the simple extrapolation

$$V_1^{n-1}(x) = U_1^{n-1}(s_{n-1}) + \frac{\partial U_1^{n-1}}{\partial x}(s_{n-1})(x-s_{n-1}), \quad x \ge s_{n-1}. \tag{20.7}$$

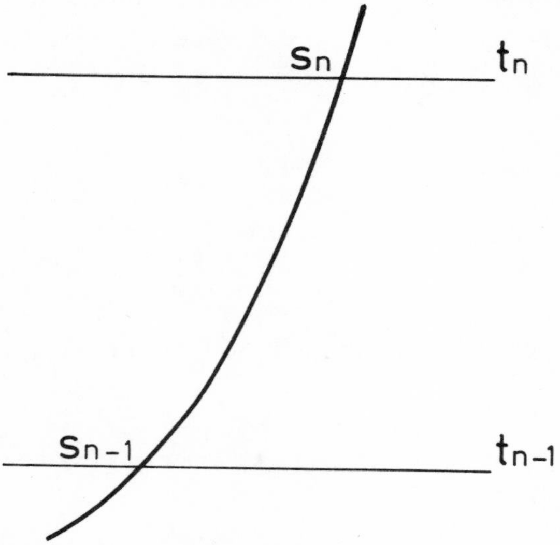

Fig. 20.2.

To the same order of accuracy he approximates to the interface condition by the equation

$$k_1 \frac{\partial U_1^n}{\partial x} - k_2 \frac{\partial U_2^n}{\partial x} + \lambda \frac{(s_n - s_{n-1})}{\Delta t} = \mu_3, \quad t = t_n, \; x = s_n. \tag{20.8}$$

The solution of the two-point boundary problem at $t = t_n$ is then effected with the use of the Riccati transformation with invariant embedding (in the modern terminology!). Full details are given in Meyer (1973), but for this particular problem the Riccati transformation is represented by the equations

$$\begin{aligned} U_1(x) &= U(x)\phi_1(x) + W(x), \\ \phi_2(x) &= R(x)U_2(x) + z(x), \end{aligned} \tag{20.9}$$

where $\phi_i(x) = k_i\ dU_i/dx$, and $U(x)$, $w(x)$, $R(x)$ and $z(x)$ are obtained in $b_1 \le x \le b_2$ as solutions of the initial-value problems

$$\frac{dU}{dx} = \frac{1}{k_1} - \frac{c_1}{\Delta t} U^2,\ U(b_1) = \beta_1/k_1 ,$$

$$\frac{dw}{dx} = -\frac{c_1}{\Delta t} U(w-V_1) - UF_1,\ w(b_1) = \alpha_1 , \tag{20.10}$$

$$\frac{dR}{dx} = \frac{c_2}{\Delta t} - \frac{1}{k_2} R^2,\ R(b_2) = \beta_2 k_2 ,$$

$$\frac{dz}{dx} = -\frac{R_2}{k_2} - \frac{c_2}{\Delta t} V_2 + F_2,\ z(b_2) = \alpha_2 k_2 ,$$

(correcting some errors in Meyer (1976)).

The free interface is then obtained as the value of x for which (20.8) is satisfied, and it is easy to see that the required value x is a root of the equation

$$\frac{\mu_1 - w}{U} - (R\mu_2 + z) + \lambda\frac{(x-s_{n-1})}{\Delta t} = \mu_3 , \tag{20.11}$$

in which we have also used the first of (20.4). Once $x = s_n$ is obtained from (20.11), then the required u_2 and ϕ_1 are obtained from the initial value problems

$$\frac{du_2}{dx} = \frac{1}{k_2}(Ru_2 + x),\ u_2 = \mu_2 \text{ at } x = s,\ s \le x \le b_2 ,$$

$$\frac{d\phi_1}{dx} = \frac{c_1}{\Delta t}(U\phi_1 + w - V_1) + F_1,\ \phi_1 = \frac{(\mu_1 - w)}{U} \text{ at } x = s, \tag{20.12}$$

$$b_1 \le x \le s ,$$

U_1 being recovered finally from the first of (20.9).

For solving the first-order differential equations Meyer recommends the trapezoidal rule, which is accurate enough in conjunction with approximations like (20.7) and (20.8), and for which we have here nothing worse than a quadratic equation to solve at each step. The fact that the computed value $x = s_n$ is not usually at a mesh point gives no serious problem with the trapezoidal rule solution of (20.12).

The advantages of this Riccati transformation are (i) its stability as distinct, for example, from a conventional shooting method, (ii) the solution only of first-order equations and (iii) the fact that s_n need not be specified in advance. This last remark would fail, for example, if we solved the two-point boundary problem by one of the modern recommended methods such as that of Pereyra (1977), and the references therein. One would then have to choose several different s_n, solve for each the boundary value problems in the two regions, and determine s_n by inverse interpolation so that the interface condition (20.8) is satisfied.

It is important to note that the problem we have discussed is linear. If the differential equations and/or boundary conditions are nonlinear the most common solution is to use some iterative method in which the equations are temporarily linearized and solved by the method outlined. Details are given in Meyer (1973) for nonlinear two-point boundary-value problems, but we know of no computations which have been performed for really significant Stefan problems.

A first extension of this method to problems in two space dimensions has been discussed by Meyer (1978), in which he uses a type of alternating-direction method in which the relevant one-dimensional equations in the coordinate directions are solved by the methods here outlined for one-dimensional problems. Only simple tests have been carried out, and it is not yet possible to give a proper evaluation of this type of method.

20.3. FRONT-TRACKING METHODS. (ii) FINITE ELEMENT METHODS

In one space dimension there are two main methods which have been used with finite elements. The first uses an isoparametric approximation (see Chapter 4 or Mitchell and Wait, 1977) with quadrilateral elements K_i^n in a space-time non-rectangular grid, arranged as shown in Figure 20.3 so that grid points lie at equal intervals on the lines t = constant, including the points at which these lines meet the interface.

For the equation

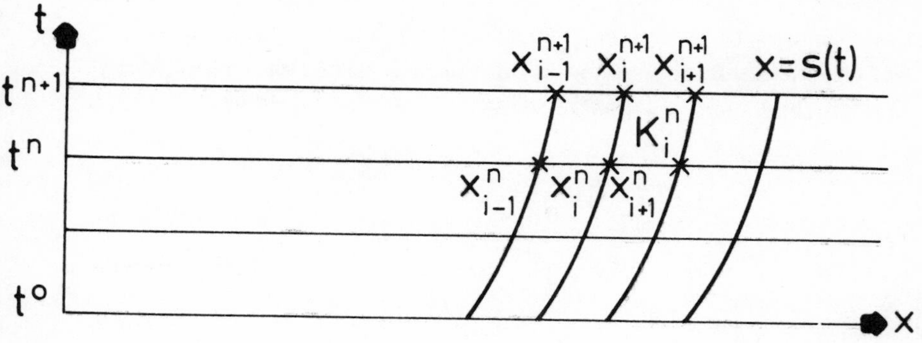

Fig. 20.3.

$$\frac{\partial u}{\partial t} = \frac{\partial^2 u}{\partial x^2}, \quad 0 < x < s(t) \ , \tag{20.13}$$

Bonnerot and Jamet (1974) use the integral identity

$$- \int_{t_1}^{t_2} \int_0^{s(t)} u \frac{\partial V}{\partial t} \, dxdt + \int_{t_1}^{t_2} \int_0^{s(t)} \frac{\partial u}{\partial x} \frac{\partial V}{\partial x} \, dxdt + $$

$$+ \int_0^{s(t_2)} u(x,t_2) \; V(x,t_2) dx - \int_0^{s(t_1)} u(x,t_1) \; V(x,t_1) dx = 0 \tag{20.14}$$

for all differentiable functions $V(x,t)$ which vanish at $x = 0$ and $x = s(t)$. They then take for V a set of bilinear iso-parametric basis functions for the quadrilateral elements of type K^n_i, and produce formulae relating the values of U and x, typically at the six labelled points of Figure 20.3, given by

$$- \tfrac{1}{4}\{(x^{n+1}_{i+1} - x^n_{i+1})(U^n_{i+1} + U^{n+1}_{i+1}) - (x^{n+1}_{i-1} - x^n_{i-1})(U^n_{i-1} + U^{n+1}_{i-1})\} +$$

$$+ \tfrac{1}{2}\{(x^{n+1}_{i+1} - x^{n+1}_{i-1}) U^{n+1}_i - (x^n_{i+1} - x^n_{i-1}) U^n_i\} - \tag{20.15}$$

$$- \tfrac{1}{2}\Delta t \left\{ \frac{U^n_{i+1} - U^n_i}{x^n_{i+1} - x^n_i} - \frac{U^n_i - U^n_{i-1}}{x^n_i - x^n_{i-1}} + \frac{U^{n+1}_{i+1} - U^{n+1}_i}{x^{n+1}_{i+1} - x^{n+1}_i} - \frac{U^{n+1}_i - U^{n+1}_{i-1}}{x^{n+1}_i - x^{n+1}_{i-1}} \right\} = 0 \, .$$

This reduces, interestingly enough, to the Crank-Nicolson formula if the mesh is rectangular.

At the interface, with a one-phase problem, they take the conditions

$$\frac{ds}{dt}(t) = -\alpha \frac{\partial u}{\partial x}(s(t),t), \quad u(s(t),t) = 0 , \qquad (20.16)$$

and use in the computation the approximate implicit replacement

$$s^{n+1} - s^{n} = -\tfrac{1}{2}\alpha\,\Delta t \frac{\partial}{\partial x}(U_N^n + U_N^{n+1}) , \qquad (20.17)$$

where the subscript N denotes a mesh point at the interface. Here the derivatives are computed using quadratic interpolation at points N, N-1 and N-2, the results being simple linear combinations of U_{N-1} and U_{N-2} since U_N is zero from the second interface condition.

The computation at time $(n+1)\Delta t$ then has the following steps:

(i) guess the interface x-derivative, and compute an approximation to s^{n+1} from (20.17);

(ii) compute the x-values at the mesh points on line $(n+1)\Delta t$, assuming equal mesh lengths;

(iii) solve for the U values at these points from the equations typified by (20.15) with suitable satisfaction of the boundary conditions on $x = 0$;

(iv) compute a better x-derivative at the interface, and return to step (i).

By this method Bonnerot and Jamet (1974) solved a simple one-phase problem with Dirichlet boundary conditions. Analysis of the rate of convergence is the subject of current research, but numerical experiments suggest that a mesh refinement with $\Delta x = \Delta t = M^{-1}$, $M = 2^p$, $p = 2,3,4,\ldots$ gives quadratic convergence.

The method is clearly applicable to two-phase problems. It could presumably be extended to the treatment of nonlinear problems, but we know of no such results. Bonnerot and Jamet (1977) used a similar method for a simple linear one-phase problem in two space dimensions, and find numerical results with first-order accuracy. Better accuracy would clearly require a more elaborate technique.

In the more usual finite element method for parabolic equations, the semi-discrete Galerkin method (see Chapter 10), a solution is obtained in the form

$$U(x,t) = \sum_{i=1}^{N} \alpha_i(t)\ \phi_i(x) \ , \qquad (20.18)$$

where the $\phi_i(x)$ are specified basis functions, and the functions $\alpha_i(t)$ are computed from a system of ordinary differential equations. Curtis (1977) extended this method to solve one-phase and two-phase Stefan problems in one space dimension and with quite complicated nonlinear differential equations and boundary conditions. The reader is recommended to consult this reference for the analytical and computational details. We note here that it was found most economic to use linear basis functions and a fairly simple discretization of the Stefan condition.

The extension of this method to two space dimensions is obviously possible but so far as we know has not yet been published.

20.4. COORDINATE TRANSFORMATION

All the other methods we shall describe bypass the treatment of the Stefan condition on a moving curved boundary. Both Crank and Ferriss, in Ockendon and Hodgkins (1975) described a "boundary-fixing" method for one space dimension based essentially on the change of the independent variable x to

$$\xi = x/s(t) \ . \qquad (20.19)$$

In (ξ,t) independent variables, equation (20.13), for example, is transformed into the nonlinear form

$$\frac{\partial^2 u}{\partial \xi^2} = s^2 \frac{\partial u}{\partial t} - \frac{\xi}{2} \frac{\partial u}{\partial \xi} \frac{d}{dt}(s^2) \ . \tag{20.20}$$

We still have to satisfy the transformed Stefan condition, but the moving boundary in the original plane becomes the fixed boundary $\xi = 1$ in the (ξ,t)-plane. Finite-difference equations are now placed on a much firmer basis, and though (20.20) is nonlinear this is commonly true of the original differential equations, as for example in (20.1) when the thermal properties vary with temperature. Höhn (1978) proves the convergence of an explicit method applied to (20.20), and that the global error is of order $(\Delta x)^2$.

More recently, similar ideas have been developed for Stefan problems in two space dimensions. Here the moving boundary is commonly expressed in a form such as

$$y = s(x,t) \ , \tag{20.21}$$

and in the associated coordinate transformation the independent variables x and t are retained but y is replaced by

$$\eta = y/s(x,t) \ . \tag{20.22}$$

Furzeland (1977c) considers a "model" one-phase problem with governing equation

$$\frac{\partial u}{\partial t} = \frac{\partial^2 u}{\partial x^2} + \frac{\partial^2 u}{\partial y^2} \ , \quad 0 < x < 1, \quad 0 < y < s(x,t) \ . \tag{20.23}$$

Boundary conditions are given on $x = 0$, $x = 1$, $y = 0$, and on $y = s(x,t)$ there are the usual two conditions, one specifying u and the other giving a relation between the velocity of the moving boundary and the normal derivative of u at this boundary.

The transformation (20.22) produces the new equation

$$\frac{\partial u}{\partial t} = \frac{\partial^2 u}{\partial x^2} + b\,\frac{\partial^2 u}{\partial x \partial \eta} + c\,\frac{\partial^2 u}{\partial \eta^2} + e\,\frac{\partial u}{\partial \eta} \ , \quad 0 < x < 1, \quad 0 < \eta < 1, \tag{20.24}$$

where

$$b = -\frac{2y}{s^2}\frac{ds}{dx}\ , \quad c = \frac{1}{s^2} + \frac{b^2}{4}\ , \quad e = \frac{\partial^2}{\partial x^2}\left(\frac{y}{s}\right) + \frac{1}{s}\frac{\partial y}{\partial t}\ . \qquad (20.25)$$

The new boundary conditions are easily obtained. For example the condition $\partial u/\partial x = 0$ on $x = 0$ becomes

$$s\frac{\partial u}{\partial x} - \frac{y}{s}\frac{\partial s}{\partial x}\frac{\partial u}{\partial \eta} = 0 \qquad (20.26)$$

on $x = 0$. The Stefan condition is also easily transformed. For example if the velocity of the moving boundary is specified to be just $-\partial u/\partial n$, then on $\eta = 1$ in the new plane we have

$$\frac{\partial y}{\partial t} = -\frac{1}{s}\left(1 + \left(\frac{\partial s}{\partial x}\right)^2\right)\frac{\partial u}{\partial \eta}\ . \qquad (20.27)$$

Furzeland's computation is essentially explicit, using forward differences in time with central differences in space. Given mesh values for the approximation U to u and the values of s on $\eta = 1$ at some time level, his steps are the following:

(i) compute the approximation Y to y from (20.22) at all grid points at this time;

(ii) compute mesh values for Y on $\eta = 1$ at the new time level from a simple explicit approximation to (20.27);

(iii) compute Y at other mesh points at the new time level from (20.22), and hence evaluate $\partial Y/\partial t$ at the old time level with a simple explicit formula;

(iv) compute U at mesh points at the new time level with a simple explicit formula for (20.24).

For this somewhat simple problem the method is quite successful, though as usual the time interval has to be rather small to avoid instabilities. In this respect the fully implicit method has better stability properties, though at each time step we have to solve an elliptic problem with mild nonlinearities. This has been achieved by Furzeland (1978) who replaced (20.27) by the approximation

$$\frac{Y^{n+1}-Y^{n}}{\Delta t} = \left[-\frac{1}{s}\left(1+\left(\frac{\partial s}{\partial x}\right)^{2}\right)\frac{\partial u}{\partial \eta}\right]^{n+1} , \qquad (20.28)$$

and solved the corresponding implicit discretization of (20.24) by the alternate line-hopscotch method of Gourlay and McKee (1977). With an estimated U^{n+1} he computes an estimated Y^{n+1} from (20.28), followed by a new estimate of U^{n+1} from the discretized form of (20.24), and the obvious iteration converges quite rapidly.

Duda *et al*. (1975) used implicit methods to solve a two-phase problem in cylindrical (r,z) coordinates, though their paper does not include too many details about the method.

The method could doubtless be extended to the fully non-linear problems of type (20.1), but we are not aware of any published work in this connectiion.

20.5. THE ISOTHERM MIGRATION METHOD

In the so-called "isotherm migration method", treated, for example, by Crank and Phahle (1973), the one-dimensional case is solved by reversing the roles of some of the dependent and independent variables, the aim as always being to "fix" the moving boundary, preferably as a straight line in the new plane.

Crank and Phahle (1973), and also Crank in Ockendon and Hodgkins (1975), discuss a linear one-phase problem in one space dimension. Instead of finding u as a function of x and t this method finds and solves equations in which x is expressed as a function of u and t, with the effect that if u is constant (say 0) on the original moving boundary this becomes the line u = 0 in the (u,t)-plane and we seek to determine x on this line.

It is easy to extend the analysis to the two-phase problem with nonlinearities at least in the differential equations. For example in the changed variables equations (20.1) become

$$k(u_i)u_i\left(\frac{\partial x}{\partial u_i}\right)^{-2}\frac{\partial^2 x}{\partial u_i^2} - \frac{dk}{du_i}\left(\frac{\partial x}{\partial u_i}\right)^{-1} = \rho(u_i)\sigma(u_i)\frac{\partial x}{\partial t} , \qquad i = 1,2, \qquad (20.29)$$

and an important point is that in the new plane we have constant u values as one set of mesh lines on which k and dk/du are known in advance. In other methods, of course, they are implied in the determination of u in, say, the (x,t)-plane. The Stefan condition in the second of (20.2) is transformed into the equation

$$k(u_2)\left(\frac{\partial x}{\partial u_2}\right)^{-1} - k(u_1)\left(\frac{\partial x}{\partial u_1}\right)^{-1} = -\rho L\,\frac{dx}{dt} \tag{20.30}$$

on u = 0, where $u_m = 0$ is the melting temperature.

The situation is then as shown in Figure 20.4, in which the relevant equations of type (20.29) are solved in regions I and II and (20.30) is satisfied on u = 0. Boundary conditions on x = 0 and x = 1 are clearly more difficult to treat unless they are of the simplest Dirichlet kind in which u(0,t) and u(1,t) are specified as constants $\alpha_1(< 0)$ and $\alpha_2(> 0)$ respectively, transforming nicely into x = 0 on one u-boundary

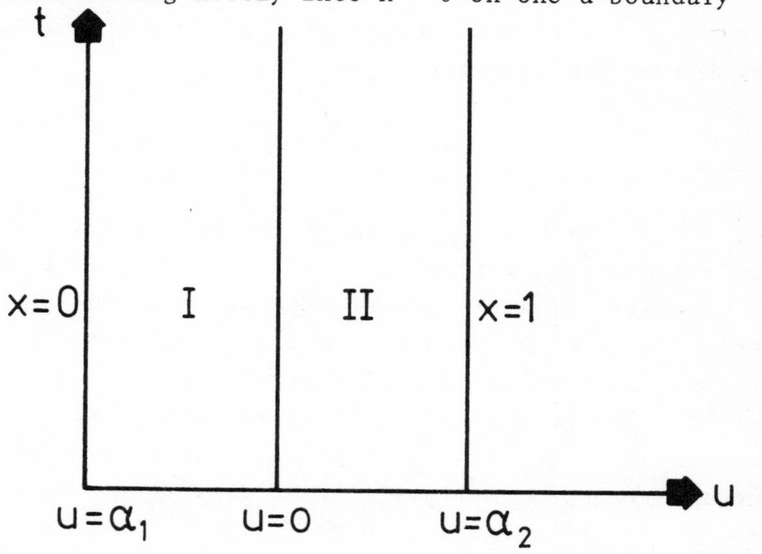

Fig. 20.4.

line and x = 1 on the other. This is the situation of Figure 20.4. A boundary condition $u(0,t) = g_1(t)$ would result in a known but curved boundary on which x = 0, and conditions involving derivatives are much more complicated to solve this way. In fact the method is essentially restricted to problems with the simplest Dirichlet boundary conditions. In particular the vanishing of $\partial u/\partial x$ at any point would appear to be almost disastrous!

Crank and Phahle (1973) used explicit methods for the finite-difference solution of (20.29) and (20.30), computing values of x at successive time intervals on discretely-spaced isotherms between $u = \alpha_1$ and $u = \alpha_2$. As usual the time step has to be rather small to guarantee stability, and Curtis (1977) found it more economic to use a Crank-Nicolson scheme, notwithstanding the need to solve nonlinear algebraic equations at each step. Turland and Wilson (1977) have also solved several problems in spherically symmetric coordinates with an implicit isotherm migration method.

In two space dimensions Crank, in Ockendon and Hodgkins (1975), proposed to solve for y (or x) in terms of t, x (or y) and u. The equation in (20.23) is transformed into

$$\frac{\partial y}{\partial t} = \frac{\partial^2 y}{\partial u^2}\left(\frac{\partial y}{\partial u}\right)^{-2} - \frac{\partial^2 u}{\partial x^2}\frac{\partial y}{\partial u}, \qquad (20.31)$$

and an interface condition equating $-\beta\ \partial u/\partial n$ to the velocity of the moving boundary becomes

$$\frac{\partial y}{\partial t} = \beta\left\{1 + \left(\frac{\partial y}{\partial x}\right)^2\right\}\left(\frac{\partial y}{\partial u}\right)^{-1}$$

on $u = 0$, which is essentially the form (20.27) used in the coordinate transformation of the preceding section 20.4.

Details of a one-phase computation which is explicit in time are given by Crank and Gupta (1974) for the problem of the solidification of an infinitely long square prism of fluid. The results are reasonably satisfactory, though a very short time step is needed for stability, and certain other necessary approximations suggest that the intervals in the space directions must also be fairly small to achieve a tolerable accuracy. To our knowledge no work has yet been published on implicit methods.

Research continues on this and analogous types of methods. Furzeland (1977c) lists in his bibliography various other coordinate transformations which have been suggested, and Crank and Crowley (1978) have obtained results with a method in which the grid effectively comprises the isotherms and the flow lines orthogonal to the isotherms.

20.6. THE ENTHALPY METHOD

In the last two sections the moving boundary has effectively been transformed by a change of variables into a fixed line in a new domain. This simplifies the numerical work to a considerable extent, but it is still necessary to satisfy the Stefan condition on this boundary. We now discuss a method in which the differential equation is modified in such a way that the Stefan condition is implied by the new equation. In reference to equations (20.1), for example, we then solve a single equation over the whole range $0 < x < 1$ and taken no explicit account of the moving boundary. The position of the latter, in fact, emerges as a by-product of the computation.

The required changes are remarkably simple. We introduce the *enthalpy*, the total heat content per unit volume, which incorporates the heat jump at the interface and is given by

$$H(u) = \int_{u_0}^{u} \rho(\bar{u})\sigma(\bar{u})\,d\bar{u}, \quad u < u_m,$$

$$H(u) = \int_{u_0}^{u} \rho(\bar{u})\sigma(\bar{u})\,d\bar{u} + \rho L, \quad u > u_m, \tag{20.33}$$

$$\int_{u_0}^{u} \rho(\bar{u})\sigma(\bar{u})\,d\bar{u} < H(u) < \int_{u_0}^{u} \rho(\bar{u})\sigma(\bar{u})\,d\bar{u} + \rho L, \; u = u_m.$$

Here u_0 is some fixed temperature, and u_m the common melting point temperature.

Equations (20.1) are now replaced by the single equation

$$\frac{\partial H}{\partial t} = \frac{\partial}{\partial x}\left\{k(u)\,\frac{\partial u}{\partial x}\right\}, \tag{20.34}$$

u retains its original boundary conditions at $x = 0$ and $x = 1$ and starting conditions at $t = 0$, and H is related to u through equations (20.33). This new differential equation is hardly meaningful at all points in a "classical" sense since H is discontinuous as a function of u at $u = u_m$. But it is meaningful in a "weak" sense in which (20.34) is cast in a form which involves no derivatives of u.

Consider, for example, boundary and initial conditions

$$u(0,t) = g_1(t), \; u(1,t) = g_2(t), \; u(x,0) = u_0(x), \quad (20.35)$$

and multiply (20.34) by a test function $\phi(x,t)$ which vanishes on $x = 0$ and $x = 1$ and $t = T$ and which has continuous first and second derivatives with respect to x and a continuous first derivative with respect to t. Then successive integration by parts produces the required weak form

$$\int_0^1 \int_0^T \left\{u \frac{\partial}{\partial x}\left(k(u) \frac{\partial \phi}{\partial x}\right) + H(u) \frac{\partial \phi}{\partial t}\right\} dxdt = \int_0^T k(g_2(t))g_2(t) \frac{\partial \phi}{\partial x}(1,t)\, dt$$

$$- \int_0^T k(g_1(t))g_1(t) \frac{\partial \phi}{\partial x}(0,t)\, dt - \int_0^1 H(u_0)\phi(x,0)\, dx, \quad (20.36)$$

an equation analogous to (20.14). Similar equations can be produced for problems with conditions more complicated than those of (20.35).

The weak solution has important properties both theoretical and numerical. First, for the theoretical results, both Taylor and Atthey, in Ockendon and Hodgkins (1975), show that any classical solution satisfies (20.36) and is therefore a weak solution; moreover (Oleinik, 1960) the weak solution is unique. It follows that the classical solution is unique. Second, for the numerical results, the discontinuities relevant to (20.34) make it difficult to prove the convergence of numerical methods of finite-difference type, but it is possible to prove that the standard finite-difference approximations to (20.34) produce results which converge to the weak solution of (20.36) and therefore to the unique classical solution of the given problem. Kamenomstaskeja (1961), and Atthey, in Ockendon and Hodgkins (1975), proved this for an obvious explicit method, and Cavialdini (1975) and Elliott (1976) proved it for a class of implicit methods for piecewise constant thermal coefficients.

The numerical work is still further simplified with

another change of variable, which might also be useful in some of the earlier methods we have described. With the Kirchoff transformation

$$C(u) = \int_{u_m}^{u} k(\bar{u})\,d\bar{u} , \qquad (20.37)$$

where u_m is taken for convenience but the lower limit could be any fixed temperature, equation (20.34) is finally reduced to the form

$$\frac{\partial H}{\partial t} = \frac{\partial^2 C}{\partial x^2} . \qquad (20.38)$$

Given a knowledge of the thermal properties of the material we can from (20.37) compile in advance a picture of C as a function of u, and then from (20.37) and (20.33) a picture of C as a function of H. They will have the general respective appearance of Figures 20.5 and 20.6.

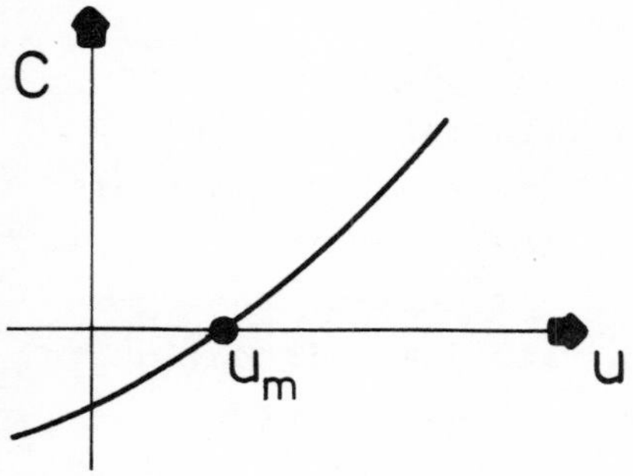

Fig. 20.5.

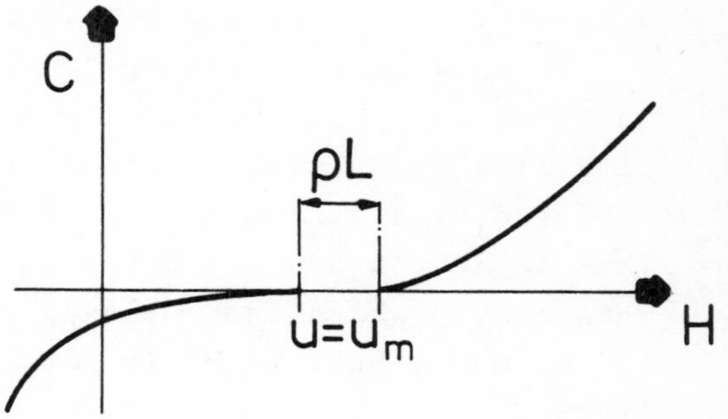

Fig. 20.6.

Following Atthey, who did not use the transformation (20.38), we can solve (20.34) simply with the explicit formula

$$H_i^{n+1} = H_i^n + \frac{\Delta t}{(\Delta x)^2} (C_{i+1}^n - 2C_i^n + C_{i-1}^n) , \qquad (20.39)$$

with suitable incorporation of the boundary conditions interpreted in terms of C. Again, however, stability considerations make it more economic to use an implicit method. The fully implicit form would replace (20.39) by

$$H_i^{n+1} = H_i^n + \frac{\Delta t}{(\Delta x)^2} (C_{i+1}^{n+1} - 2C_i^{n+1} + C_{i-1}^{n+1}) ,(20.40)$$

giving rise to the solution of nonlinear equations at each time step. Here C is a known function of H, and solution proceeds by a combination of Newton's method and a Gauss-Seidel process, perhaps with the introduction of an SOR-type parameter to accelerate the convergence (see chapter 18 for a discussion of these methods). For piecewise constant thermal coefficients, Elliott (1976) proved the uniqueness of the solution of (20.40) and the convergence of the Gauss-Seidel iteration.

In this formulation we have allowed a discontinuity in the enthalpy. Several authors, however, have thought it necessary to apply a smoothed form for H, with rapid but continuous change between temperatures $u_m - \Delta u_m$, and $u_m + \Delta u_m$ where Δu_m is small. This simplifies the analysis of uniqueness and of the convergence of finite-difference methods, but it is also claimed that it is physically more meaningful, since in practice the change of state probably does occur over a small range of critical temperatures.

It is also interesting to observe that the presence of an extra body-heating term in the differential equation may give rise to a "mushy" region in which the sharp boundary is replaced by a finite region at the critical temperature u_m. This region, again, is revealed as a by-product of the computation in which the enthalpy retains its jump discontinuity, and examples have been given by Atthey (1974) and Crowley (1975).

The extension to two or more space dimensions is quite straightforward. In general equation (20.38) will become

$$\frac{\partial H}{\partial t} = \nabla^2 U \ , \qquad (20.41)$$

associated with the usual types of boundary conditions. For this problem Crowley (1978) used an explicit method, and various other authors, for example Meyer (1973, 1976), Elliott (1976) and Furzeland (1977a) have used implicit methods both with finite differences and finite elements, and both with and also without the smoothing of the enthalpy. In this connection Furzeland (1977b) gives a useful bibliography of analytical and numerical results and methods.

Other authors, notably Bonacina *et al.* (1973), have smoothed the enthalpy and, with finite-difference methods in one space dimension, have used a three-level implicit formula originally proposed by Lees (1966) which has the numerical attributes but avoids the stability disabilities of explicit formulae. In particular this method avoids the solution of nonlinear algebraic equations. Comini *et al.* (1974) have extended this method to more difficult problems in three space dimensions, using the Lees formula together with a Galerkin-type finite element method.

20.7. METHODS USING QUADRATIC PROGRAMMING AND VARIATIONAL INEQUALITIES

Another technique which avoids explicit use of the Stefan interface condition formulates the problem in terms of certain inequalities. This is more interesting for the one-phase problem, and we shall therefore consider this in connection with equations (20.1) in which $u_2 \equiv 0 = u_m$. Corresponding to the enthalpy formulation of section 20.6 we introduce the variables

$$\overline{H}(u) = \int_0^u \rho(\overline{u})\sigma(\overline{u})\,d\overline{u}\ , \quad C(u) = \int_0^u k(\overline{u})\,d\overline{u}\ , \tag{20.42}$$

and the new variable w by $\partial w/\partial t = C(u)$ or, equivalently, by

$$w(x,t) = \int_0^t C(u)\,d\overline{t}\ . \tag{20.43}$$

From the relation (20.38) between $\overline{H}(u)$ and $C(u)$ we can then define H^* by

$$\overline{H}(u) = H^*\left(\frac{\partial w}{\partial t}\right)\ . \tag{20.44}$$

We now use the enthalpy differential equation (20.38), integrate it formally with respect to t, and produce for w the differential equation

$$\begin{aligned} H^*\left(\frac{\partial w}{\partial t}\right) &= \frac{\partial^2 w}{\partial x^2} + \overline{H}(u_0), \quad t > 0, \quad 0 < x < s(0)\ , \\ H^*\left(\frac{\partial w}{\partial t}\right) &= \frac{\partial^2 w}{\partial x^2} + \rho L, \quad t > \ell(x), \quad s(0) < x < 1, \end{aligned} \tag{20.45}$$

where $t = \ell(x)$ is the interface so that (20.45) refers to region I, and

$$w = 0, \quad t > 0, \quad s(t) < x < 1 \tag{20.46}$$

in region II. Any boundary condition for $u(0,t)$ can be expressed in terms of $w(0,t)$, any initial condition on $u(x,0)$ gives

$$w(x,0) = 0, \quad 0 < x < 1 , \tag{20.47}$$

and from the initial condition the term $\overline{H}(u_0)$ in (20.45) is easily calculable.

The next important observation is that these various equations combine to give the inequalities

$$H^*\left(\frac{\partial w}{\partial t}\right) - \frac{\partial^2 w}{\partial x^2} - g(x) \le 0, \quad w \le 0, \tag{20.48}$$

where

$$g(x) = \overline{H}(u_0), \quad 0 < x < s(0) ,$$
$$g(x) = \rho L, \quad s(0) < x < 1, \tag{20.49}$$

and these are supported by the equality

$$\left\{H^* \left(\frac{\partial w}{\partial t}\right) - \frac{\partial^2 w}{\partial x^2} - g(x)\right\} w = 0 . \tag{20.50}$$

The first of (20.48) is satisfied with equality in region I, and the inequality holds in region II since the first two terms vanish and ρL is positive. The second of (20.48) follows since $u < 0$ in region I so that $C(u)$ is everywhere negative, and (20.50) follows since one of the factors vanishes in each region.

If the thermal coefficients are independent of temperature so that

$$H^*\left(\frac{\partial w}{\partial t}\right) = h \frac{\partial w}{\partial t} , \tag{20.51}$$

where h is a known constant, then the fully implicit method for (20.48) produces the inequality

$$h\,\frac{W_i^{n+1} - W_i^n}{\Delta t} - \frac{W_{i+1}^{n+1} - 2W_i^{n+1} + W_{i-1}^{n+1}}{(\Delta x)^2} - g(x_i) \le 0,\ i = 0,1,\ldots,N-1, \tag{20.52}$$

where $N\Delta x = 1$, and with some adjustment for the boundary condition at $x = 0$, also assumed here to be linear in w. Equation (20.52) can then be expressed in the form

$$A\underset{\sim}{w} + \underset{\sim}{b} \le \underset{\sim}{0}\ , \tag{20.53}$$

where A is a known matrix and $\underset{\sim}{b}$ a known vector, and the vector $\underset{\sim}{w} = [W_0^{n+1}, W_1^{n+1}, \ldots, W_{N-1}^{n+1}]^T$. The other equations in (20.48) and (20.50) give

$$\underset{\sim}{w} \le \underset{\sim}{0},\ (A\underset{\sim}{w} + \underset{\sim}{b})^T\underset{\sim}{w} = 0\ . \tag{20.54}$$

This algebraic problem can now be solved by known methods in quadratic programming. Defining

$$\underset{\sim}{z} = A\underset{\sim}{w} + \underset{\sim}{b} \tag{20.55}$$

we have to solve

$$A\underset{\sim}{w} = \underset{\sim}{z} - \underset{\sim}{b},\ \underset{\sim}{w}^T\underset{\sim}{z} = 0 \quad \text{such that} \quad \underset{\sim}{w} \le \underset{\sim}{0},\ \underset{\sim}{z} \le \underset{\sim}{0}.$$

If A is symmetric and positive definite this is equivalent to the quadratic programming problem:

$$\text{minimize} \quad \underset{\sim}{b}^T\underset{\sim}{w} + \tfrac{1}{2}\underset{\sim}{w}^T A\underset{\sim}{w} \quad \text{such that} \quad \underset{\sim}{w} \le \underset{\sim}{0}\ . \tag{20.57}$$

Cryer (1971) solved this problem by an SOR-type iterative method, generating successive vectors $\underset{\sim}{z}^{(k+1)}$ and $\underset{\sim}{w}^{(k+1)}$ with components

$$z_i^{(k+1)} = b_i + \sum_{j=1}^{i-1} A_{ij}\, w_j^{(k+1)} + \sum_{j=i}^{n} A_{ij}\, w_j^{(k)} ,$$

$$w_i^{(k+1)} = \min \{ 0, w_i^{(k)} + \omega z_i^{(k+1)}/A_{ii} \} , \tag{20.58}$$

where ω is a suitable parameter in the range $0 < \omega < 2$. Cryer proved the convergence of the method, which is essentially the sort of technique which early hand-relaxation operators would use without compunction! Elliott (1976, 1978a, 1978b) applied it to several one-phase Stefan-type problems.

Just as in the enthalpy method, so here the rigorous justification of the use of (20.45) depends on the analysis of the corresponding weak formulation. The weak formulation gives rise, corresponding to (20.48) - (20.50) for the "classical solution", to a parabolic variational inequality given here by

$$\left(H^*\left(\frac{\partial w}{\partial t}\right), v - w\right) + a(w, v - w) \geq (g, v - w) ,$$

$$(w,v) = \int_0^1 wv\,dx, \quad a(w,v) = \int_0^1 \frac{\partial w}{\partial x}\frac{\partial v}{\partial x}\,dx \tag{20.59}$$

in connection with the boundary condition

$$u(0,t) = f(t) , \tag{20.60}$$

for all functions v which are non-positive in $0 < x < 1$ and agree with w on $x = 0$, and for which $\int_0^1 v^2 dx$ and $\int_0^1 (\partial v/\partial x)^2\, dx$ are bounded.

This variational inequality can now be used to produce a relevant algebraic problem of type (20.56) obtained by a finite element discretization. Elliott (1978b) investigated the stability and convergence of finite element explicit and implicit discretizations in the case of constant thermal coefficients, proving convergence to the relevant weak solution and obtaining results for stability similar to those for the standard parabolic problem.

The method is obviously applicable to problems with more

than one space dimension, and Elliott (1976, 1978b) gives examples with both explicit and implicit discretizations. Other literature on variational inequality techniques is listed in Elliott (1976) and Ockendon and Hodgkins (1975), and this is an area of continuing current research. An important part of this is the extension of the quadratic programming technique to the more general nonlinear case.

20.8. EVALUATION OF METHODS

Like all previous relative evaluations this must be open ended, since almost all the methods listed here are still developing both in theory and in practice. In the present state of the art it is likely that the enthalpy method is most useful, for the nonlinear one- or two-phase problems and in one or more dimensions, whenever the Stefan conditions are similar to those of equation (20.2) which can be absorbed in the enthalpy formulation. Interface conditions like (20.4) cannot be absorbed in this way, and such problems cannot be treated directly by any of the boundary-fixing methods including the isotherm migration method, or by variational inequalities. In such cases something like the method of lines is needed.

For linear one-phase problems the quadratic programming technique is quite attractive, particularly if it can be developed for economic computation in the nonlinear case. The isotherm migration method, in one space dimension, is useful for some specially suited problems and has no special difficulties with nonconstant thermal coefficients, but it has too many other difficulties to qualify as a general purpose technique.

Finally, as always, the choice of finite-difference or finite element discretization seems at present to be largely a question of taste, and much more numerical experimentation is needed before firm decisions can be made. This, of course, is true with respect to all the different formulations we have discussed.

Another important problem, on which more research would be useful, is in the production of analytical solutions valid for small time. These are particularly needed in situations, such as the "sudden quenching" problem, when there is a significant singularity at $x = 0$, $t = 0$ or when, as noted by Parker and Crank (1964) in the simple heat conduction problem without a phase change, nothing is gained by using very small time intervals in a numerical method since with some types of boundary condition truncation errors persist rather than disappear. Some analytical solutions appear in the literature, such as those of Carslaw and Jaeger (1959), and Boley in Ockendon and Hodgkins (1975), but they are rather scattered and it would be useful to collect and collate them in one coherent account.

PART V

HYPERBOLIC EQUATIONS

This part contains a brief introduction to the theory of well-posedness in hyperbolic equations. Characteristic and finite-difference methods are discussed in some detail with emphasis on the relations between well-posedness, stability and convergence. In the final chapter, there is a discussion of numerical methods for the important practical problem of elliptic-hyperbolic equations.

Contents

Chapter	Title	Contributor
21	Hyperbolic Equations and Characteristics.	J.E. Walsh
22	Difference Methods for Hyperbolic Equations.	J.M. Watt
23	Numerical Solution of Mixed Elliptic-Hyperbolic Equations	J.A. Hendry

21

HYPERBOLIC EQUATIONS AND CHARACTERISTICS

21.1. NATURE OF THE CHARACTERISTICS

Partial differential equations of hyperbolic type are often formulated as first-order systems, and we start by considering such a system in the form

$$\sum_{j=1}^{n} \alpha_{ij} \frac{\partial v_j}{\partial x} + \sum_{j=1}^{n} \beta_{ij} \frac{\partial v_j}{\partial y} = \gamma_i \;, \quad i = 1,2,\ldots,n \;. \qquad (21.1)$$

We assume initially that the coefficients α_{ij}, β_{ij} are constants, and that γ_i does not depend upon any v_j. The system (21.1) may be written as follows in matrix form

$$A \frac{\partial \underset{\sim}{v}}{\partial x} + B \frac{\partial \underset{\sim}{v}}{\partial y} = \underset{\sim}{\gamma} \;, \qquad (21.2)$$

where $A = [\alpha_{ij}]$, $B = [\beta_{ij}]$, $\underset{\sim}{v} = [v_1,v_2,\ldots,v_n]^T$, $\underset{\sim}{\gamma} = [\gamma_1,\gamma_2,\ldots,\gamma_n]^T$. The equations are hyperbolic if the system can be reduced by a linear transformation to n equations each of which specifies the derivatives of $v_1,v_2,\ldots,v_n$ in one direction only. Suppose the ith direction is given by

$$dx - \lambda_i dy = 0 \;; \qquad (21.3)$$

then the derivative of any function w in this direction is $\lambda_i\, \partial w/\partial x + \partial w/\partial y$. The equations (21.2) are transformed by premultiplying by a matrix M, and we want the result to be of the following form

$$M\left(A \frac{\partial \underset{\sim}{v}}{\partial x} + B \frac{\partial \underset{\sim}{v}}{\partial y}\right) = D \frac{\partial \underset{\sim}{w}}{\partial x} + \frac{\partial \underset{\sim}{w}}{\partial y} \;, \qquad (21.4)$$

where $D = \mathrm{diag}(\lambda_i)$ and $\underset{\sim}{w} = MB\underset{\sim}{v}$. To obtain this form, the ith row $\underset{\sim}{m}_i^T$ of M must satisfy

$$\underset{\sim}{m}_i^T A = \lambda_i \underset{\sim}{m}_i^T B \;, \qquad (21.5)$$

showing that the λ_i are the eigenvalues of the generalized problem

$$\det(A - \lambda B) = 0 \; . \tag{21.6}$$

The condition for a hyperbolic system is that (21.6) has n real roots, corresponding to n real directions in the (x,y)-plane. We assume in general that the roots are distinct.

If the system is hyperbolic, it may be reduced to n uncoupled first-order differential equations for the variables w_i, in the directions given by λ_i. The problem is very simple when A and B are constant, so that the λ_i are fixed. Usually, however, the coefficients and right-hand side of (21.1) depend on x,y, and $\underset{\sim}{v}$, so that the directions λ_i are not constant throughout the integration, but vary from point to point. In either case, provided the roots of (21.6) are real and distinct at all points in the domain of interest, we can define a network of curves (the characteristics) along which simplified equations of the form (21.4) hold, and which can be used for approximate solution. Special cases arise when some of the roots coincide (the parabolic case) or when the roots are partly real and partly complex; we return to these later.

Let us now consider a second-order partial differential equation, and find the condition for real characteristics. The general quasi-linear equation has the form

$$a\,\frac{\partial^2 u}{\partial x^2} + 2b\,\frac{\partial^2 u}{\partial x \partial y} + c\,\frac{\partial^2 u}{\partial y^2} + p\,\frac{\partial u}{\partial x} + q\,\frac{\partial u}{\partial y} + su = f. \tag{21.7}$$

Putting $v_1 = u$, $v_2 = \partial u/\partial x$, $v_3 = \partial u/\partial y$, equation (21.7) may be written as the following first-order system

$$\frac{\partial v_1}{\partial x} = v_2 \; , \quad \frac{\partial v_2}{\partial y} = \frac{\partial v_3}{\partial x} \; ,$$
$$a\,\frac{\partial v_2}{\partial x} + b\left(\frac{\partial v_2}{\partial y} + \frac{\partial v_3}{\partial x}\right) + c\,\frac{\partial v_3}{\partial y} + pv_2 + qv_3 + sv_1 = f \; . \tag{21.8}$$

Comparing with (21.2), we see that the matrices A and B are

$$A = \begin{bmatrix} 1 & 0 & 0 \\ 0 & 0 & -1 \\ 0 & a & b \end{bmatrix} , \qquad B = \begin{bmatrix} 0 & 0 & 0 \\ 0 & 1 & 0 \\ 0 & b & c \end{bmatrix} . \tag{21.9}$$

Equation (21.6) for the characteristic directions is therefore

$$a - 2\lambda b + \lambda^2 c = 0 , \tag{21.10}$$

which has distinct real roots if

$$b^2 - ac > 0 . \tag{21.11}$$

This is the condition for (21.7) to be hyperbolic.

The existence of real characteristics may be taken as the defining property of a hyperbolic system, and it is important both in the theory and in some methods of calculating the solution. In particular, the characteristics show what boundary conditions are needed to define the solution in a certain region. We illustrate this for the case $n = 2$, where there are two characteristics given by λ_1, λ_2 through any point of space. In general they are curved, but because characteristics of the same family do not intersect, they can be used as the basis for a coordinate grid.

In Figure 21.1(a), suppose we have values of v_1, v_2 specified on the line Q_1Q_2, which is not a characteristic. Then we can

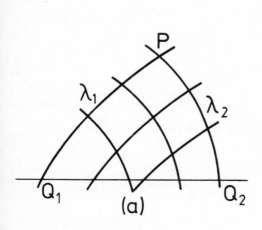

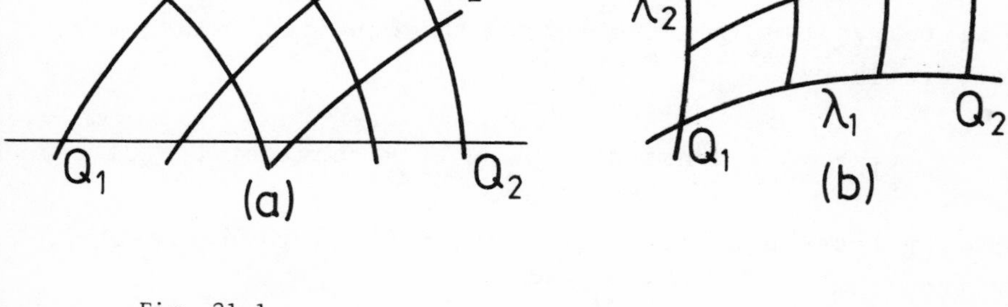

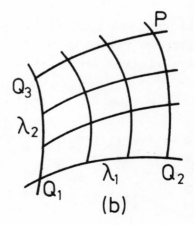

Fig. 21.1

integrate along characteristic curves to find v_1 and v_2 at all points in the region PQ_1Q_2. The solution at P is dependent only on the boundary data given on the segment Q_1Q_2, and any data outside this segment will not affect it. In Figure 21.1(b), suppose the boundary data is given on the two characteristic curves Q_1Q_2, Q_1Q_3. If one variable is specified on each of Q_1Q_2, Q_1Q_3, and both variables at Q_1, we can use the differential relation along characteristics to find values of the other variable along these curves. The complete solution may then be obtained in the region $Q_1Q_2PQ_3$. (See Courant and Hilbert, 1962, section 5.6 for a full discussion.)

For a system of n equations there are n characteristics through any point P, and the solution at P is determined by the data on a segment of the initial line cut off by the extreme characteristics through P. We return to the question of boundary conditions for a well-posed problem at the end of the next section.

For a second-order equation such as (21.7), we might enquire whether it can be solved in a closed region, with boundary values specified at all points of the boundary. Consider the following equation

$$\frac{\partial^2 u}{\partial x^2} - \frac{\partial^2 u}{\partial y^2} = 0 \qquad (21.12)$$

in the unit square with vertices (0,0), (0,1), (1,0), (1,1), and suppose values of u are given on the four sides. The characteristics for equation (21.12) are

$$x + y = \text{constant}, \quad x - y = \text{constant}. \qquad (21.13)$$

If we write $v_1 = \partial u/\partial x$, $v_2 = \partial u/\partial y$, the equations along the characteristics are

$$v_1 + v_2 = \text{constant}, \quad v_1 - v_2 = \text{constant}\ . \qquad (21.14)$$

By applying these equations along AB, BC, CD, DA in Figure 21.2, we can show that

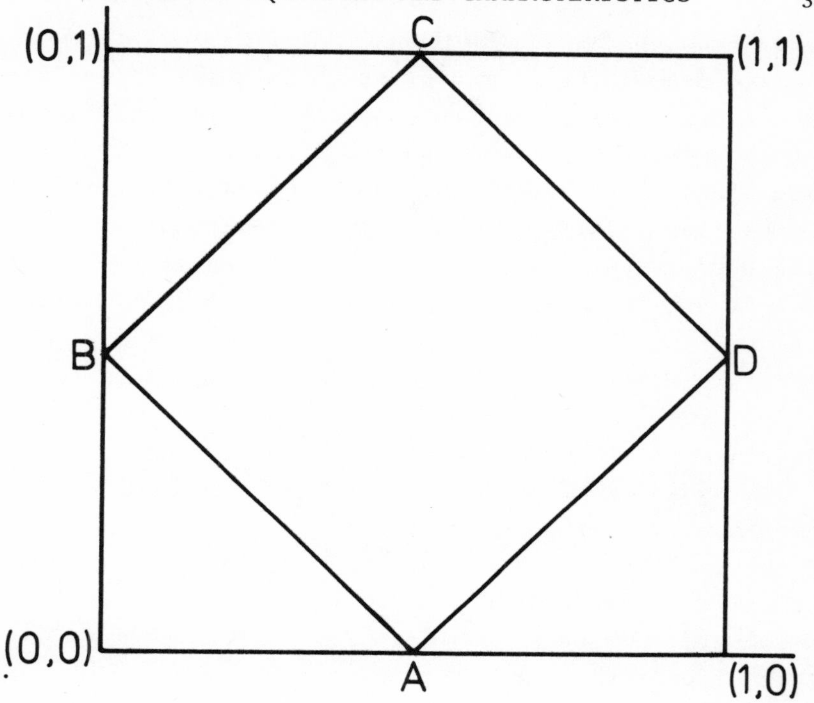

Fig. 21.2.

$$\left.\begin{aligned} v_1(C) - v_2(C) &= v_1(A) + v_2(A) - 2v_2(B), \\ v_1(C) + v_2(C) &= v_1(A) - v_2(A) + 2v_2(D). \end{aligned}\right\} \quad (21.15)$$

Adding these two equations, we see that $v_1(C)$ is determined by $v_1(A)$, $v_2(B)$, and $v_2(D)$, and it follows that the values of u at A, B, C, D cannot be specified independently. However, we can solve the problem if suitable boundary conditions are given on three sides of the square, e.g. v_1 and v_2 on $y = 0$, and one each of v_1 and v_2 on $x = 0$ and $x = 1$.

An important consequence of the relation between the characteristics and the boundary conditions is that discontinuities on the boundary are propagated through the solution along characteristic curves. In cases where such discontinuities occur, it is almost essential to calculate the solution on the characteristic mesh, which gives a natural way of tracing the discontinuity and maintaining accuracy. This will be discussed further in section 21.3.

21.2. SOLUTION ALONG THE CHARACTERISTICS

To make use of the characteristics in numerical solution, we have to integrate simultaneously the differential equations for the characteristics, and the special relations satisfied along them. Returning to the first-order system (21.1), suppose the values of $v_1, v_2, \ldots, v_n$ are specified on an initial line Q_1Q_2 (Figure 21.3) which is not a characteristic, and that we want to integrate the equations to a neighbouring point P

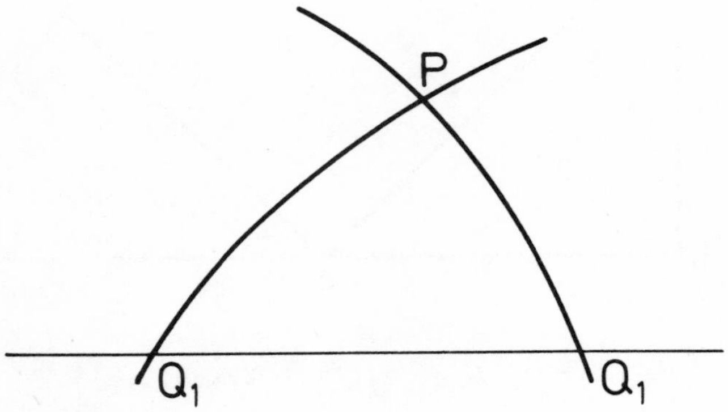

Fig. 21.3.

of the characteristic mesh. The procedure is relatively simple if $n > 2$, because we can select the two points Q_1, Q_2, and find the point of intersection P of the characteristics through them (see Ames, 1969, section 4.3).

For general $n > 2$, it is more difficult to find the points of the mesh. To advance the solution to P, we need to know the n points $Q_1, Q_2, \ldots, Q_n$ in which the characteristics through P cut the initial line. One way of proceeding is to estimate the position of P, then fix its coordinates and solve the characteristic equations for the points $Q_1, Q_2, \ldots, Q_n$. In general the characteristics depend on $\underset{\sim}{v}$ as well as on x and y, and so the values of $\underset{\sim}{v}$ at P have to be found at the same time.

Suppose the point P(x,y) has been fixed, and we have estimated the values of $v_i(P)$, $i = 1,2,\ldots,n$. Then we evaluate the coefficients α_{ij}, β_{ij}, γ_i in (21.1) at P, and solve (21.6) to find the characteristic directions. This gives n differential relations of the form

$$dx - \lambda_i(\underset{\sim}{v},x,y)\,dy = 0\ . \tag{21.16}$$

Along the ith characteristic, the variables v_i satisfy the relation

$$\sum_{j=1}^{n} \sigma_{ij}dv_j = \tau_i dy\ , \tag{21.17}$$

where

$$\sigma_{ij} = \sum_{k=1}^{n} m_{ik}\beta_{kj}\ ,\ \tau_i = \sum_{k=1}^{n} m_{ik}\gamma_k\ . \tag{21.18}$$

The coefficients are all dependent on x, y, and $\underset{\sim}{v}$, and so the equations (21.16), (21.17) must be solved iteratively. The usual procedure is to approximate (21.16) and (21.17) by the trapezium rule, giving

$$x(P) - x(Q_i) = \tfrac{1}{2}\{\lambda_i(P) + \lambda_i(Q_i)\}(y(P) - y(Q_i)) \tag{21.19}$$

for (21.16), and similarly for (21.17). There are effectively n unknown quantities to be found in order to determine $Q_1,Q_2,\ldots,Q_n$, since they lie on a fixed line, and n unknown function values of $v_1,v_2,\ldots,v_n$ at P. From (21.16) and (21.17) we have 2n equations, and we can show that the simple iteration converges provided the step-length is small enough (Courant and Hilbert, 1962, section 5.6).

It is clear that the geometrical problem of keeping track of the mesh-lines can be very complicated, particularly in an automatic program. Even for n = 2, we may find that the mesh gets very distorted as the integration proceeds, so that points have to be inserted or deleted in order to maintain a reasonable distribution of mesh-lines.

A method of simplifying the calculation is to use the characteristic relations on a regular (x,y) mesh. This was first suggested by Hartree, and it is described in detail by Ames (1969, section 4.12). Assume the solution is known

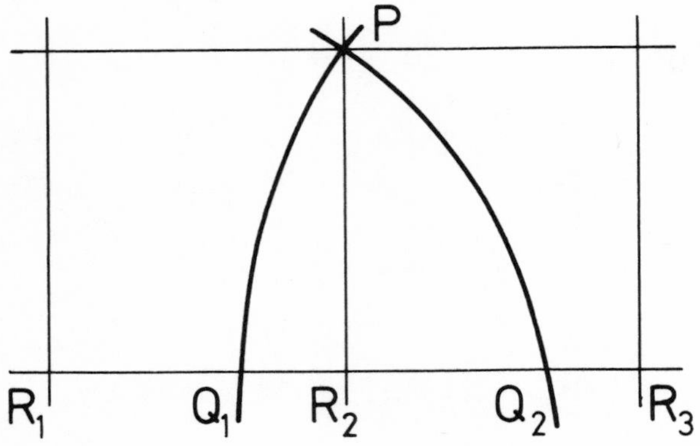

Fig. 21.4.

on a regular mesh up to the line which includes the segment R_1R_3 (Figure 21.4). The values at a new mesh-point such as P are obtained by tracing the characteristics through P back to the previous mesh-line, which they intersect in $Q_1, Q_2, \ldots, Q_n$ say. The function values at these points are obtained by interpolation between solution values at the mesh points R_i. The integration is then carried out as before by using the relations (21.16) and (21.17) in an approximate form. A formula such as (21.19) is relatively low-order, and so the interpolation presents no difficulty, because it need only be as accurate as the integration. This method is easier to program than one using a pure characteristic mesh, and it extends more readily to problems in three dimensions (see section 21.4).

We now consider the boundary conditions required for a well-posed problem. Usually the region of interest for an equation such as (21.2) is bounded on at least two sides, say by the lines $x = 0$, $y = 0$, so that we are considering the positive quadrant. Often there is a further boundary in the x-direction, say at $x = 1$, and the region is open in the y-direction (Figure 21.5). Suppose the value of $\underset{\sim}{v}$ is given on $y = 0$, with auxiliary conditions on the other boundaries of the form

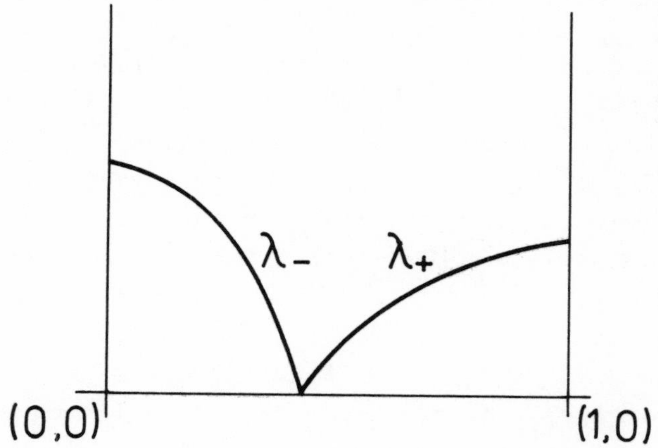

Fig. 21.5.

$$\left.\begin{aligned} S\underset{\sim}{v} &= \underset{\sim}{b}, \quad \text{on } x = 0, \\ T\underset{\sim}{v} &= \underset{\sim}{c}, \quad \text{on } x = 1. \end{aligned}\right\} \tag{21.20}$$

If this information is enough to define the solution everywhere, we must be able to find the value $\underset{\sim}{v}$ on x = 0 and x = 1. Suppose the equation (21.6) for the characteristic directions has k positive and (n-k) negative roots. Then we need (n-k) additional conditions on x = 1 and k additional conditions on x = 0, and these conditions together with the characteristic relations must determine all the values of v_i on the boundaries. The characteristic form of equation (21.2) is

$$D\,\frac{\partial \underset{\sim}{w}}{\partial x} + \frac{\partial \underset{\sim}{w}}{\partial y} = M\underset{\sim}{\gamma}\ , \tag{21.21}$$

where $D = \mathrm{diag}(\lambda_i)$, $\underset{\sim}{w} = MB\underset{\sim}{v}$. Assuming that the first k elements of D are positive, it follows that the first k components of $\underset{\sim}{w}$ together with the condition $T\underset{\sim}{v} = \underset{\sim}{c}$ must determine all the variables v_i on x = 1. Suppose the first k rows of M form a matrix M_1, then if T is (n-k)×n the necessary condition is

$$\begin{pmatrix} M_1 B \\ T \end{pmatrix} \text{ is non-singular for } x = 1, \tag{21.22}$$

and a similar condition may be obtained for x = 0. (The rows of M are the left eigenvectors of the generalized eigenvalue problem (21.6).) A fuller discussion is given by Morton (1977, p.701). It is assumed here that the differential equations (21.2) are linear; if they are nonlinear, we have to assume that each λ_i (which depends on $\underset{\sim}{v}$) is of constant sign.

21.3. DISCONTINUITIES AND SHOCK WAVES

As noted earlier, the solution of a hyperbolic equation may be discontinuous across a characteristic. Let us now look at the situation in more detail for the case of a linear system. We have seen that the first-order equations (21.1) may be reduced to a system of ordinary differential equations along characteristics, of the form (21.17). If the variables are transformed by writing $\underset{\sim}{w} = MB\underset{\sim}{v}$, we obtain the equations in the simpler form (21.21). Suppose one of the w_i is discontinuous at a point Q on the initial line, say

$$w_i(Q+) = w_i(Q-) + \delta_i \quad . \tag{21.23}$$

Then because the equations are linear, the jump δ_i satisfies the usual relation in the direction given by λ_i, that is

$$d\delta_i = (M\underset{\sim}{\gamma})_i dy \ , \tag{21.24}$$

corresponding to (21.21). So the solution may be computed on the characteristic mesh, with all variables except w_i continuous along the corresponding mesh-lines. Thus the complete solution is continuous everywhere in the plane except across the λ_i characteristic QP (Figure 21.6).

In the nonlinear case the coefficients of M and B are dependent on $\underset{\sim}{v}$, and consequently we cannot write the system in the form (21.21), except locally. The lines of discontinuity do not necessarily lie along characteristics,

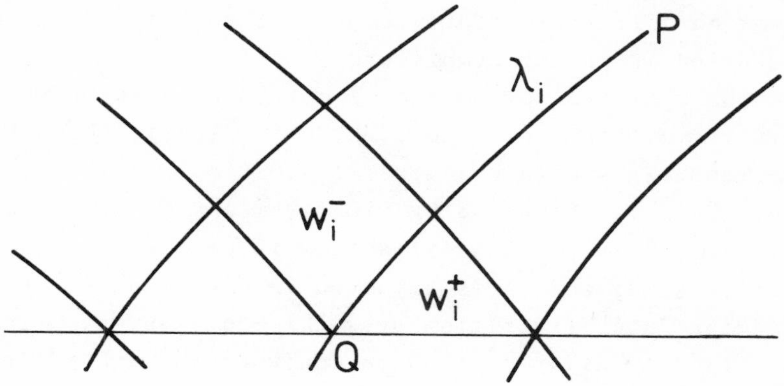

Fig. 21.6.

and furthermore they can arise from points within the region, without being propagated from the boundary conditions.

The treatment of nonlinear discontinuities (or shocks) depends very much on the form of the equations and on any further conditions satisfied by the variables. In practical problems, a shock corresponds to some physical phenomenon about which we may have further information (for example, a conservation law). A well-known example of a nonlinear system is the following (the equations for compressible flow)

$$\frac{\partial \rho}{\partial t} + \frac{\partial}{\partial x}(\rho u) = 0,$$

$$\frac{\partial}{\partial t}(\rho u) + \frac{\partial}{\partial x}(\rho u^2 + p) = 0, \qquad (21.25)$$

$$\frac{\partial p}{\partial t} + \rho c^2 \frac{\partial u}{\partial x} + u\frac{\partial p}{\partial x} = 0 .$$

If a shock wave arises in the course of integrating these equations, we have a curve across which p, ρ, and u are discontinuous. But from the general properties of fluid flow, the mass, momentum, and energy are conserved across the shock, so there are three special relations which apply on the line of discontinuity. This provides enough information to integrate the equations, either on a characteristic mesh or on a finite-difference mesh, though the detailed calculation is rather complicated.

A somewhat simpler method is obtained by introducing an artificial term into the equations (21.25), which has the effect of "smoothing out" the shock over a short distance. This is discussed by Richtmyer and Morton (1967, section 12.10 ff.). The extra term is described as a pseudo-viscosity term, and it is adjusted in magnitude so as to spread the shock transition over three or four mesh intervals. It is usually applied in conjunction with a finite-difference method on a regular mesh. An alternative formulation of a similar type introduces a dissipative term directly into the finite-difference approximation (Lax's method). This approach has been very successful, and is easy to program.

21.4. MORE DIFFICULT PROBLEMS

It is possible to extend the theory of characteristics to equations in three dimensions, but the geometry is more complex, and methods of integration based on characteristics have not been widely used. Butler (1960) discusses this problem and gives a computational procedure. Taking the third variable to be the time t, suppose we have the linear first-order system

$$A\,\frac{\partial \underset{\sim}{v}}{\partial x} + B\,\frac{\partial \underset{\sim}{v}}{\partial y} + C\,\frac{\partial \underset{\sim}{v}}{\partial t} = \underset{\sim}{\gamma}\,, \tag{21.26}$$

representing n equations in the n variables $v_1, v_2, \ldots, v_n$. We define a characteristic as a surface on which specified values of the variables do not define a unique solution of (21.26). Let $\phi(x,y,t) = 0$ be such a surface; then if (ξ,η,τ) is normal to the surface, we must have

$$\det(A\xi + B\eta + C\tau) = 0. \tag{21.27}$$

This homogeneous equation represents a cone, which is nondegenerate for a hyperbolic system. Thus there is a family of characteristic surfaces for the system.

We can find a curve called the bicharacteristic in $\phi = 0$ such that the differential equations (21.26) may be reduced to a relation between the derivatives along the curve and the derivatives in some other direction in the surface. The actual method of integration is somewhat similar to Hartree's method mentioned earlier. From a forward point in time we find bicharacteristic curves which intersect a fixed surface (which may be a plane) at an earlier time, and the values of v_i at the points of intersection are found by interpolation. The characteristic conditions are then used to advance the solution, but the details are too lengthy to give here. The specification of boundary conditions for a well-posed problem in three dimensions is discussed by Morton (1977, p. 704). For another method, see Johnston and Pal (1972).

So far we have assumed that the differential equations are purely hyperbolic, but there are many problems which give rise to equations or systems of an intermediate character, and the conditions for well-posedness and for stable solution methods need special consideration. There is no general theory which covers all cases, but certain classes of problems have been analysed in detail. For example, Courant and Hilbert (1962, section 6.16) consider equations of mixed elliptic and hyperbolic type. The most difficult problems arise with nonlinear equations, where the type of the equation may change in different parts of the region.

Some simple examples which illustrate developments of hyperbolic problems are given by Carrier and Pearson (1976), for example, Burgers' equation which is

$$\frac{\partial u}{\partial t} = \varepsilon \frac{\partial^2 u}{\partial x^2} - u\frac{\partial u}{\partial x} . \qquad (21.28)$$

This is a parabolic equation, but for small ε it approaches a hyperbolic form for which a shock wave may occur. Another example of a mixed problem of this type is the diffusion-convection equation, which is discussed further in chapter 11.

22

DIFFERENCE METHODS FOR HYPERBOLIC EQUATIONS

22.1. INTRODUCTION

The general treatment of hyperbolic equations is complicated by the great variety of physical phenomena that are important in different situations. For instance, in meteorology low speed compressible flow is considered and nonlinear interactions over long time periods are of interest; whereas, in aeronautics the phenomena of interest are mainly steady state, but the flow is high speed with shocks and turbulence, and boundaries (of wings or turbine blades) are very important. This great diversity means that it is usually necessary to construct special methods which are suitable for the particular problem concerned and complete books have been written about particular problems (see for instance Teman (1977) for the solution of the Navier-Stokes equations).

Another complication is that the analytic theory of the equations we wish to solve is not complete and it is not always clear what boundary conditions must be imposed to ensure a well-posed problem (see also chapter 21). To illustrate this problem, which is more severe in multidimensional equations, we condense some remarks from Richtmyer and Morton (1967, p.352). "The linearised problem of Taylor instability for compressible fluids is not properly posed; it is possible that these problems were properly posed before linearisation, rather likely that they are properly posed when compressibility is taken into account and almost certain that they are properly posed if effects like viscosity, surface tension and heat conduction are included".

In this brief chapter we are unable to do more than give an introduction to the treatment of problems with smooth solutions, that is shocks are not treated, and we concentrate on systems of first order equations.

22.2. SIMPLE DISCRETIZATIONS

Consider as a model the simple first-order constant coefficient equation

$$u_t + au_x = f(x,t), \quad a > 0 . \tag{22.1}$$

Table 22.1 lists a number of simple discretizations that have been used for this equation, and also lists the von Neumann stability conditions, and the order of the local discretization error. The diagrams show the arrangement of points used, and the cross indicates the centre of the approximation.

The von Neumann stability criterion for these discretizations is obtained by trying to obtain a solution to the equations of the form

$$U_j^n = \lambda^n e^{ij\xi} , \quad |\xi| < \pi .$$

Substitution into the difference equation gives an equation for λ, and for stability we require that all such λ satisfy

$$|\lambda| \le 1 + O(\Delta t) , \text{ for all } \xi : |\xi| \le \pi .$$

This follows from the fact that any x-variation can be decomposed by Fourier analysis into a sum of terms of the form $e^{ij\xi}$. (See chapter 9 for a justification of a similar analysis for parabolic equations.)

It will be seen that the explicit schemes all have a stability requirement $0 \le pa \le 1$ or $|pa| \le 1$ (where $p = \tau/h = \Delta t/\Delta x$). That a requirement of this form is necessary was first pointed out by Courant, Friedrichs and Lewy (1928). The *region of dependence* of a point $(\bar{x},\bar{t})$ is the set of points of the $t = 0$ plane, the initial values, which affect the solution at $(\bar{x},\bar{t})$. It is clear that for convergence the region of dependence of the difference solution must *include* that of the analytic solution; for otherwise a change in initial conditions could change the analytic but not the difference solution at $(\bar{x},\bar{t})$.

Now the characteristics of (22.1) are the lines $x\text{-}at$ = constant, so the region of dependence of $(\bar{x},\bar{t})$ is the point $(\bar{x}-a\bar{t},0)$. For method A of the table the region is the set of points

$$(\bar{x}-\bar{t}/p,0),\ldots,(\bar{x}-h,0), \ (\bar{x},0).$$

TABLE 22.1

Finite Difference Schemes for $u_t + au_x = f(x,t)$

$$(x_j, t_n) = (jh, n\tau) \qquad p = \tau/h$$

A 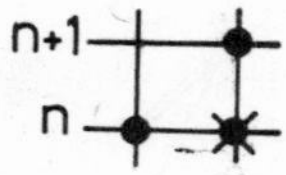

$$(U_j^{n+1} - U_j^n) + pa(U_j^n - U_{j-1}^n) = \tau f_j^n$$

stable for $0 \le pa \le 1$ l.d.e. = $0(h+\tau)$

B 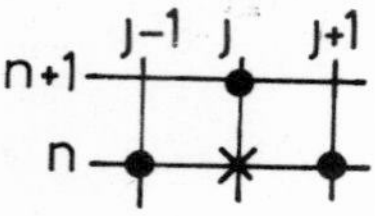

$$\{U_j^{n+1} - \tfrac{1}{2}(U_{j+1}^n + U_{j-1}^n)\} + \tfrac{1}{2}\, pa(U_{j+1}^n - U_{j-1}^n) = \tau f_j^n$$

stable for $|pa| \le 1$ l.d.e. = $0(h^2+\tau)$

C 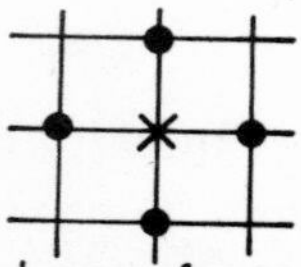
Leap frog

$$(U_j^{n+1} - U_j^{n-1}) + pa(U_{j+1}^n - U_{j-1}^n) = 2\tau f_j^n$$

stable for $|pa| \le 1$ l.d.e. = $0(h^2+\tau^2)$

D 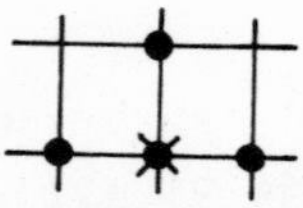
Lax-Wendroff

$$(U_j^{n+1} - U_j^n) + pa(U_{j+1}^n - U_{j-1}^n)/2$$

$$- \frac{pa^2}{2}(U_{j+1}^n - 2U_j^n + U_{j-1}^n) = \tau f_j^n + \frac{\tau^2}{2}\left(\frac{\partial f^n}{\partial t}{}_j - a\frac{\partial f^n}{\partial x}{}_j\right)$$

stable for $|pa| \le 1$ l.d.e. = $0(h^2+\tau^2)$

E 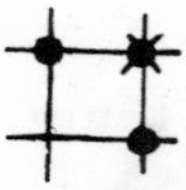

$$(U_j^{n+1} - U_j^n) + pa(U_j^{n+1} - U_{j-1}^{n+1}) = \tau f_j^n$$

stable for $pa \ge 0$ l.d.e. = $0(h+\tau)$

F 
Box scheme

$$\{(1+pa)U_{j+1}^{n+1} + (1-pa)U_j^{n+1} - (1-pa)U_{j+1}^n - (1+pa)U_j^n\}$$

$$= 2\tau f_{j+\frac{1}{2}}^{n+\frac{1}{2}}$$

unconditionally stable   l.d.e. = $0(h^2+\tau^2)$

G 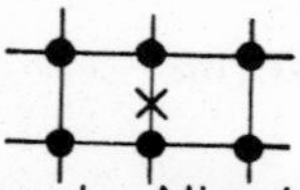
Crank Nicolson type.

$$(U_j^{n+1} - U_j^n) + pa(U_{j+1}^{n+1} + U_{j+1}^n - U_{j-1}^{n+1} - U_{j-1}^n)/2 = \tau f_j^{n+\frac{1}{2}}$$

unconditionally stable   l.d.e. = $0(h^2+\tau^2)$

Hence for convergence $1/p \geq a$, that is $pa \leq 1$, and since the method is consistent, $pa \leq 1$ is a stability condition.

The methods in the table all generalize easily to the variable coefficient case for systems of equations

$$\underset{\sim}{u}_t + A(x,t)\underset{\sim}{u}_x = \underset{\sim}{f}(x,t) \tag{22.2}$$

where $\underset{\sim}{u}$ is a vector of q components and $A(x,t)$ is a square $q \times q$ matrix. For example, the leapfrog method can now be written as

$$(\underset{\sim}{U}_j^{n+1} - \underset{\sim}{U}_j^{n-1}) + pA_j^n(\underset{\sim}{U}_{j+1}^n - \underset{\sim}{U}_{j-1}^n) = 2\tau\, \underset{\sim}{f}_j^n , \tag{22.3}$$

and the Lax Wendroff method becomes, when $\underset{\sim}{f} = \underset{\sim}{0}$,

$$\underset{\sim}{U}_j^{n+1} = \left\{ \begin{aligned} &1 + \tfrac{1}{2}pA_j^{n+\frac{1}{2}}(\Delta_x + \nabla_x) \\ &+ \tfrac{1}{4}p^2(A_j^{n+\frac{1}{2}}\Delta_x A_j^{n+\frac{1}{2}}\nabla_x + A_j^{n+\frac{1}{2}}\nabla_x A_j^{n+\frac{1}{2}}\Delta_x) \end{aligned} \right\} \underset{\sim}{U}_j^n . \tag{22.4}$$

The von Neumann stability criterion must now be applied to the problem with "frozen" coefficients, that is with $A(x,t)$ replaced by $A(\bar{x},\bar{t})$, and for stability, in each case, we require

$$|p\, \lambda_s(\bar{x},\bar{t})| \leq 1, \quad s = 1,2,\ldots,q , \tag{22.5}$$

for $\lambda_1, \lambda_2, \ldots, \lambda_q$ eigenvalues of $A(\bar{x},\bar{t})$. For stability of the variable coefficient problem we demand that (22.5) holds at all $(\bar{x},\bar{t})$ in the region of interest.

Discretizations can be divided into two classes, *dissipative* and *non-dissipative*. A discretization is dissipative if each λ of the von Neumann stability criterion satisfies, for some $\delta > 0$ and positive integer m (usually unity),

$$|\lambda| \leq 1 - \delta|\xi|^{2m}, \quad \text{for all } |\xi| \leq \pi$$

If this criterion is satisfied, approximate solutions have norm gradually decreasing with time, and any corresponding 'energy' decays gradually, instead of being conserved.

Alternatively if $|\lambda| = 1$ for all s and ξ, the discretization is *non-dissipative* and 'energy' is strictly conserved. Such discretizations are best used for problems (such as those occurring in meteorology) which involve a long time scale and where the energy of the system is not fed in from the boundary conditions and forcing terms. On the other hand for high-speed phenomena with much energy input across boundaries and from forcing terms, dissipative approximations are more suitable, and provide some smoothing of the solution which is often useful if shock waves are being treated in a simple way.

In practical computation, the methods used are at least second-order, commonly explicit, but sometimes implicit as in the Box or Crank-Nicolson schemes. Recent work (Roberts and Weiss (1966), Fromm (1968), Grammelvedt (1969), Kreiss and Oliger (1972)) has shown that if the solution is reasonably smooth it is advantageous to use methods which approximate the space derivatives to fourth or sixth order accuracy; this has led to a 50% saving of work when solving the shallow water equations.

22.3. NONLINEAR EQUATIONS IN CONSERVATION FORM

An important class of nonlinear equations can be written conservation law form. We therefore consider the system of equations

$$\underset{\sim}{u}_t + \{\underset{\sim}{f}(\underset{\sim}{u})\}_x = \underset{\sim}{0} \,, \qquad (22.5)$$

where $\underset{\sim}{f}$ is a vector function with q components. Denoting $\underset{\sim}{f}(\underset{\sim}{U}_j^n)$ by $\underset{\sim}{F}_j^n$, method A (in Table 22.1) can be generalized to

$$(\underset{\sim}{U}_j^{n+1} - \underset{\sim}{U}_j^n) + p(\underset{\sim}{F}_j^n - \underset{\sim}{F}_{j-1}^n) = \underset{\sim}{0} \,,$$

with similar generalizations for most of the other methods.

The appropriate generalization of the Lax-Wendroff scheme is more difficult and takes on a Runge-Kutta form with intermediate approximate values. For an S_β^α scheme these intermediate values approximate u at the points $(x_{j+\beta}, t_{n+\alpha})$. All the schemes reduce to the familiar Lax-Wendroff scheme when the equation is linear. The usual $S^{\frac{1}{2}}_{\frac{1}{2}}$ scheme is

$$\begin{aligned} \bar{\underset{\sim}{U}}^{n+\frac{1}{2}}_{j+\frac{1}{2}} &= \tfrac{1}{2}(\underset{\sim}{U}^n_{j+1}+\underset{\sim}{U}^n_j) - \tfrac{1}{2}p(\underset{\sim}{F}^n_{j+1}-\underset{\sim}{F}^n_j) \ , \\ \underset{\sim}{U}^{n+1}_j &= \underset{\sim}{U}^n_j - p(\bar{\underset{\sim}{F}}^{n+\frac{1}{2}}_{j+\frac{1}{2}}-\bar{\underset{\sim}{F}}^{n+\frac{1}{2}}_{j-\frac{1}{2}}) \ , \end{aligned} \tag{22.6}$$

using an obvious notation. In aeronautics, the schemes S^1_1 and S^1_0 (McCormack, 1971) are very popular and successful. The first of these is

$$\begin{aligned} \bar{\underset{\sim}{U}}^{n+1}_j &= \underset{\sim}{U}^n_j - p(\underset{\sim}{F}^n_j-\underset{\sim}{F}^n_{j-1}) \ , \\ \underset{\sim}{U}^{n+1}_j &= \underset{\sim}{U}^n_j - \tfrac{1}{2}p(\underset{\sim}{F}^n_j-\underset{\sim}{F}^n_{j-1}+\bar{\underset{\sim}{F}}^{n+1}_{j+1}-\bar{\underset{\sim}{F}}^{n+1}_j) \ . \end{aligned} \tag{22.7}$$

S^1_0 is similar and can be derived from this by changing the x-direction. The advantage of S^1_0 and S^1_1 over $S^{\frac{1}{2}}_{\frac{1}{2}}$ lies in the ease of adding a dissapative term $-b\underset{\sim}{u}_{xx}$ to $\{\underset{\sim}{f}(\underset{\sim}{u})\}_x$, suitable differences of $\underset{\sim}{U}^n$ and $\bar{\underset{\sim}{U}}^n$ being added to $\underset{\sim}{F}^n$ and $\bar{\underset{\sim}{F}}^n$ (Morton, 1977). The choice between S^1_0 and S^1_1 depends on the propagation direction of the solution.

The form $S^{\frac{1}{2}}_{\frac{1}{2}}$ due to Richtmyer was generalised to $S^{\alpha}_{\frac{1}{2}}$ by McGuire and Morris (1974). Further work has been done by Shankar and Anderson (1975) and Lerat and Peyret (1975), and the method has been extended to third order (Warming, Kutler and Lomax, 1973).

The generalization of the basic Lax-Wendroff method to two space dimensions leads to a wide variety of different methods which usually involve nine space points. For a survey of these see Turkel (1974); for locally one-dimensional methods see Gourlay and Morris (1968).

22.4. BOUNDARY CONDITIONS

For the simple problem

$$u_t + a\,u_x = 0, \quad x \geq 0,\ t \geq 0 \ , \tag{22.8}$$

the characteristics, on which the solution is constant, are the lines

$$x - at = \text{constant},$$

and it is clear that if $a > 0$, for the solution to be unique, it must be given on both the positive x-axis and t-axis. If $a < 0$ however the values of u can only be given on the positive x-axis.

Now consider the system of q equations

$$\underset{\sim}{u}_t - A\underset{\sim}{u}_x = \underset{\sim}{0} \quad 0 \le x \le 1, \quad t \ge 0 . \tag{22.9}$$

If A is diagonal with the first s components positive and remaining (q-s) components negative, it follows that $\underset{\sim}{u}(x,0)$ must be given for all x, $0 \le x \le 1$, and that the first s components of $\underset{\sim}{u}(0,t)$ and the last (q-s) components of $\underset{\sim}{u}(1,t)$ must be given for all $t > 0$. If A is not diagonal similar considerations apply to the eigenvectors corresponding to positive and negative eigenvalues, but it is more difficult to apply the correct conditions. The condition at $x = 0$ may however involve *all* eigenvectors of A provided that, if the components corresponding to negative eigenvalues are known, those corresponding to positive eigenvalues are determined uniquely. Hence there must be s linearly independent conditions at $x = 0$ and q-s at $x = 1$ (see chapter 21). When we attempt to approximate boundary conditions by differences, similar criteria must be adhered to, in order that the complete system of difference equations have a unique stable solution.

Rigorous analyses of the effect of boundary conditions on the stability of discretizations have only recently become available, see Kreiss (1968), Osher (1969a, 1969b), Gustafsson, Kreiss and Sundström (1972). These results are restricted to the case of linear equations, sometimes with constant coefficients only.

The idea of the results is an extension of the von Neumann Fourier analysis in which allowance is made for the boundary conditions. Consider difference equations reduced to the following form, which is always possible (Richtmyer and Morton, 1967), (again $\underset{\sim}{U}$ has q components and the coefficients are constant matrices):

$$B_0 \underset{\sim}{U}_j^{n+1} + B_1 \underset{\sim}{U}_{j+1}^{n+1} = A_0 \underset{\sim}{U}_j^n + A_1 \underset{\sim}{U}_{j+1}^n + \underset{\sim}{f}_j, \quad n,j = 0,1,2,\ldots, \tag{22.10}$$

with boundary conditions

$$B_0 \underset{\sim}{U}_0^{n+1} + B_1 \underset{\sim}{U}_1^{n+1} = \alpha_0 \underset{\sim}{U}_0^n + \alpha_1 \underset{\sim}{U}_1^n + \underset{\sim}{g}, \quad n = 0,1,2,\ldots . \tag{22.11}$$

Now seek solutions of the form

$$\underset{\sim}{U}_j^n = \lambda^n \mu^j \underset{\sim}{V}, \tag{22.12}$$

to the homogeneous form of (22.10, 22.11), which are bounded in space; we pursue a non-rigorous argument.

First think of the solution at a point P (corresponding to j=0) a long way from the boundary, so that with h small the solution extends (almost) to infinity in both directions (j increasing and decreasing). If $|\mu| > 1$, as x increases, that is j tends to infinity, $\| \underset{\sim}{U}_j^n \| \to \infty$. Similarly, if $|\mu| < 1$ and the solution is finite and non-zero at P, then $\| \underset{\sim}{U}_j^n \| \to \infty$ as j decreases. Hence for a bounded solution, non-zero at P we must have $|\mu| = 1$. Now if $|\lambda| > 1$ the solution (22.12) will increase without limit as t increases. On requiring that these solutions, bounded in space, be also bounded in time we get the following stability criterion: if (22.12) is a solution of (22.10) with $\underset{\sim}{f}_j = \underset{\sim}{0}$ and if $|\mu| = 1$, we must have $|\lambda| \leq 1$ for $\Delta t \to 0$.

Substitution of (22.12) into (22.10) gives

$$C(\lambda,\mu)\underset{\sim}{V} \equiv \{\lambda(B_0+\mu B_1) - (A_0+\mu A_1)\}\underset{\sim}{V} = \underset{\sim}{0}, \tag{22.13}$$

and for a given value of λ, there are q corresponding values of μ found from det $C(\lambda,\mu) = 0$. The above criterion is applied to these q roots and shows that for all λ such that $|\lambda| > 1$ the corresponding roots μ_r must satisfy $|\mu_r| \neq 1$. Hence for all values of λ with modulus greater than unity there are a fixed number s of the μ_r with modulus less than unity, and the remaining (q-s) have modulus greater than unity.

We turn now to the boundary conditions, and seek a solution of (22.10) which is finite, non-zero at the boundary $x = 0$, and is bounded for all x, or equivalently for all j. This solution can be written as a sum over those μ_r with modulus less than unity,

$$\underset{\sim}{U}_j^n = \lambda^n \sum_{r=1}^{s} C_r \, \mu_r^j \, \underset{\sim}{V}_r \, , \tag{22.14}$$

if we assume that the eigenvalues μ_r are distinct and that $\underset{\sim}{V}_r$ is the eigensolution of (22.13) corresponding to eigenvalue μ_r. We want (22.14) to be a solution of the complete system (22.10, 22.11), so we substitute (22.14) into (22.11) to get the condition

$$\sum_{r=1}^{s} C_r \{\lambda(\beta_0+\mu_r\beta_1) - (\alpha_0+\mu_r\alpha_1)\} \, \underset{\sim}{V}_r = \underset{\sim}{g} \, . \tag{22.15}$$

For stability it must not be possible for small perturbations to stimulate a solution which increases indefinitely with t. So the homogeneous form of this equation must not have a non-zero solution for $|\lambda| > 1$.

Hence a second stability requirement, coming from the boundary condition is that the matrix $E(\lambda)$ of the equations (22.15) satisfies

$$\det E(\lambda) \neq 0 \, , \quad \text{for } |\lambda| > 1. \tag{22.16}$$

Boundary conditions at the other end of the range give rise to a similar condition.

The work by Kreiss and others mentioned above has extended this analysis to give necessary and sufficient conditions for stability, which we quote in the form given by Morton (1977, p.711).

Theorem Under a number of reasonable assumptions a difference scheme which is either non-dissipative or dissipative is stable if and only if

$$\det E(\lambda) \neq 0 \quad \text{for all } \lambda \text{ such that } |\lambda| \geq 1.$$

(Note: If the scheme is dissipative or non-dissipative it satisfies the von Neumann stability test).

These results for one space variable have been extended to two space variables by Kreiss (1970, 1973) for the analytic problem and in unpublished work by Kreiss and Oliger for the difference problem. In this case the above theorem is still true if slightly modified; stability still holds if det $E(\lambda_0) = 0$ for $|\lambda_0| = 1$ provided λ_0 is a multiple root of det $E(\lambda) = 0$ (Sundström, 1978).

Some results of this theory are that for the correct condition of reflection at a smooth boundary, when using either the leapfrog or Lax - Wendroff scheme in the interior of the region, either method A of Table 22.1 or method F or the box scheme can be used at the boundary to complete a stable discretization (Gustafsson, Kreiss and Sundström (1972).

22.5. CONCLUSIONS

The solution of hyperbolic partial differential equations is a very large subject calling for great skill analytically in setting up the equations to be solved in a well-posed way, and also knowledge of the latest theoretical results and practical methods for the numerical solution of problems in the particular field of interest.

For more information the standard text books, and particularly Richtmyer and Morton (1967), should be consulted. In preparing this chapter, we have drawn heavily on the survey article by Morton (1977) which should certainly be read before consulting individual papers, a large number of which are contained in its list of references.

23
NUMERICAL SOLUTION OF MIXED ELLIPTIC–HYPERBOLIC EQUATIONS

23.1. INTRODUCTION

A general second order (linear) partial differential equation, in two dimensions, is

$$au_{xx} + 2bu_{xy} + cu_{yy} + pu_x + qu_y + su = f, \quad (x,y) \in \Omega, \qquad (23.1)$$

where a,b,...,f are functions of x,y. Over Ω, the discriminant

$$\Delta = b^2 - ac$$

determines the type of equation (23.1); that is parabolic, hyperbolic or elliptic, see chapter 21. If Δ changes sign in Ω then equation (23.1) is a mixed elliptic-hyperbolic equation. The simplest example of a mixed equation is the Tricomi equation

$$yu_{xx} - u_{yy} = f, \quad (x,y) \in \Omega \equiv \{-1 \leq x, y \leq 1\}. \qquad (23.2)$$

For $y < 0$ (23.2) is elliptic (Δ negative), for $y > 0$ it is hyperbolic (Δ positive) while on the line $y = 0$ it is parabolic ($\Delta = 0$).

For mixed equations it is usually difficult, in general, to specify appropriate boundary conditions to ensure the existence and uniqueness of solutions. (Some results for equation (23.2) are given in Garabedian (1964).) However, some progress can be made from the theory of positive symmetric differential equations (Friedrichs, 1958) outlined in section 23.2.

Section 23.3 contains some recent numerical work (based on the finite element method) for the solution of mixed equations while section 23.4 briefly reviews earlier work using the finite-difference method. Finally section 23.5 describes an approach to the solution of the nonlinear mixed elliptic-hyperbolic partial differential equations that arise in transonic flow.

23.2. SYMMETRIC POSITIVE DIFFERENTIAL EQUATIONS

A unified treatment of elliptic, hyperbolic and mixed elliptic-hyperbolic equations is possible (up to a point) from the theory of symmetric positive differential equations introduced by Friedrichs (1958). For simplicity, the description is here confined to a pair of equations in two space variables, although the formalism is more generally applicable.

Consider the pair of first order equations (E, F, G 2×2 matrices and functions of x,y)

$$K\underset{\sim}{v} \equiv 2E\underset{\sim}{v}_x + 2F\underset{\sim}{v}_y + G\underset{\sim}{v} = \underset{\sim}{f}\ (x,y) \in \Omega \tag{23.3}$$

where $\underset{\sim}{v} = (v_1, v_2)^T$, $\underset{\sim}{v}_x = (v_{1x}, v_{2x})^T$ and $\underset{\sim}{v}_y = (v_{1y}, v_{2y})^T$. Then the operator K is symmetric positive if

$$\begin{aligned} &\text{(i)} \quad E = E^T,\ F = F^T, \\ &\text{(ii)} \quad Z + Z^T \text{ is positive definite } (Z = G - E_x - F_y). \end{aligned} \tag{23.4}$$

(By E_x is meant the matrix with elements $(E_{ij})_x$.)

For a symmetric positive operator, an appropriate boundary condition operator M can be specified such that (23.3) is well posed. Introduce the boundary matrix $B = n_1E + n_2F$, where $\underset{\sim}{n}$ = unit vector along the outward normal to Γ. Then, if $M = N - B$ where

$$\begin{aligned} &\text{(i)} \quad N + N^T \text{ is positive semidefinite,} \\ &\text{(ii)} \quad \ker(N - B) \oplus \ker(N + B) = R^2, \end{aligned} \tag{23.5}$$

equation (23.3) will be well-posed with the boundary conditions

$$M\underset{\sim}{v} = \underset{\sim}{0} \tag{23.6}$$

on Γ. For future reference note that

$$K\underset{\sim}{v} \equiv E\underset{\sim}{v}_x + F\underset{\sim}{v}_y + (E\underset{\sim}{v})_x + (F\underset{\sim}{v})_y + Z\underset{\sim}{v} .$$

The adjoint operator K^* is defined by

$$K^*\underset{\sim}{v} = - E\underset{\sim}{v}_x - F\underset{\sim}{v}_y - (E\underset{\sim}{v})_x - (F\underset{\sim}{v})_y + Z^T\underset{\sim}{v} , \qquad (23.7)$$

while the adjoint boundary operation M^* is

$$M^* = N^T + B .$$

Introducing the vector $\underset{\sim}{v} = (v_x, v_y)^T$, the Tricomi equation (23.2) can be rewritten as the pair of equations

$$\begin{pmatrix} y & 0 \\ 0 & 1 \end{pmatrix} \underset{\sim}{v}_x + \begin{pmatrix} 0 & -1 \\ -1 & 0 \end{pmatrix} \underset{\sim}{v}_y = \underset{\sim}{f} \qquad (23.8)$$

in Ω, with $\underset{\sim}{f} \equiv (f,0)^T$. Equation (23.8) satisfies the first condition of (23.4), but not the second. Premultiplying (23.8) by

$$\begin{pmatrix} b & cy \\ c & b \end{pmatrix}$$

leads to a first order system with (in the notation of (23.4))

$$Z \equiv \tfrac{1}{2} \begin{pmatrix} (c_y - b_x)y + c & b_y - c_x y \\ b_y - c_x y & c_y - b_x \end{pmatrix}$$

By an appropriate choice of b, c, Z can be made positive definite. For example,

$$b = - b_0 - b_1 x, \; c = c_0, \; b_0, c_0, b_1 \text{ constants}$$

gives

$$Z \equiv \tfrac{1}{2} \begin{pmatrix} b_1 y + c_0 & 0 \\ 0 & b_1 \end{pmatrix}$$

which is positive definite provided $b_1 > 0$ and $b_1 y + c_0 > 0$

over Ω.

The quadratic form for B in this case is

$$\underset{\sim}{v}B\underset{\sim}{v} = \frac{(b^2-c^2y)(n_2v_1-n_1v_2)^2-(n_2^2-yn_1^2)(bv_1+cv_2)^2}{2(bn_1+cn_2)} .$$

For this case it is easy to construct a matrix N which is positive semidefinite with

$$\underset{\sim}{v}N\underset{\sim}{v} = \frac{|b^2-c^2y|(n_2v_1-n_1v_2)^2+|n_2^2-yn_1^2|(bv_1+cv_2)^2}{2|bn_1+cn_2|} .$$

It can also be shown that N satisfies (ii) of (23.5). Thus the boundary condition $(N - B)\underset{\sim}{v} = \underset{\sim}{0}$ on Γ is admissible.

Katsansis (1969) has given a detailed discussion of the wide range of admissible boundary conditions for a mixed elliptic-hyperbolic equation (slightly more general than the Tricomi equation). However, the difficulty in practical situations is the determination of suitable admissible boundary conditions from the given physical information on the boundary. Similar manipulations can be used to recast an elliptic or hyperbolic equation as a symmetric positive differential operator (Friedrichs, 1958).

23.3. FINITE ELEMENT METHOD APPROACHES TO THE SOLUTION OF MIXED EQUATIONS

The finite element method has been used in two ways to obtain the numerical solution of mixed equations. In one approach, the mixed equation is first reformulated, along the lines of section 23.2, the resulting symmetric positive equation then being suitably discretized. Alternatively, the given mixed equation can be discretized directly.

An appropriate starting point for a finite element method uses the concept of the weak solution of a differential equation. Such a starting point is available for a symmetric positive differential equation, but the resulting finite element method would require the trial function space to explicitly satisfy the adjoint boundary condition (Friedrichs, 1958; Aziz and Leventhal, 1976; Aziz, Fix and Leventhal, 1976).

Instead a method based on an integral relation (the "firs identity", Friedrichs, 1958) has been proposed. By an integration by parts, using the definitions of (23.7),

$$(K\,\underset{\sim}{v},\ \underset{\sim}{w})_{\Omega} + (M\,\underset{\sim}{v},\ \underset{\sim}{w})_{\Gamma} = (\underset{\sim}{v},\ K^{*}\,\underset{\sim}{w})_{\Omega} + (\underset{\sim}{v},\ M^{*}\,\underset{\sim}{w})_{\Gamma}$$

where

$$(\underset{\sim}{w},\underset{\sim}{v})_{\Omega} = \iint_{\Omega} (w_1 v_1 + w_2 v_2)\ d\Omega$$

with a similar definition for $(\underset{\sim}{w},\underset{\sim}{v})_{\Gamma}$. Then, if $\underset{\sim}{v} \in \mathcal{H}_1(\Omega)$ is a solution of equation (23.3) subject to the admissible boundary condition (23.5), it satisfies

$$(K\,\underset{\sim}{v},\ \underset{\sim}{w})_{\Omega} + (M\,\underset{\sim}{v},\ \underset{\sim}{w})_{\Gamma} = (\underset{\sim}{f},\underset{\sim}{w})_{\Omega} \quad \text{for all } \underset{\sim}{w} \in \mathcal{H}_1(\Omega) \qquad (23.9)$$

To numerically estimate the solution $\underset{\sim}{v}$, introduce as usual a suitable expansion set $\underset{\sim}{\phi}^{(i)}$ and estimate the solution as

$$\underset{\sim}{v} \simeq \sum_{i=1}^{N} \alpha_i\, \underset{\sim}{\phi}^{(i)}$$

by finding the vector $\underset{\sim}{\alpha}$ from the linear equations

$$L\,\underset{\sim}{\alpha} = \underset{\sim}{g}$$

where

$$L_{ij} \equiv (K\,\underset{\sim}{\phi}^{(i)},\ \underset{\sim}{\phi}^{(j)})_{\Omega} + (M\,\underset{\sim}{\phi}^{(i)},\underset{\sim}{\phi}^{(j)})_{\Gamma}\ , \quad i,j = 1,2,\ldots,N$$

$$g_i = (\underset{\sim}{f},\ \underset{\sim}{\phi}^{(i)})_{\Omega}\ . \qquad i=1,2,\ldots,N$$

In Aziz and Leventhal (1976), and Aziz, Fix and Leventhal (1976) this method is applied to the partial differential equation

$$[G(x,y)u_x]_x + u_{yy} = f \quad \text{in } \Omega\ ,$$

for various G (which change sign in Ω). After appropriate manipulations to rewrite the partial differential equation as a symmetric positive equation, piecewise linear approximating functions on a suitable triangulation of Ω were used. The numerical results indicate an $O(h^2)$ convergence rate (in the $\mathcal{L}_2$ norm) of the computed results to the exact solution. (Note that the rate effectively refers to the components of the derivatives u_x, u_y.)

We now consider some direct approaches to finite element methods. Recently Fix and Gurtin (1977) have proposed a "patched variational approach" to the solution of a mixed equation. The region Ω is split into elliptic and hyperboic regions (Ω_E, Ω_H, $\Omega = \Omega_E \cup \Omega_H$) with boundary conditions given on the elliptic boundary Γ_E and the hyperbolic boundary Γ_H, together with an interface condition along the parabolic line ℓ.

Mathematically the original problem is

$$\mathrm{div}(A\,\underset{\sim}{\nabla}\underset{\sim}{u}) = f \text{ in } \Omega,$$

subject to the boundary conditions

$$u = 0 \quad \text{on } \Gamma_E ,$$

$$C_1\, \underset{\sim}{\nabla}\underset{\sim}{u} = \underset{\sim}{0} \quad \text{on } \Gamma_H \quad (\text{for a suitable } C_1).$$

In Ω_H, this is reformulated (via the transformation $\underset{\sim}{v} = T\underset{\sim}{\nabla}\underset{\sim}{u}$) and the problem considered as the coupled system:

$$\mathrm{div}(A\underset{\sim}{\nabla}\underset{\sim}{u}) = f \quad \text{in } \Omega_E ,$$

$$u = 0 \quad \text{on } \Gamma_E ,$$

$$(A\underset{\sim}{\nabla}\underset{\sim}{u})^T\underset{\sim}{n} = (A\,T^{-1}\,\underset{\sim}{v})^T\underset{\sim}{n} \quad \text{on } \ell ,$$

$$E\,\underset{\sim}{v}_x + F\,\underset{\sim}{v}_y + G\,\underset{\sim}{v} = \underset{\sim}{g} \quad \text{in } \Omega_H ,$$

$$C\,\underset{\sim}{v} = \underset{\sim}{0} \quad \text{on } \Gamma_H ,$$

$$C\,\underset{\sim}{v} = C\,T\,\underset{\sim}{\nabla}\underset{\sim}{u} \quad \text{on } \ell \text{ where } C_1 = CT .$$

This coupled problem is then solved using a standard elliptic finite element functional in Ω_E, together with a functional (similar to (23.9)) in Ω_H. Again the numerical results indicate $O(h^2)$ convergence for the components of the derivatives when linear trial functions are used.

In a similar vein, Trangenstein (1977) reformulates the Tricomi equation as an elliptic problem (with a non-local boundary condition along the parabolic line) and a hyperbolic problem. The underlying intention is to solve in the elliptic region and use the solution on the parabolic line as initial data in the hyperbolic region. No numerical results are quoted.

Fix and Gunzberger (1978) have proposed a least squares finite element approach for the solution of mixed equations. This has the attractive feature of requiring no information about the location of the parabolic line, but it does require rewriting a second order equation as a system of first order equations (or using a sufficiently continuous basis in the finite element method).

23.4. FINITE-DIFFERENCE APPROACHES TO THE SOLUTION OF MIXED EQUATIONS

In the symmetric positive differential equation approach, the equation is discretised in a manner which produces a system of difference equations with a positive definite matrix. In Schecter (1960) and Friedrichs (1958), such a differencing is applied directly to a symmetric positive equation while Katsansis (1969) first integrates the symmetric positive equation over a suitable mesh and then differences this integrated form. In this latter case, the theory would predict an $\mathcal{L}_2$ convergence rate (for the derivatives) of $O(h^\mu)$, $\mu \leq \frac{1}{2}$, while practical results (see Aziz, Fix and Leventhal, 1976) indicate $\mu = 0.2$.

In a direct approach, the mixed second order partial differential equation is discretised directly using suitable difference schemes in the elliptic and hyperbolic regions and an appropriate matching scheme across the parabolic line.

In Ogawa (1961) a difference scheme, based on the knowledge of the characteristics in the hyperbolic region, is proposed together with convergence results. However, the scheme is only applicable to a restricted class of mixed equations and domains.

A more general direct scheme has been borrowed from non-linear transonic calculations (Murman and Cole, 1971) see section 23.5. Central differences are used in the elliptic region in both x- and y-directions, while in the hyperbolic region one sided differencing is used in one direction with central differences in the other. This effectively compromises with accuracy to achieve stability. Results in Fix and Gurtin (1977) indicate an O(h) convergence rate for the $\mathcal{L}_2$ norm of the error in the derivatives.

23.5. NONLINEAR MIXED EQUATIONS

An important practical application of a nonlinear mixed elliptic-hyperbolic equation arises in transonic flow over an aerofoil surface. Initially the flow is subsonic (local mach number $M < 1$, see Figure 23.1), a supersonic patch ($M > 1$)

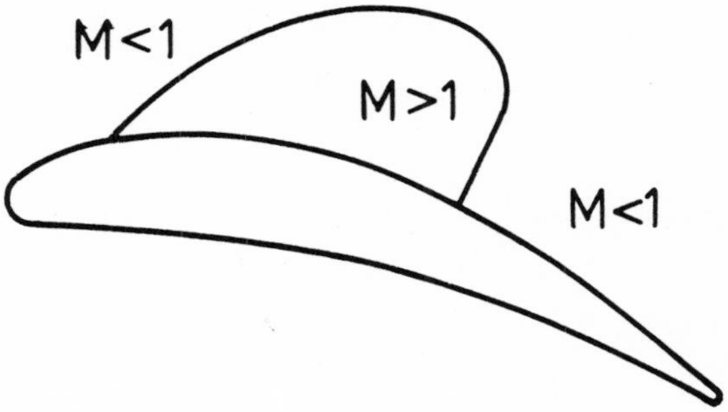

Fig. 23.1.

develops on the aerofoil surface and finally the fluid flows subsonically.

For a suitable fluid, the above situation (in terms of the velocity potential ϕ) is described by the potential equation,

$$(a^2-u^2)\ \phi_{xx} - 2uv\ \phi_{xy} + (a^2-v^2)\ \phi_{yy} = 0 \qquad (23.10)$$

where $u = \phi_x$, $v = \phi_y$ and a is the local speed of sound. The local mach number $M = q/a$ $(q^2 = u^2 + v^2)$. Computationally a continuous solution ϕ with (possibly) discontinuous derivatives is sought.

A useful, approximation, retaining the essential features of (23.10), is the transonic small perturbation equation

$$A\phi_{xx} + \phi_{yy} = 0, \quad A = K - (\gamma+1)\,\phi_x\,, \quad K,\ \gamma \text{ constants.} \qquad (23.11)$$

For details of the derivation of equations (23.10), (23.11) see Jameson (1976a,b) and for a review of transonic flow, see Hall (1976).

Numerical schemes to handle shocks include an artificial viscosity term (to dampen out instabilities which otherwise might occur, see Potter, 1973). This is the essential idea of the Murman-Cole scheme mentioned earlier.

Consider first the transonic small perturbation equation. Ignoring, for the present, nonlinear aspects, then suitable difference approximations at the point $(i\Delta x,\ j\Delta y)$ are

$$p_{ij} = A_{ij}\,\delta_x^2\,\phi_{ij}/\Delta x^2, \quad A_{ij} = K - (\gamma+1)\,\frac{\phi_{i+1,j}-\phi_{i-1,j}}{2\Delta x}\,,$$

$$q_{ij} = \delta_y^2\,\phi_{ij}/\Delta_y^2\,.$$

The Murman-Cole scheme introduces a switching function μ, such that

$$\mu_{ij} = \begin{cases} 0 & A_{ij} > 0 \text{ (elliptic)} \\ 1 & A_{ij} < 0 \text{ (hyperbolic)} \end{cases}$$

and replaces equation (23.11) by the difference scheme

$$p_{ij} + q_{ij} - [\mu_{ij}\,(p_{ij} - p_{i-1,j})] = 0\,. \qquad (23.12a)$$

In the supersonic region, the bracketed term can be considered as an approximation to

$$-\ \mu\Delta x\,\frac{\partial}{\partial x}\,(A\phi_{xx}) = -\ \mu\Delta x\,(A\phi_{xxx} - (\gamma+1)\,\phi_{xx}^2)$$

The first term here acts like an artificial viscosity (of order Δx).

Alternatively, equation (23.12a) can be interpreted (as in section 23.3) as using central differences everywhere in the y-direction and one sided differences in the x-direction for supersonic flow (with central differences in the x-direction for subsonic flow). While these ideas (with a suitable iterative scheme for the nonlinear features) can be used for transonic computations, the physically correct type of shocks are not obtained. The remedy in this case is straightforward, and merely requires the addition of the artificial viscosity in conservative form resulting in the scheme

$$p_{ij} + q_{ij} - \mu_{ij}\, p_{ij} + \mu_{i-1,j}\, p_{i-1,j} = 0 . \qquad (23.12b)$$

This effectively smears the shock over two adjacent mesh points. The development of suitable iterative schemes for use with equation (23.12b) and the results obtained can be found in Jameson (1976 a,b).

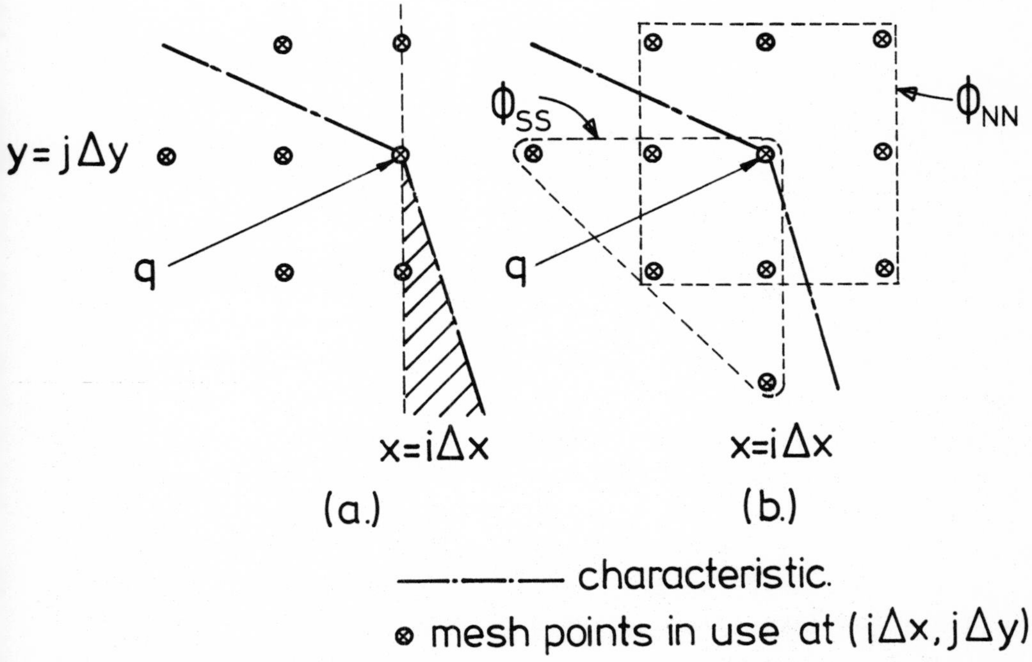

Fig. 23.2.

The construction of appropriate difference schemes for the potential equation (23.10) is more difficult since the characteristics are no longer (locally) bisected by the direction of the x-axis (see Figure 23.2a). If the direct analogue of the Murman-Cole scheme were used at a supersonic point, then a computational mesh as in Figure 23.2a would result with the domain of dependence of the numerical scheme (i.e. $x \leq i\Delta x$) not including that of the differential equation (the shaded area is missing). Jameson (1976) proposes a locally rotated difference scheme, based on coordinates s,n locally aligned with and normal to the flow direction respectively. In this system, equation (23.10) becomes

$$(a^2-q^2)\,\phi_{ss} + a^2\,\phi_{nn} = 0,$$

where

$$\phi_{ss} = \frac{1}{q^2}\,(u^2\,\phi_{xx} + 2uv\,\phi_{xy} + v^2\,\phi_{yy}),$$

$$\phi_{nn} = \frac{1}{q^2}\,(v^2\,\phi_{xx} - 2uv\,\phi_{xy} + u^2\,\phi_{yy}).$$

All the derivatives in ϕ_{ss} are replaced by backward differences, while those in ϕ_{nn} are replaced by central differences giving the computational molecule of Figure 23.2b which now has a numerical domain of dependence including that of the equation. This scheme is again equivalent to the introduction of an artificial viscosity, and again it is necessary to add the viscosity in conservative form to capture the correct type of shock. The ideas of this rotated scheme are used to model the appropriate form of the viscosity terms, and the implementation details are in Jameson (1976).

To date, it appears that the finite element method has been little used in transonic flow computations. Carey (1978) has formulated variational principles for equations (23.10), (23.11) and reports the development of a related implementation.

An optimal control formalism based on the finite element method has also been proposed for transonic flow (Glowinski, 1976, Hunt, 1979).

REFERENCES

ABBOTT, J.P. (1977). Numerical continuation methods for non-linear equations and bifurcation problems, Ph.D. thesis, Computer Centre, A.N.U. Canberra.

ADAMS, G.K. & COOK, G.B. (1960). The effect of pressure on the mechanism and speed of the hydrazine decomposition flame. *Comb. and Flame*, 4, 9-18.

ADAMS, R.A. (1975). *Sobolev Spaces*, Academic Press, New York.

AGMON, S. (1965). *Lectures on Elliptic Boundary Value Problems*, Van Nostrand, New Jersey.

AHLFORS, L.V. (1953). *Complex Analysis*. McGraw-Hill, New York.

AITCHISON, J.M. (1972). Numerical Treatment of the Singularity in a Free Boundary Problem, *Proc.Roy.Soc.A*, 330, pp. 573-580.

——— (1977). The Numerical Solution of a Minimisation Problem Associated with a Free Surface Flow, *J.Inst. Math.Applics.*, 20, pp. 34-44.

ALBRECHT, J., COLLATZ, L. & HAMMERLIN, G. (eds.) (1978). *Numerische Behandlung von Differentialgleichungen mit besonderer Berücksichtigung Greier Randwertaufgaben*, Int. Ser. Num. Math., 39, Birkhauser Verlag, Basel.

ALEXANDER, R. (1977). Diagonally Implicit Runge-Kutta Methods for Stiff O.D.E.'s, *SIAM J.Numer.Anal.* 14, pp. 1006-1021.

ALLAMI, K. & THOMAS, D.P. (1975). A numerical investigation of acoustic diffraction by a circular aperture in a rigid screen and some related problems. N.A. Rept. 10, Dundee University.

AMES, W.F. (1969). *Numerical Methods for Partial Differential Equations*, Nelson, London.

ANSELONE, P.M.(1971). *Collectively compact operator approximation theory and applications to integral equations*. Prentice-Hall, Englewood Cliffs, N.J.

ARMIJO, L. (1966). Minimisation of functions having Lipschitz continuous first partial derivatives, *Pacific J. Math.*, 16, pp. 1-3.

ARTHURS, A.M. (1970). *Complementary Variational Principles*, Clarendon Press, Oxford.

ATKINSON, K.E. (1966). Extension of the Nyström method for the numerical solution of linear integral equations of the second kind. MRC Tech. Summ. Rept. 686, University of Wisconsin, Madison.

——— (1976). *A survey of numerical methods for the solution of Fredholm integral equations of the second kind.* SIAM Publications, Philadelphia.

ATTHEY, D.R. (1974). A finite-difference scheme for melting problems. *J.Inst.Math.Applics.* 13, pp. 353-366.

——— (1975). A finite-difference scheme for melting problems based on the method of weak solutions. pp. 182-191 of Ockendon, J.R. and Hodgkins, W.R. (eds.) 1975.

AZIZ, A.K. ed. (1972). *The Mathematical Foundations of the Finite Element Method with Applications to Partial Differential Equations.* Academic, New York.

———, FIX, G. & LEVENTHAL, S. (1976). On the Numerical Solution of Linear Mixed Problems, in Glowinski, R. and Lions, J.L. (eds.) (1976).

——— & LEVENTHAL, S. (1976). Numerical Solution of Linear Partial Differential Equations of Elliptic-Hyperbolic Type, in Hubbard, B.E. (ed.) (1976).

BABUSKA, I. & RHEINBOLDT, W.C. (1977a). Error Estimates for Adaptive Finite Element Computations, *SIAM J.Numer.Anal.*, 15, pp. 736-754.

——— ——— (1977b). Computational Aspects of the Finite Element Method, in Rice, J.R. (ed.) (1977).

BAIOCCHI, C., COMINCIOLI, V., MAGENES, V. & POZZI, G.A. (1973). Free Boundary Problems in the Theory of Fluid Flow through Porous Media: Existence and Uniqueness Theorems, *Ann.Mat. Pure et Appl.*, 96, pp. 1-82.

——— ——— ——— ——— (1974). Fluid Flow through Porous Media: a New Theoretical and Numerical Approach. Publication No.69. Laboratorio di Analisi Numerica. Pavia.

BAKER, C.T.H. (1977). *The numerical treatment of integral equations.* Clarendon Press, Oxford.

BARKER, V.A. (Ed.) (1976). *Sparse Matrix Techniques.* Springer-Verlag Lecture Notes in Mathematics, 572, Berlin.

BARNHILL, R.E. (1976). Blending Function Finite Elements for Curved Boundaries, in Whiteman, J.R. (ed.) (1976a).

BARNHILL, R.E. & WHITEMAN, J.R. (1973). Error Analysis of Finite Element Methods with Triangles for Elliptic Boundary Value Problems, pp. 83-111 in Whiteman, J.R. (ed.) (1973).

BARNSLEY, M.F. & ROBINSON, P.D. (1977). Bivariational bounds for nonlinear problems. *J.Inst.Math.Applics*. 20, pp. 485-504.

BARRODALE, I. & YOUNG, A. (1970). Computational experience in solving linear operator equations using the Chebyshev norm, pp. 115-142 of Hayes, J.G. (ed.) (1970).

BARTELS, R. & DANIEL, J.W. (1974). A conjugate gradient approach to nonlinear elliptic boundary value problems in irregular regions, pp. 1-11 of Watson G.A. (ed.) (1974).

BATES, R.H.J. (1969). Theory of the Point Matching method for perfectly conducting waveguides and transmission lines, *I.E.E.E. Trans*. MIT. 17, p. 194.

BATHE, K-J. & KHOSHGOFTAAR, M.R. (1978). Finite Element Free Surface Seepage Analysis without Mesh Iteration, *Int.J. Num.Meth. and Anal.Meth. in Geomechanics*, 1, (in press).

BEAR, J. (1972). *Dynamics of Fluids in Porous Media*. Elsevier, New York.

BERGMAN, S. & SCHIFFER, M. (1953). *Kernel functions and elliptic differential equations in mathematical physics*. Academic, New York.

BERNAL, M.J.M. & WHITEMAN, J.R. (1970). Numerical treatment of biharmonic boundary value problems with re-entrant boundaries, *Comput.J.*, 13, pp. 87-91.

BIRKHOFF, G. & VARGA, R.S. (1959). Implicit alternating direction methods, *Trans.Amer.Math.Soc*. 92, pp.13-24.

BLEDJIAN, L. (1973). Computation of time-dependent laminar flame structure. *Comb. and Flame*, 20, pp. 5-17.

BLUE, J.L. (1977). Boundary integral solutions of Laplace's equation. Comp. Sci. Tech. Rept. No. 60, Bell Laboratories, Murray Hill, N.J.

BOLEY, B.A. (1975). The embedding technique in melting and solidification problems. pp. 150-172 of Ockendon, J.R. and Hodgkins, W.R. (eds.) (1975).

BONACINA, C., COMINI, G., FASENO, A. & PRIMICERIO, M. (1973). Numerical solution of phase-change problems. *Int.J.Heat Mass Transfer* 16, pp. 1825-1832.

BONNEROT, R. & JAMET, P. (1974). A second-order finite-element method for the one-dimensional Stefan problem. *Int.J. Num.Meth.Engng*. 8, pp. 811-820.

BONNERTOT, R. & JAMET, P. (1977). Numerical computation of the free boundary for the two-dimensional Stefan problem by space-time finite elements. *J.Comp.Phys.* 25, pp. 163-181.

BOURGAT, J.F. (1976). Numerical study of a dual iterative method for solving a finite element approximation of the biharmonic equation, *Comp.Math.Appl.Mech.Eng.*, 9, pp. 203-218.

BRAMBLE, J.H. (ed.) (1966). *Numerical Solution of Partial Differential Equations*, Academic, New York.

——— & HILBERT, S.R. (1970). Estimation of linear functionals on Sobolev spaces with application to Fourier transforms and spline interpolation. *SIAM J. Numer.Anal.*, 7, pp. 113-124.

——— & HUBBARD, B.E. (1968). Effects of boundary regularity on the discretization error in the fixed membrane eigenvalue problem. *SIAM J.Numer.Anal.*, 5, pp. 835-863.

BRANDT, A. (1973). Multi-level Adaptive Technique (MLAT) for Fast Numerical Solution to Boundary Value Problems, in Cabannes, H. & Temam, R. (eds.) (1973).

——— (1977). Multi-level Adaptive Solutions to Boundary Value Problems, *Math.Comp.*, 31, pp. 333-390.

BRIGHAM, E.O. (1974). *The Fast Fourier Transform*, Prentice Hall, Englewood Cliffs, N.J.

BROWN, J.H. (1976). Conforming and Non-conforming Finite Element Methods. Ph.D. Thesis, University of Dundee.

BROYDEN, C.G. (1965). A class of methods for solving non-linear simultaneous equations, *Math.Comp.*, 19, pp. 577-593.

——— (1971). The convergence of an algorithm for solving sparse non-linear systems, *Math.Comp.*, 25, pp. 285-294.

BULL, D.C. & QUINN, C.P. (1975). An interferometric study of autoignition at hot surfaces. Proc. of 2nd European Symposium on Combustion. (Published by) Sect. Francaise du Combustion Institut, pp. 551-555.

BUNCH, J.R. & PARLETT, B.N. (1971). Direct Methods for Solving Symmetric Indefinite Systems of Linear Equations, *SIAM.J.Numer.Anal.*, 8, pp. 639-655.

BUNEMAN, O. (1969). A compact non-iterative Poisson solver, Report SUIPR-294, Inst. Plasma Research, Stanford University.

BURTON, A.J. (1974). Numerical solution of scalar diffraction problems, chapter 21 of Delves, L.M. and Walsh, J.E. (eds.) 1974.

BUTLER, D.S. (1960). The numerical solution of hyperbolic systems of partial differential equations in three independent variables. *Proc.Roy.Soc.A*, 255, pp. 232-252.

BUTCHER, J.C. (1964). Implicit Runge-Kutta Processes, *Math. Comp*. 18, pp. 50-64.

BUZBEE, B.L. & DORR, F.W. (1974). The direct solution of the biharmonic equation on rectangular regions and the Poisson equation on irregular regions, *SIAM J.Numer.Anal.*, 11, pp. 753-763.

——— ———, GEORGE, J.A. & GOLUB, G.H. (1971). The direct solution of the discrete Poisson equation on irregular regions, *SIAM J.Numer.Anal.*, 8, pp. 722-736.

———, GOLUB, G.H. & NIELSON, C.W. (1970). On direct methods for solving Poisson's equations, *SIAM J.Numer. Anal.*, 7, pp. 627-656.

CABANNES, H. & TEMAN, R.(eds.) (1973). *Third International Conference in Numerical Methods in Fluid Mechanics*, Lecture Notes in Physics, 18, Springer, Berlin.

CANNON, J.R. (1967). Determination of the unknown coefficient k(u) in the equation $\nabla k(u)\nabla u = 0$ from overspecified boundary data. *J.Math.Anal.Appl*. 18, pp. 112-114.

——— (1968). Determination of an unknown heat source from over-specified boundary data. *SIAM.J.Numer.Anal*. 5, pp. 275-286.

CAREY, G.F. (1978). Variational Principles for the Transonic Airfoil Problem, *Comp.Meth.Appl.Mech.Engng.*, 13, pp. 129-140.

CARRÉ, B.A. (1961). The determination of the optimum accelerating factor for successive over-relaxation. *Comput.J*. 4, pp. 73-78.

CARRIER, G.F. & PEARSON, C.E. (1976). *Partial Differential Equations, Theory and Technique*. Wiley, New York.

CARSLAW, H. & JAEGER, J. (1959). *The conduction of heat in solids*. Oxford University Press, Oxford.

CAVENDISH, J.C., PRICE, H.S. & VARGA, R.S. (1969). Galerkin methods for the numerical solution of boundary value problems, *Soc.Petrol.Eng.J.*, 246, pp. 204-220.

CAVIALDINI, J.F. (1975). Analyse numerique d'un probleme de Stefan a deux phases par une méthode l'élements finis. *SIAM J.Numer.Anal.* 12, pp. 464-487.

CHENG, R.T., Finite element solution of a class of problems involving a corner singularity. Manuscript.

CHRISTIE, I., GRIFFITHS, D.F., MITCHELL, A.R. & ZIENKIEWICZ, O.C. (1976). Finite element methods for second order differential equations with significant first derivatives, *Int.J.Num.Meth.Engng.*, 10, pp.1389-1396.

CHURCHILL, R.V. (1960). *Complex Variables and Applications*, (3rd ed.) McGraw-Hill, New York.

CIARLET, P.G. (1974). Conforming and nonconforming finite element methods for solving the plate problem, pp. 21-32 of G.A. Watson (ed.) (1974).

——— (1976). *Numerical Analysis of the Finite Element Method*, Presse de la Universite de Montreal, Montreal.

——— (1977). *Introduction to the Numerical Analysis of the Finite Element Method*, North Holland, Amsterdam.

——— (1978). *The Finite Element Method for Elliptic Problems*, North Holland, Amsterdam.

——— & RAVIART, P.A. (1972). General Lagrange and Hermite interpolation in R^n with applications to finite element methods. *Arch.Rat.Mech.Anal.*, 46, pp. 177-199.

COLE, P. (1977). A new Green's function for solving the exterior Neumann problem of accoustics for an open arc. Ph.D. thesis, Manchester University.

COLLATZ, L. (1960). *The Numerical Treatment of Differential Equations*. (3rd ed.), Springer, Berlin.

——— (1977). Private communication to Rosser and Whiteman.

COLTON, D.L. (1976). *Solution of boundary value problems by the method of integral operators*. Pitman, London.

COMINI, G., DEL GUIDICE, S., LEWIS, R.W. & ZIENKIEWICZ, O.C. (1975). Finite element solution of non-linear heat conduction problems with special reference to phase change *Int.J.Num.Math.Engng.* 8, pp. 613-624.

CONCUS, P. & GOLUB, G.H. (1973). Use of fast direct methods for efficient numerical solution of nonseparable elliptic equations, *SIAM J.Numer.Anal.*, 10, pp. 1103-1120.

COOLEY, J.W. & TUKEY, J.W. (1965). An algorithm for machine calculation of complex Fourier series, *Math.Comp.*, 19, pp. 297-301.

COPSON, E.T. (1955). *An Introduction to the Theory of Functions of a Complex Variable*, Oxford University Press, Oxford.

COURANT, R., FRIEDRICHS, K.O. & LEWY, H. (1928). Uber die Partiellen Differenzengleichungen der Mathematischen Physik, *Math.Ann.*, 100, pp. 32-74.

——— & HILBERT, D. (1953). *Methods of Mathematical Physics*. Vol. I. Interscience, London.

——— ——— (1962). *Methods of Mathematical Physics*. Vol. II. Interscience, London.

COX, M.G. (1978). The Numerical Evaluation of a Spline from its B-spline Representation, *J.Inst.Math.Applics*. 21, pp. 135-143.

CRANK, J. (1975). Finite-difference methods. pp. 192-207 of Ockendon J.R. and Hodgkins, W.R. (eds.) (1975).

——— & CROWLEY, A.B. (1978). Isotherm migration along orthogonal lines in two dimensions. *Int.J.Heat Mass Transfer*, 21, pp. 393-398.

——— & FURZELAND, R.M. (1976). Modified finite-difference approximations near the singularity in the Motz problem, Technical Report TR/60, Dept. of Maths., Brunel University.

——— & GUPTA, R.S. (1974). Isotherm migration method in two dimensions. Technical Report TR/42, Dept. of Maths. Brunel University.

——— & PHAHLE, R.D. (1973). Melting ice by the isotherm migration method. *Bull.Inst.Maths. Applics.*, 9, pp. 12-14.

CROWLEY, A.B. (1975). Differential equations with industrial applications. D.Phil. thesis, University of Oxford.

——— (1978). Numerical solution of Stefan problems. *Int.J.Heat Mass Transfer*, 21, pp. 215-219.

CRYER, C.W. (1970a). The solution of the Dirichlet problem for Laplace's equation when the boundary data is discontinuous and the domain has a boundary which is of bounded variation by means of the Lebesque-Stieltjes integral equation for the double-layer potential. Comp. Sci.Tech.Rep. No. 99, University of Wisconsin, Madison.

——— (1970b). On the Approximate Solution of Free Boundary Problems using Finite Differences. *J.Assoc. Comput.Mach.*, 17, pp. 397-411.

——— (1971). The Solution of a Quadratic Programming Problem using Systematic Overrelaxation. *SIAM J. Control*, 9, pp. 385-392.

——— (1977). A Bibliography of Free Boundary Problems, MRC. Tech.Summ.Rept. 1793. Mathematics Research Center, University of Wisconsin, Madison.

CULHAM, W.E. & VARGA, R.S. (1971). Numerical methods for time dependent, non-linear boundary value problems, *Soc.Petrol.Eng.J.*, 248, pp. 374-388.

CURTIS, A.R., POWELL, M.J.D. & REID, J.K. (1974). On the estimation of sparse Jacobian matrices, *J.Inst.Maths. Applics.*, 13, pp. 117-119.

——— & REID, J.K. (1974). The choice of step lengths when using differences to approximate Jacobian matrices. *J.Inst.Math.Applics.*, 13, pp. 121-126.

CURTIS, P.E.M. (1977). Numerical Analysis of Stefan problems. D.Phil.thesis, University of Oxford.

CUTHILL, E. (1972). Several Strategies for Reducing the Band-width of Matrices, pp. 157-166 of Rose, D.J. & Willoughby, R.J. (eds.) (1972).

——— & McKEE, J. (1969). Reducing the bandwidth of sparse symmetric matrices. *Proc.24th National Conf. ACM*, Brandon Systems Press, N.J. pp. 157-172.

DAVIES, B. & HENDRY, J.A. (1977). The indirect design of fluid channels using a Global Variational method with trial functions not satisfying the prescribed boundary conditions. *Int.J.Num.Meth.Engng.*, 11, pp. 579-591.

DEKKER, K., VAN DER HOUWEN, P.J., VERWER, J.G. & WOLKENFELT, P.H.M. (1977). Comparing Stabilized Runge-Kutta Methods for Semi-Discretized Parabolic and Hyperbolic Equations. Report NW 45/77. Mathematische Centrum, Amsterdam.

DELVES, L.M. (1977). On the Solution of sets of linear equations arising from Galerkin methods. *J.Inst.Math.Applics.* 10, pp. 163-171.

——— & HALL, C.A. (1976). An implicit matching principle for Global Element Calculations, Research Rept. CSS/76/10/3, University of Liverpool; submitted for publication.

——— & PHILLIPS, C. (1977). A Fast Implementation of the Global Element Method, Research Rept. CSS/76/11/2, University of Liverpool; submitted for publication.

——— & PHILLIPS, C. (1978). Numerical Experience with the Elliptic PDE program GEM2, Research Report CSS/78/9/1, University of Liverpool.

——— & WALSH, J.E. (Eds.) (1974). *Numerical solution of integral equations*. Clarendon Press, Oxford.

DEWEY, C.F. & GROSS, J.F. (1967). Exact similar solutions of the laminar boundary-layer equations. *Advances in Heat Transfer*, 4, pp. 317-446. Academic Press, New York.

D'JAKANOV, E.G. (1969). On certain iterative methods for solving nonlinear difference equations, pp. 7-22 of Morris, J.L. (ed.) (1969).

DONNELLY, J.D.P. (1969). Eigenvalues of membranes with re-entrant corners. *SIAM.J.Numer.Anal.* 6, pp. 47-61.

DORR, F.W. (1970). The direct solution of the Poisson equation on a rectangle, *SIAM Rev.*, 12, pp. 248-263.

DOUGLAS, J. Jnr. (1962). Alternating direction methods for three space variables, *Numer.Math.*, 4, pp. 41-63.

——— (1972). A Superconvergence Result for the approximate solution of the Heat Equation by a Collocation Method, pp. 475-490 of Aziz, A.K. (ed.) (1972).

——— & DUPONT, T. (1971). Alternating Direction Galerkin Methods on Rectangles, pp. 133-214 of Hubbard, B. E. (ed.) (1971).

——— & GUNN, J.E. (1964). A general formulation of alternating direction methods. *Numer.Math.* 6, pp. 428-453.

——— & RACHFORD, H.H. Jnr. (1956). On the numerical solution of heat conduction problems in two and three space variables, *Trans.Amer.Math.Soc.* 82, pp. 421-439.

DUDA, J.L., MALONE, M.F. & NOTTER, R.H. (1975). Analysis of two-dimensional diffusion-controlled moving boundary problems. *Int.J.Heat Mass Transfer* 18, pp. 901-910.

ELLIOTT, C.M. (1976). Some applications of the finite-element method in numerical analysis. D.Phil.thesis, University of Oxford.

——— (1978a). Moving Boundary Problems and Linear Complementarity, in Albrecht J. *et al.* (eds.) (1978).

——— (1978b). On the numerical solution of a class of moving boundary problems posed as parabolic variational inequalities. *J.Inst.Math.Applics.*, to appear.

ELVIUS, T. (1971). Numerical solution of a problem in potential theory using integral equations. Uppsala University, Dept. of Computer Science, Report 32.

ENRIGHT, W.H., HULL, T.E. & LINDBERG, B. (1975). Comparing Numerical Methods for Stiff Systems of O.E.E:s, *BIT*, 15, pp. 10-48.

EVANS, H.L. (1968). *Laminar boundary-layer theory*, Addison-Wesley, Reading, Mass.

FAIRWEATHER, G. (1978). *Finite Element Galerkin Methods for Differential Equations*, Lecture Notes in Pure and Applied Mathematics, 34, Marcel Dekker Inc., Basel.

FALK, R.S. (1974). Error Estimates for the Approximation of a Class of Variational Inequalities, *Math.Comp.* 28, pp. 963-971.

FALKNER, V.M. & SKAN, S.W. (1931). Some approximate solutions of the boundary-layer equations. *Phil.Mag.*, 12, pp. 865-896.

FERRIS, D.H. (1975). Fixation of a moving boundary by means of a change of independent variable. pp. 251-255 of Ockendon, J.R. and Hodgkins, W.R. (eds.) (1975).

FEDERENKO, R.P. (1962). A relaxation method for solving elliptic difference equations. *USSR Comp.Math. and Math.Phys.* 1, pp. 1092-1096.

FISCHER, D., GOLUB, G., HALD, O., LEIVA, C. & WIDLUND, O. (1974). On Fourier-Toeplitz methods for separable elliptic problems, *Math.Comp* ., 28, pp. 349-368.

FIX, G.J., GULATI, S. & WAKOFF, G.I., (1973). On the Use of Singular Functions with Finite Element Approximations. *J.Comp.Phys.*, 13, pp. 209-

——— & GUNZBERGER, M.D. (1978). On Least Squares Approximations to Indefinite Problems of Mixed Type, *Int.J.Num.Methods Engng.*, 12, pp. 453-470.

——— & GURTIN, M.E. (1977). On Patched Variational Principles with Application to Elliptic and Mixed Elliptic-Hyperbolic Problems, *Num.Math.*, 28, pp. 259-271.

FORSYTHE, G.E. & WASOW, W.R. (1960). *Finite difference methods for partial differential equations*. Wiley, New York.

FOX, L. (1950). The numerical solution of elliptic differential equations when boundary conditions involve a derivative, *Phil.Trans.Roy.Soc.A*, 242, pp. 345-378.

——— (ed.) (1962). *Numerical solution of ordinary and partial differential equations*. Pergamon Press, Oxford.

——— (1971). Some experiments with singularities in linear elliptic partial differential equations. *Proc.Roy.Soc.A*, 323, pp. 179-190.

——— (1975). What are the best numerical methods? pp. 210-214 of Ockendon, J.R. and Hodgkins, W.R. (eds.) (1975).

——— (1977). Finite-difference Methods for Elliptic Boundary-Value problems, pp. 799-881 of Jacobs, D.A.H. (ed.) (1977).

FOX, L., HENRICI, P. & MOLER, C. (1967). Approximations and bounds for eigenvalues of elliptic operators. *SIAM J. Numer. Anal.*, 4, pp. 89-102.

——— & MAYERS, D.F. (1968). *Computing methods for scientists and engineers*. Clarendon Press, Oxford.

——— & SANKAR, R. (1969). Boundary singularities in linear elliptic partial differential equations. *J.Inst. Maths.Applics.*, 5, pp. 340-350.

——— ——— (1973). The Regula-Falsi Method for Free-Boundary Problems, *J.Inst.Math.Applics.*, 12, pp. 49-54.

——— & SOUTHWELL, R.V. (1945). Biharmonic analysis applied to the flexure and extension of flat elastic plates. *Phil.Trans.Roy.Soc.A*, 239, pp. 419-460.

FREDERICKSON, P.O. (1975). Fast approximate inversion of large sparse linear systems. Maths. report 7-75, Lakehead University, Thunder Bay, Ontario.

FRIEDMAN, A. (1964). *Partial Differential Equations of Parabolic Type*, Prentice-Hall, Englewood Cliffs, N.J.

FREIDRICHS, K.O. (1958). Symmetric Positive Differential Equations, *Comm.Pure Appl.Math.*, 11, pp. 333-418.

FROMM, J.E. (1968). A method for reducing dispersion in convective difference schemes, *J.Comp.Phys.* 3, pp. 176-189.

FURZELAND, R.M. (1977a). The numerical solution of partial differential equations with boundary singularities and moving boundaries. Ph.D. thesis, Brunel University.

——— (1977b). A Survey of the Formulation and Solution of Free and Moving Boundary (Stefan) Problems. Tech.Rept. TR/76. Dept. of Mathematics, Brunel University.

——— (1977c). The numerical solution of two-dimensional moving boundary problems using curvilinear coordinate transformations. Tech. Rept. TR/77. Dept. of Mathematics, Brunel University.

FUSSELL, D.D. & HELLUMS, J.D. (1965). The numerical solution of boundary-layer problems. *A.I.Ch.E. Journal*, 11, pp. 733-739.

GAKHOV, F.D. (1966). *Boundary value problems*. Pergamon, Oxford.

GALLAGHER, R.H. (1975). *Finite Element Analysis*. Prentice Hall, New Jersey.

GARABEDIAN, P.R. (1964). *Partial Differential Equations*, Wiley, New York.

GAWLEY, R. & DELVES, L.M. (1973). A Global Variational Algorithm for Boundary Value Problems Over Simply Connected Regions. Research Report, University of Liverpool, unpublished.

GAY, D.M. (1977). Some convergence properties of Broyden's method, Working paper no. 175, N.B.E.R., Cambridge, Massachusetts.

GEAR, C.W. (1971). *Numerical Initial Value Problems in Ordinary Differential EQuations*. Prentice-Hall, Englewood Cliffs, N.J.

GEORGE, J.A. (1973). Nested dissection of a regular finite element mesh. *SIAM J.Numer.Anal.* 10, pp. 345-363.

——— (1976). Solution of linear systems of equations: direct methods for finite element problems. pp. 52-101 of Barker V.A. (ed.) (1976).

——— & LIU, J.W.H. (1975a). An automatic partitioning and solution scheme for solving large sparse positive definite systems of linear algebraic equations. University of Waterloo, Ontario, Report CS-75-17.

——— (1975b). Some results on fill-in for sparse matrices. *SIAM J.Numer. Anal.* 12, pp. 452-455.

——— (1976a). On the application of the minimum degree algorithm to finite element systems. University of Waterloo, Ontario, Report CS-76-16.

——— ——— (1976b). An Automatic Nested Dissection Algorithm for Irregular Finite Element Problems. University of Waterloo Report CS-76-38. To appear in *SIAM J. Numer. Anal.*

GERDES, W. (1978). Die Lösung des Angangs-Randwertproblems für die Wärmeleitungsgleichung in $\mathbb{R}^3$ mit einer Integralgleichungsmethode nach dem Rotheversfahren. *Computing*, 19, pp. 251-268.

GIBBS, N.E., POOLE, W.G. Jnr. & STOCKMEYER, P.K. (1976). An algorithm for reducing the bandwidth and profile of a sparse matrix. *SIAM J.Numer.Anal.* 13, pp. 236-250.

GILBERT, R.P. (1970). Integral operator methods for approximate solutions of Dirichlet problems. *ISNM* 15 (Iterationsverfahren, Numerische Mathematik, Approximations-theorie), Birkhauser Verlag, Basel.

GILL, P.E. & MURRAY, W. (1974). Safeguarded steplength algorithms for optimisation using descent methods, NPL report NAC 37, National Physical Laboratory, Teddington, Middlesex.

GLADWELL, I. (1978). The Development of the Boundary Value Codes in the Ordinary Differential Chapter of the NAG Library. Numerical Analysis Report No. 30, Department of Mathematics, University of Manchester. To be published.

GLOWINSKI, R. (1973). Approximations externes, par éléments finis de Lagrange d'ordre un et deux, du problème de Dirichlet pour l'óperateur biharmonique. Méthode iterative de resolution des problemes approches. pp. 123-171 of J.J.H. Miller (ed.) (1973).

GLOWINSKI, R. & LIONS, J.L. (eds.) (1976). *Computing Methods in Applied Sciences*, Lecture Notes in Physics, 58, Springer, Berlin.

——— PERIAUX, J. & PIRONNEAU, O. (1976). Use of Optimal Control Theory for the Numerical Simulation of Transonic Flow by Finite Elements, in van de Vouren, A.I. & Zandbergen, P.J. (eds.) (1976).

GODUNOV, S.K. & RYABENKII, V.S. (1964). *Theory of Difference Schemes*, North Holland, Amsterdam.

GÖRTLER, H. (1957). A new series for the calculation of steady laminar boundary-layer flows. *J.Math.Mech.*, 6, pp. 1-66.

GOTTLIEB, D. & GUSTAFSSON, B. (1976). Generalised du Fort-Frankel Methods for Parabolic Initial-Boundary Value Problem, *Siam.J.Numer.Anal.*, 13, pp. 129-144.

GOURLAY, A.R. (1970). Hopscotch, a Fast Second Order Partial Differential Equation Solver, *J.Inst.Math.Applics.*, 6, pp. 375-390.

——— (1977). Splitting Methods for Time-Dependent Partial Differential Equations, pp. 757-796 of Jacobs, D.A.H. (ed.) (1977).

——— & McKEE, S. (1977). The construction of hopscotch methods for parabolic and elliptic equations in two space dimensions with a mixed derivative, *J.Comp. Appl.Math.*, 3, pp. 201-206.

——— & MORRIS, J., Ll (1968). A Multistep Formulation of the Optimised Lax-Wendroff Method for Nonlinear Hyperbolic Systems in Two Space Variables, *Math.Comp.*, 22, pp. 715-720.

GOURSAT, E. (1964). *A course in mathematical analysis* 3, part 2, *Integral equations, calculus of variations*. Dover, New York.

GRAM, C. (Ed.) (1962). *Selected numerical methods*. Regencentralen, Copenhagen.

GRAMMELTREDT, A. (1969). A survey of finite difference schemes for the primitive equations for a barotropic fluid, *Mon.Weath.Rev.* 97, pp. 384-404.

GRAY, W.G. & PINDER, G.F. (1976). An analysis of the numerical solution of the transport equation, *Water Resour.Res.*, 12, pp. 547-555.

——— ——— & BREBBIA, C.A. (1977) (eds.) *Finite elements in water resources*, Pentech Press, Plymouth.

GREEN, C.D. (1969). *Integral equation methods*. Nelson, London.

GREGORY, J.A., FISHELOV, D., SCHIFF, B. & WHITEMAN, J.R. (1978). Local mesh refinement with finite elements for elliptic problems, *J.Comp.Phys.* 28, pp.

GUSTAFSSON, B., KREISS, H.O. & SUNDSTRÖM, A. (1972). Stability theory for difference approximations of Mixed Initial Boundary Value Problems II, *Math.Comp.* 26, pp. 649-686.

——— & SUNDSTRÖM, A. (1978). Incompletely Parabolic Systems in Fluid Mechanics, *SIAM J.Appl.Math.* To appear.

GUYMON, G.L. (1970). A finite element solution of the one-dimensional diffusion-convection equation, *Water Resour. Res.*, 6, pp. 204-210.

HAGEMAN, L.A. (1972). The estimation of acceleration parameters for the Chebyshev polynomial and the successive overrelaxation iteration methods. AEC Res. and Dev. report WAPD-TM-1038.

HALL, G. & WATT, J.M. (1976). *Modern Numerical Methods for Ordinary Differential Equations*. Clarendon Press, Oxford.

HALL, M.G. (1976). Methods and Problems in the Calculation of Transonic Flow, in van de Vooren, A.I. & Zandbergen, P.J. (eds.) (1976).

HANSEN, E. & HOUGAARD, P. (1974). On a moving boundary problem. *J.Inst.Maths.Applics.* 13, pp. 385-398.

HARTREE, D.R. (1937). On an equation occurring in Falkner and Skan's approximate treatment of the equations of the boundary layer. *Proc.Camb.Phil.Soc.*, 33, pp. 223-239.

——— & WOMERSLEY, J.R. (1937). A method for the numerical or mechanical solution of certain types of partial differential equations. *Proc.Roy.Soc.London*, A, 161, pp. 353-366.

HAYES, J.G. (ed.) (1970). *Numerical Approximation to Functions and Data*. Athlone Press, London.

HAYES, J.K. (1968). The Laplace Fortran code. Los Alamos Scientific Lab.Rep. LASL-4004. Los Alamos, New Mexico.

———, KAHANER, D.K. & KELLNER, R.G. (1972). An improved method for numerical conformal mapping. *Math. Comp.*, 26, pp. 327-334.

——— ——— ——— (1975). A numerical comparison of integral equations of the first and second kind for conformal mapping. *Math.Comp.* 29, pp. 512-521.

HEINRICH, J.C., HUYAKORN, P.S. & ZIENKIEWICZ, O.C. (1977). An 'upwind' finite element scheme for two-dimensional convective transport equations, *Int.J.Num.Meth.Engng.* 11, pp. 131-143.

——— & ZIENKIEWICZ, O.C. (1977). Quadratic finite element schemes for two-dimensionsal convective transport problems, *Int.J.Num.Meth.Engng.* 11, pp. 1831-1844.

HENDRY, J.A. (1978). Singular Problems and the Global Element Method. *Int.J.Num.Meth. Engng.* To appear.

——— & DELVES, L, M. (1978). The Global Element Method Applied to a Harmonic Mixed Boundary Value Problem, Research Rept. CSS/77/9/2, University of Liverpool. Submitted for publication.

HENRICI, P. (1957). A survey of I.N. Vekua's theory of elliptic partial differential equations with analytical coefficients. *ZAMP*, 3, pp. 169-203.

——— (1974). *Applied and Computational Complex Analysis*, Vol.I, Wiley, New York.

HENRY, A.F. & YANG, S.T. (1976). The Finite Element Synthesis Method. *Nuc.Sc. and Engng.* 59, pp. 63-67.

HINDMARSH, A.C. (1973). GEARB: solution of ordinary differential equations having banded Jacobian. Report UCID-30059, Lawrence Livermore Laboratory, Livermore, California.

——— (1974). GEAR: Ordinary Differential Equation Solver. UCID 30001 Rev. 3. Lawrence Livermore Laboratory, Livermore, California.

——— & BYRNE, G.D. (1975). EPISODE: An Experimental Package for the Integration of Systems of Ordinary Differential Equations. UCID 30112. Lawrence Livermore Laboratory, Livermore, California.

HOCKNEY, R.W. (1965). A fast direct solution of Poisson's equation using Fourier analysis. *J.Assoc.Comput.Mach.*, 12, pp. 95-113.

——— (1970). The Potential Calculation and Some Applications, *Methods in Computational Physics*, 9, Academic Press, New York.

HOFMANN, P. (1967). Asymptotic expansions of the discretization error of boundary-value problems of the Laplace equation in rectangular domains. *Numer.Math.* 9, pp. 302-322.

HOHN, W. (1978). Konvergenzordnung bei einen expliziten Differenzenfahren zur numerischen Losung des Stefan-Problems, in Albrecht J. *et al.* (eds.) (1978).

HOLT, M. (ed.) (1971). *2nd International Conference on Numerical Methods in Fluid Dynamics*, Lecture Notes in Physics, 8, Springer, Berlin.

HOOD, P. (1976). Frontal solution program for unsymmetric matrices. *Int.J.Num.Meth.Engng.* 10, pp. 379-400.

HOWARTH, L. (1938). On the solution of the laminar boundary-layer equations. *Proc.Roy.Soc. A.* 164, pp. 547-579.

HUBBARD, B.E. (1966). Remarks on the Order of Convergence in the Discrete Dirichlet Problem, in Bramble, J.H. (ed.) (1966).

——— (ed.) (1971). *Numerical Solution of Partial Differential Equations, II*, Academic Press, New York.

——— (ed.) (1976). *Numerical Solution of Partial Differential Equations, III*, Academic Press, New York.

HUNT, B. (ed.) (1979). *Numerical Methods in Applied Fluid Dynamics*, Academic Press, London.

HUYAKORN, P.S. & TAYLOR, C. (1977). Solution of the transient convection-diffusion equation using asymmetric weighting functions and extended upwind finite element scheme, Univ. of Wales, Swansea, Civil Engineering Report.

IRONS, B.M. (1970). A frontal solution program for finite element analysis. *Int.J.Numer.Meth.Engng.* 2, pp. 5-32.

——— & RAZZAQUE, A. (1972). Experience with the Patch Test, pp. 557-587 in Aziz, A.K. (ed.) (1972).

ISAACSON, E. & KELLER, H.B. (1966). *Analysis of Numerical Methods*, Wiley, New York.

IVANOV, V.V. (1976). *The theory of approximate methods and their application to the numerical solution of singular integral equations*. (Translation) Noordhoff International, Leyden.

JACOBS, D.A.H. (1973a). E3DES: A subprogram for solving the difference equations for three-dimensional linear elliptic or parabolic partial differential equations on a rectangular mesh. CERL report RD/L/P/10/83.

——— (1973b). The strongly implicit procedure for biharmonic problems. *J.Comp.Phys.* 13, pp. 303-315.

——— (eds.) (1977). *The State of the Art in Numerical Analysis*. Academic, London.

JAMESON, A. (1976a). Numerical Solution of Non-Linear Partial Differential Equations of Mixed Type, in Hubbard, B.E. (ed.) (1976).

——— (1976b). *Transonic Flow Calculations*, Von Karman Institute Lecture Series on Computational Fluid Dynamics, 87.

JAMET, R. (1977). Estimation of the Interpolation Error for Quadrilateral Finite Elements which can Degenerate into Triangles, *SIAM J.Numer.Anal.*, 14, pp. 925-930.

JANOVSKY, V. & PROCHARKA, P. (1976). The nonconforming finite element method for a clamped plate with ribs, *Aplikace Matematiky* 21, pp. 273-289.

JASWON, M.A. (1963). Integral equation methods in potential theory - I. *Proc.Roy.Soc.A*, 275, pp. 23-32.

——— & SYMM, G.T. (1977). *Integral Equation Methods in Potential Theory and Elastostatics*, Academic Press, London.

JENNINGS, A. (1966). A compact storage scheme for the solution of symmetric linear simultaneous equations. *Comput. J.* 9, pp. 281-285.

JEPPSON, R.W. (1972). Inverse Solution to Three-Dimensional Potential Flows. *J.Engng.Mech.Div. ASCE*, 98, pp. 789-812.

JOHNSTON, R.L. & PAL, S.K. (1972). The numerical solution of hyperbolic systems using bicharacteristics, *Math.Comp.*, 26, pp. 377-392.

JORDAN, W.B. (1970). A.E.C. Res. and Dev. Rept. KAPL-M-7112.

KAMENOMOSTSKAJA, S.L. (1961). On the Stefan problem. *Mat.Sb.* 53, pp. 489-514.

KANTOROVICH, L.V. & AKHILOV, G.P. (1964). *Functional Analysis in Normed Spaces*, Pergamon Press, Oxford.

——— & KRYLOV, V.I. (1964). *Approximate Methods of Higher Analysis*, Noordhoff, Groningen.

KAO, T. & ELROD, H.G. (1974). Laminar shear-stress pattern for nonsimilar incompressible boundary layers. *AIAA Journal*, 12, pp. 1401-1408.

——— ——— (1976). Rapid calculation of heat transfer in nonsimilar laminar incompressible boundary layers. *AIAA Journal*, 14, pp. 1746-1751.

KATSANSIS, T. (1969). Numerical Solution of the Tricomi Equation using Theory of Symmetric Positive Definite Equations, *SIAM J.Numer.Anal.*, 6, pp. 236-253.

KELLER, H.B. (1971). A new difference scheme for parabolic problems, pp. 327-350 of Hubbard, B.E. (ed.) (1971).

——— & CEBECI, T. (1971). Accurate numerical methods for boundary-layer flows. I. Two-dimensional laminar flows, pp. 92-100 of Holt, M. (ed.) (1971).

——— ——— (1972). Accurate numerical methods for boundary-layer flows. II. Two-dimensional turbulent flows. *AIAA Journal*, 10, pp. 1193-1199.

KERSHAW, D.S. (1978). The incomplete Cholesky-conjugate gradient method for the iterative solution of systems of linear equations. *J.Comp.Phys*. 26, pp. 43-65.

KIKUCHI, N. (1977). An Analysis of the Variational Inequalities of Seepage Flow by Finite-Element Methods. *Q.App.Math*., 35, pp. 149-163.

KLEINMAN, R.E. & WENLAND, W.L. (1977). On Neumann's method for the exterior problem for the Helmholtz equation. *J.Math.Anal.Appl*., 27, pp. 170-202.

KONDRAT'EV, V.A. (1968). Boundary problems for elliptic equations with conical or angular points, *Trans.Moscow Math.Soc*. 17, pp.

KREISS, H.O. (1968). Stability theory for difference approximations of Mixed Initial Boundary Value Problems I, *Math. Comp*, 22, pp. 703-714.

——— & OLIGER, J. (1972). Comparison of accurate methods for the integration of Hyperbolic Equations, *Tellus* 24, pp. 199-215.

KUPRAZDE, V.D. (1965). *Potential methods in the theory of elasticity*. Israel Program for Scientific Transl., Jerusalem.

LAMBERT, J.D. (1973). *Computational Methods in Ordinary Differential Equations*. Wiley, London.

LANDAU, L. & LIFCHITZ, E. (1967). *Theorie de l'Elasticité*, Mir, Moscow.

LAROCK, B.E. & TAYLOR, C. (1976). Computing Three-Dimensional Free Surface Flows. *Int.J.Num.Meth.Engng*., 10, pp. 1143-1152.

LEENDERTSE, J.J. (1977). Comment on "An analysis of the numerical solution of the transport equation", *Water Resour.Res*., 13, pg. 219.

LEES, M. (1966). A linear three-level difference scheme for quasi-linear parabolic equations. *Math.Comp*. 20, pp. 516-522.

LEHMAN, R.S. (1959). Development at an analytic corner of solutions of elliptic partial differential equations, *J.Math.Mech*. 8, pp. 727-760.

LERAT, A. & PEYRET, R. (1975). Propriétés dispersives et dissipatives d'une classe de schemes aux differences pour les systemes hyperboliques nonlineares. *Rech. Aerosp*., 2, pp. 61-79.

LEVIN, D., PAPAMICHAEL, N. & SIDERIDIS, A. (1977). On the use of conformal transformations for the numerical solution of harmonic boundary value problems, *Comp.Meth. Appl.Mech.Engng*. 12, pp. 201-218.

LEVIN, D., PAPAMICHAEL, N. & SIDERIDIS, A. (1978). The Bergman kernel method for the numerical conformal mapping of simply connected domains, *J.Inst.Math. Applics*. (to appear).

LIONS, J.L. and STAMPACCHIA, G. (1967). Variational Inequalities. *Comm.Pure Appl.Math*. 22, pp. 493-519.

LONSETH, A.T. (1977). Sources and applications of integral equations. *SIAM Review* 19, pp. 241-278.

LOVE, A.E.H. (1944). *Mathematical Theory of Elasticity*, Dover, New York.

LOZIUK, L.A., ANDERSON, J.C. & BELYTSCHKO, T. (1972). Hydrothermal analysis by finite element method, *Proc. ASCE J. of the Hydraulics Div. HY* 11, pp. 1983-1998.

LUXMORE, A.R. & OWEN, D.R.J. (eds.) (1978). *Numerical Methods for Fracture Mechanics*.

LYNN, M.S. & TIMLAKE, W.P. (1967). The use of multiple deflations in the numerical solution of singular systems of equations with applications to potential theory, IBM Publication 37.017, Houston Scientific Center.

_____ _____ (1968). The numerical solution of singular integral equations of potential theory. *Num.Math*., 11, pp. 77-98.

McCORMACK, R.W. (1971). Numerical solution of the interaction of a shock wave with a laminar boundary layer, pp. 151-163 of Holt, M. (ed.) (1971).

McGUIRE, G.R. & MORRIS, J. Ll. (1974). A class of implicit second order accurate dissipative schemes for solving systems of conservation laws. *J.Comp.Phys*. 14, pp. 126-147.

MANGASARIAN, O.L. (1963). Numerical solution of the first biharmonic problem by linear programming, *Int.J.Engng. Sci*. 1, pp. 231-240.

MARKOWITZ, H.M. (1957). The elimination form of the inverse and its application to linear programming. *Management Science* 3, pp. 255-269.

MEIJERINK, J.A. & VAN DER VORST, H.A. (1977). An iterative solution method for linear systems of which the coefficient matrix is a symmetric M-matrix. *Math.Comp*. 31, pp. 148-162.

MEINARDUS, G. (1967). *Approximation of Functions; Theory and Numerical Methods*. Springer Tracts. 13, Springer, Berlin.

MERK, H.J. (1959). Rapid calculations for boundary-layer transfer using wedge solutions and asymptotic expansions. *J.Fluid Mechs*., 5, pp. 460-480.

MEYER, G.H. (1970). On a free interface problem for linear ordinary differential equations and the one-phase Stefan problem. *Numer.Math.* 16, pp. 248-267.

——— (1973). Multidimensional Stefan problems. *SIAM J. Numer.Anal.* 10, pp. 522-538.

——— (1976). The numerical solution of Stefan problems with front-tracking and smoothing methods. Technical Rept. TR/62, Maths.Dept. Brunel University.

——— (1977a). An Application of the Method of Lines to Multi-Dimensional Free Boundary Problems. *J.Inst. Math.Applics.*, 20, pp. 317-329.

——— (1977b). One-dimensional parabolic free boundary problems. *SIAM Rev.* 19, pp. 17-34.

——— (1977c). An alternating direction method for multi-dimensional parabolic free surface problems. *Int. J.Num.Meth.Engng.*, 11, pp. 741-752.

——— (1978). The Method of Lines for Poisson's Equation with Nonlinear or Free Boundary Conditions. *Numer. Math.*, 29, pp. 329-344.

MILLER, G.F. (1974). Fredholm equations of the first kind. Ch. 13 of Delves L.M. and Walsh J.E. (Eds.) (1974).

MILLER, J.J.H. (ed.) (1973). *Topics in Numerical Analysis*, Academic Press, London.

MIRANDA, C. (1948-49). Formule di maggiorazione e teorema di esistenza per le funzioni biarmoniche di due variabili, *Giornale di Matematiche Battaglini* 78, pp. 97-118.

MITCHELL, A.R. (1969). *Computational Methods in Partial Differential EQuations*, Wiley, London.

——— & GRIFFITHS, D.F. (1978). Semi-Discrete Generalised Galerkin Methods for Time-Dependent Conduction-Convection Problems. Report NA/24. Department of Mathematics, University of Dundee.

——— & WAIT, R. (1977). *The Finite Element Method in Partial Differential Equations*, Wiley, London.

MOGEL, T.R. & STREET, R.L. (1974). A Numerical Method for Steady-State Cavity Flows. *J.Ship Res.*, 18, pp. 22-31.

MORLEY, L.S.D. (1973). Finite Element Solution of Boundary-Value Problems with Non-Removable Singularities. *Phil. Trans.Roy.Soc. London A*, 275, pp. 463-488.

MORRIS, J.Ll. (ed.) (1969). *Numerical Solution of Differential Equations*, Lecture Notes in Mathematics, 109, Springer, Berlin.

MORRIS, J.Ll. (ed.) (1971). *Applications of Numerical Analysis*, Lecture Notes in Mathematics, 228, Springer, Berlin.

MORRIS, J.Ll. & WAIT, R. (1978). Crack-tip Elements with Curved Boundaries and Variable Nodes. *App.Math.Mod.* (to appear

MORTON, K.W. (1977). Initial-value problems by finite-difference and other methods, in Jacobs, D.A.H. (ed.) (1977).

MOTZ, H. (1946). Treatment of singularities of partial differential equations by relaxation methods, *Q.Appl. Math.*, 4, pp. 371-377.

MURMAN, E.M. & COLE, J.D. (1971). Calculation of Plane Steady Transonic Flows, *AIAA J.*, 9, pp. 114-121.

MUSA, F.A. & DELVES, L.M. (1977). Weakly Asymptotically Diagonal Systems, submitted for publication.

MUSKHELISHVILI, N.I. (1953). *Singular integral equations* (Translation) Noordhoff, Gröningen.

NEUMANN, S.P. & WITHERSPOON, P.A. (1970). Finite Element Method of Analysing Steady Seepage with a Free Surface. *Water Resour.Res.*, 6, pp. 889-897.

NICHOL, K. (1967). Extension of a recent paper by Fox, Henrici and Moler on eigenvalues of elliptic operators. *SIAM J.Numer.Anal.* 4, pp. 483-488.

NICOLAIDES, R.A. (1975). On multiple grid and related techniques for solving discrete elliptic systems. *J.Comp. Phys.*, 19, pp. 418-431.

NOBLE, B. (1971). A bibliography on methods for solving integral equations - Subject listing. MRC Tech.Summ. Rep.No. 1177. University of Wisconsin, Madison.

——— (1977). The Numerical Solution of Integral Equations, pp. 915-966 of Jacobs, D.A.H. (ed.) (1977).

NØRSETT, S.P. (1974). Semi-Explicit Runge-Kutta Methods. Mathematics Computation Rept. No. 6/74. University of Trondheim.

OCKENDON, J.R. and HODGKINS, W.R. (Eds.) (1975). *Moving boundary problems in heat flow and diffusion*. Clarendon Press, Oxford.

OGAWA, H. (1961). On Difference Methods for the Solution of a Tricomi Problem, *Trans.Amer.Math.Soc.*, 100, pp. 404-424.

O'LEARY, D.P. (1977). A Generalized Conjugate Gradient Method for Solving a Class of Quadratic Programming Problems. Stanford Rept. CS-77-638.

OLEINIK, O.A. (1960). A method of solution of the general Stefan problem. *Soviet Math.Dokl.* 1, pp. 1350-1354.

OLIVER, J. (1972). A doubly-adaptive Clenshaw-Curtis quadrature method. *Comput.J.* 15, pp. 141-147.

ORTEGA, J.M.& RHEINBOLDT, W.C. (1970). *Iterative Solution of Nonlinear Equations in Several Variables*, Academic Press, New York.

OSHER, S. (1969a). Stability of difference approximations of the dissapative type for mixed initial boundary problems I. *Math.Comp.* 23, pp. 335-340.

——— (1969b). Systems of difference equations with general homogeneous boundary conditions, *Trans.Amer. Math.Soc.* 137, pp. 177-201.

——— (1972). Stability of Parabolic Difference Approximations to certain Mixed Initial-Boundary Value Problems, *Math.Comp.*, 26, pp. 13-39.

OVERLEY, J.R., OVERHOLSER, K.A. & REDDIEN, G.W. (1978). Calculation of minimum ignition energy and time-dependent laminar flame profiles. *Comb. and Flame*, 31, pp. 69-83.

PAPAMICHAEL, N. & SIDERIDIS, A. (1978). The use of conformal transformations for the numerical solution of elliptic boundary value problems with boundary singularities. *J.Inst.Math.Applics.* (to appear).

——— & SYMM, G.T. (1978). Numerical techniques for two dimensional Laplacian problems. *Comp.Meth.Appl. Mech.Eng.* 6, pp. 175-194.

——— & WHITEMAN, J.R. (1973). A numerical conformal transformation method for harmonic mixed boundary value problems in polygonal domains, *J.Appl.Math.Phys. (ZAMP)* 24, pp. 304-316.

PARKER, I.B. & CRANK, J. (1964). Persistent discretization errors in partial differential equations of parabolic type. *Comput.J.* 7, pp. 163-167.

PATANKAR, S.V. & SPALDING, D.B. (1967). *Heat and Mass Transfer in Boundary Layers*, Morgan-Grampian, London.

PEACEMAN, D.W. & RACHFORD, H.H. Jnr. (1955). The numerical solution of parabolic and elliptic differential equations, *SIAM J.Appl.Math.*, 3, pp. 28-41.

PEACEMAN, D.W. & RACHFORD, H.H. Jnr. (1962). Numerical calculation of multi-dimensional miscible displacement, *Soc.Petrol. Eng.J.*, 237, pp. 327-338.

PEREYRA, V. (1977). Algorithms for solving two-point boundary-value problems (in Symposium on High Speed Computer and Algorithm Organisation, Urbana, Illinois).

——— , PROSKUROWSKI, W. & WIDLUND, O. (1977). High order fast Laplace solvers for the Dirichlet Problem on general regions, *Math.Comp.*, 31, pp. 1-16.

POTTER, D. (1973). *Computational Physics*, Wiley, London.

PRENTER, P.M. (1975). *Splines and Variational Methods.* Wiley, New York.

PRICE, H.S., CAVENDISH, J.C. & VARGA, R.S. (1968). Numerical methods of high-order accuracy for diffusion-convection equations, *Soc.Petrol.Eng.J.*, 243, pp. 293-303.

——— & VARGA, R.S. (1962). Recent numerical experiments comparing successive overrelaxation iterative methods with implicit alternating direction methods, Gulf Research and Development Co. report, Harmarville, Pa. U.S.A.

PROSKUROWSKI, W. & WIDLUND, O. (1976). On the numerical solution of Helmholtz's equation by the capitance matrix method, *Math.Comp.*, 30, pp. 433-468.

PROTTER, M.H. & WEINBERGER, H.F. (1967). *Maximum Principles in Differential Equations.* Prentice-Hall, Englewood Cliffs, N.J.

REID, J.K. (1966). A method for finding the optimum successive over-relaxation parameter. *Comput.J.* 9, pp. 200-204.

——— (ed.) (1971a) *Large Sparse Sets of Linear Equations*, Academic Press, London.

——— (1971b). On the Method of Conjugate Gradients for the Solution of Large Sparse Systems of Equations, pp. 231-254 of Reid, J.K. (ed.) (1971a).

——— (1972). Two Fortran subroutines for direct solution of linear equations whose matrix is sparse, symmetric and positive definite. AERE Report R. 7119. HMSO, London.

——— & WALSH, J.E. (1965). An elliptic eigenvalue problem for a re-entrant region. *SIAM J.Appl.Math.* 13, pp. 837-850.

RHEINBOLDT, W.C. (1974). *Methods for Solving Systems of Nonlinear Equations*, Regional Conference Series in Applied Mathematics, 14, SIAM, Philadelphia.

RICE, J.R. (ed.) (1977). *Mathematical Software III*, Academic Press, New York.

RICHTMYER, R.D. & MORTON, K.W. (1967). *Difference Methods for Initial Value Problems*. Wiley, Interscience, New York.

RILEY, A. (1977). Complementary variational principles and the finite element method, M.Sc. thesis, Department of Mathematics, University of Manchester.

ROACH, G.F. (1970). *Green's functions: introductory theory with applications*. Van Nostrand, London.

ROBERTS, K.V. & WEISS, N.O. (1966). Convective difference schemes, *Math.Comp.*, 20, pp. 272-297.

ROBINSON, A. (1976). Thermochemical modelling of a parallel-plate reactor. Conf. on Partial Differential Equations in Industry, Department of Mathematics, Manchester (unpublished).

ROSE, D.J. & BUNCH, J.R. The role of partitioning in the numerical solution of sparse systems, pp. 177-187 of Rose, D.J. & Willoughby, R.A. (eds.) (1972).

——— & WILLOUGHBY, R.A. (Eds.) (1972). *Sparse matrices and their applications*. Plenum Press, New York.

ROSSER, J.B. (1978). Majorisation formulas for a biharmonic function of two variables, Tech.Rept. BICOM 78-10, Institute of Computational Mathematics, Brunel University.

——— & PAPAMICHAEL, N. (1973). A power series solution of a harmonic mixed boundary value problem, Technical Report TR/35, Dept. of Maths., Brunel University, or MRC Tech. Survey Rept. 1405, Univ. of Wisconsin, Madison.

SANKAR, R. (1967). Numerical solution of differential equations. D.Phil. Thesis, University of Oxford.

SAYLOR, P.E. (1974). Second order strongly implicit symmetric factorization methods for the solution of elliptic difference equations. *SIAM J.Numer.Anal.*, 11, pp. 894-908.

SCHECTER, S. (1960). Quasi-Tridiagonal Matrices and Type-Insensitive Difference Equations, *Q.Appl.Math.* 18, pp. 285-295.

SCHEFFLER, R.S. & WHITEMAN, J.R. (1978). Institute of Computational Mathematics Rept., Brunel University.

SCHIFF, B., WHITEMAN, J.R. & FISHELOV, D. (1979). Determination of a Stress Intensity Factor Using Local Mesh Refinement, in Whiteman, J.R. (ed.) (1979).

SCHLICHTING, H. (1960). *Boundary-layer theory. 4th Ed.* McGraw-Hill, New York.

SCHREM, E. (1971). Computer implementation of the finite element procedure. ONR Symposium on numerical and computer methods in structural mechanics, Urbana, Illinois.

SCHUBERT, L.K. (1970). Modification of a quasi-Newton method for nonlinear equations with a sparse Jacobian, *Math. Comp.*, 24, pp. 27-30.

SCHULTZ, M.H.(1973). *Spline Analysis*. Prentice-Hall, Englewood Cliffs, N.J.

SCOTT, L.R. (1976). A survey of displacements for the plate bending problem, Proc. US-Germany Symposium on Formulations and Computational Algorithms in Finite Element Analysis, M.I.T.

SHANKAR, V. & ANDERSON, D. (1975). Numerical Solutions of Supersonic Corner Flow, *J.Comp.Phys.*, 17, pp. 160-180.

SIEMIENIUCH, J.L. & GLADWELL, I. (1974). On Time-Discretizations for Linear Time-Dependent Partial Differential Equations. Numerical Analysis Rept. 5, Mathematics Dept., University of Manchester.

——— ——— (1978). Analysis of Explicit Difference Methods for a Diffusion-Convection Equation. *Int.J.Num.Meth.in Engng.*, 12, pp. 899-916.

SINCOVEC, R.F. & MADSEN, N.K. (1975). Software for nonlinear partial differential equations. *ACM Trans. on Math. Software*, 1, pp. 233-260.

SMITH, A.M.O., & CLUTTER, D.W. (1963). Solution of the incompressible laminar boundary-layer equations. *AIAA J*, 1, pp. 2062-2071.

SMITH, G.D. (2nd ed.) (1977). *Numerical Solution of Partial Differential Equations*, Oxford University Press, Oxford.

SMITH, I.M. (1977). Integration in time of diffusion and diffusion-convection equations, pp. 1.3-1.20 of Gray, W.G., Pinder, G.F. & Brebbia, C.A. (eds.) (1977).

——— (1978). Transient phenomena of offshore foundations, Ch. 15 of Zienkiewicz, O.C., Lewis, R.W. & Stagg, K.G. (eds.) (1978).

———, FARRADAY, R.V. & O'CONNOR, B.A. (1973). Rayleigh-Ritz and Galerkin finite elements for diffusion-convection problems, *Water Resour.Res*, 9, pp. 593-606.

SMITH, I.M., SIEMIENIUCH, J.L. & GLADWELL, I. (1977). Evaluation of Nørsett Methods for Integrating Differential Equations in Time, *Int.J.Num. & Anal.Meth. in Geomechanics*, 1, pp. 57-74.

SMITH, J. (1968). The coupled equation approach to the numerical solution of the biharmonic equation by finite differences, I, II, *SIAM J.Numer.Anal.* 5, pp. 323-339.

SMITHIES, F. (1958). (2nd ed.) (1967). *Integral equations.* Cambridge Univ. Press, Cambridge.

SOKOLNIKOFF, I.S. (1956). *Mathematical Theory of Elasticity*, McGraw-Hill, New York.

SOUTHWELL, R.V. (1956). *Relaxation Methods in Engineering Science*, Oxford University Press, Oxford.

SPALDING, D.B., STEPHENSON, P.L. & TAYLOR, R.G. (1971). A calculation procedure for the prediction of laminar flame speeds. *Comb. and Flame*, 17, pp. 55-64.

——— (1972). A novel finite difference formulation for differential expressions involving both first and second derivatives, *Int.J.Num.Meth.Engng.* 4, pp. 551-559.

SPARROW, E.M., QUACK, H. & BOERNER, C.J. (1970). Local nonsimilarity boundary-layer solutions. *AIAA J.* 8, pp. 1936-1942.

——— & YU, H.S. (1971). Local nonsimilarity thermal boundary-layer solutions. *Trans ASME, J.Heat Transfer*, 93, pp. 328-334.

SPEELPENNING, B. (1973). The generalized element method. Private communication.

STEPHENSON, P.L. & TAYLOR, R.G. (1973). Laminar flame propagation in hydrogen, oxygen, nitrogen mixtures. *Comb. and Flame*, 20, pp. 231-244.

STONE, H.L. (1968). Iterative solution of implicit approximations of multi-dimensional partial differential equations. *SIAM J.Numer.Anal.*, 5, pp. 530-558.

——— & BRIAN, P.L.T. (1963). Numerical solution of convective transport problems, *Amer.Inst.Chem.Eng.J.*, 9, pp. 681-688.

STRANG, G. & FIX, G.J. (1973). *An Analysis of the Finite Element Method.* Prentice-Hall. Englewoods Cliffs, N.J.

SWARZTRAUBER, P.N. (1974). A direct method for the discrete solution of separable elliptic equations, *SIAM J. Numer.Anal.*, 11, pp. 1136-1150.

SYMM, G.T. (1963). Integral equation methods in potential theory - II, *Proc.Roy.Soc.A*, 275, pp. 33-46.

SYMM, G.T. (1964). Integral equation methods in elasticity and potential theory. NPL Report Ma 51, Teddington, Middlesex.

——— (1966). An integral equation method in conformal mapping, *Numer.Math.*, 9, pp. 250-258.

——— (1973). Treatment of singularities in the solution of Laplace's equation by an integral equation technique. NPL Report NAC 31, Teddington, Middlesex.

——— & PITFIELD, R.A. (1974). Solution of Laplace's equation in two dimensions. NPL Report NAC 44, Teddington, Middlesex.

SZABO, B.A. & MEHTA, A.K. (1978). P-Convergent Finite Element Approximations in Fracture Mechanics. *Int.J.Num.Math. Engng.* (to appear).

TAYLER, A.B. (1975). The mathematical formulation of Stefan problems. pp. 120-137 of Ockendon, J.R. and Hodgkins, W.R. (eds.) (1975).

TEE, G.J. (1963). A novel finite-difference approximation to the biharmonic operator, *Comput.J.* 6, pp. 177-192.

TEMAN, R. (1973). *Numerical Analysis*, D. Reidel, Dordrecht.

——— (1977). *Navier-Stokes Equations: Theory and Numerical Analysis*, North-Holland, Amsterdam.

TERRILL, R.M. (1960). Laminar boundary-layer flow near separation with and without suction. *Phil.Trans.Roy.Soc. London A*, 253, pp. 55-100.

TERRY, F.H. (1967). Iterative solution of Neumann boundary value problems with application to electrocardiography, Ph.D. thesis, Case Western Researve University.

THATCHER, R.W. (1975). Singularities in the Solution of Laplace's Equations in Two Dimensions, *J.Inst.Maths. Applics.*, 16 pp. 303-319.

——— (1976). The Use of Infinite Refinement of Singularities in the Solution of Laplace's Equation. *Numer.Math.*, 25, pp. 163-168.

THOMÉE, V. (1972). Spline Approximation and Difference Schemes for the Heat Equation, pp. 711-746 of Aziz, A.K. (ed.) (1972).

TIMOSHENKO, S. & GOODIER, J.N. (1951). *Theory of Elasticity*, McGraw-Hill, New York.

TRANGENSTEIN, J.A. (1977). A Finite Element Method for the Tricomi Problem in the Elliptic Region, *SIAM J.Numer. Anal.* 14. pp. 1066-1078.

TURKEL, E. (1974). Phase error and stability of second order methods for hyperbolic problems I, *J.Comp.Phys.* 15, pp. 226-250.

TURLAND, B.D. & WILSON, G. (1977). An implicit isotherm migration method. U.K.A.E.A. (Culham) report. RS(77)N42.

URSELL, F. (1973). On the exterior problems of acoustics. *Proc.Camb.Phil.Soc.* 74, 117-125.

VAN DE VOOREN, A.I. & ZANDBERGEN, P.J. (eds.) (1976). *Fifth International Conference in Fluid Dynamics*. Lecture Notes in Physics, 59, Springer, Berlin.

VAN GENUCHTEN, M.T. & GRAY, W.G. (1978). Analysis of some dispersion-corrected numerical schemes for solution of the transport equation, *Int.J.Num.Meth.Engng.* 12, pp. 387-404.

VARAH, J.M. (1971). Stability of Difference Approximations to the Mixed Initial Boundary Value Problems for Parabolic Systems. *SIAM.J.Numer.Anal.*, 8, pp. 598-615.

VARAH, J.M.(1972). On the solution of block-tridiagonal systems arising from certain finite-difference equations. *Math. Comp.*, 26, pp. 859-868.

VARGA, R.S. (1962). *Matrix Iterative Analysis*, Prentice-Hall, Englewood Cliffs, New Jersey.

VENIT, S. (1973). Mesh Refinements for Parabolic Equations of Second Order, *Math.Comp.*, 27, pp. 745-754.

VERWER, J.G. (1977a). A Class of Stabilized Three-Step Runge-Kutta Methods for the Numerical Integration of Parabolic Problems. Rept. NW 39/77. Mathematische Centrum, Amsterdam.

——— (1977b). An Implementation of a Class of Stabilized Explicit Methods for the Time Integration of Parabolic Equations. Rept. NW 38/77, Mathematische Centrum, Amsterdam.

VOLKOV, E.A. (1963). The removal of singularities in the solution of boundary-value problems for the Laplace equation in a region with a smooth boundary. *USSR Comp. Math. & Math.Phys.* 3, pp. 139-152.

VON MISES, R. & FRIEDRICHS, K.O. (1971). *Fluid Dynamics*, Applied Mathematical Sciences 5, Springer Verlag, New York.

WACHSPRESS, E.L. (1966). *Iterative Solution of Elliptic Systems and Applications to the Neutron Diffusion Equations in Reactor Physics*, Prentice-Hall, Englewood Cliffs, N.J.

——— & BECKER, M. (1965). Variational Synthesis with Discontinuous Trial Functions. Proc.Conf.Appl.Comp. Methods in Reactor Problems, Rept. ANL-7050, Argonne National Laboratory.

——— & HABELTER, G.J. (1960). An alternating direction implicit iteration technique, *J.SIAM Appl.Math.* 8, pp. 403-424.

WAIT, R. (1978a). Finite Element Methods for Elliptic Problems with Singularities. *Comp.Meth.Appl..Mech.Engng.* 13, pp. 141-150.

——— (1978b). A Note on Quarter Point Triangular Elements. *Int.J.Num.Meth.Engng* ., 12, pp. 1333-1336.

——— (1978c). A Curved Macro-Element for Cracks and Corners (to appear).

——— (1979). *The Numerical Solution of Algebraic Equations*. Wiley, Chichester.

——— & MITCHELL, A.R. (1971). Corner singularities in elliptic problems by finite element methods, *J.Comp. Phys*. 8, pp. 45-52.

WALSH, J.E. (1960). Numerical solution of partial differential equations. D.Phil. thesis, Oxford.

WANG, G.T. (1953). *Applied Elasticity*, McGraw-Hill, New York.

WARMING, R.F., KUTLER, P. & LOMAX, H. (1973). Second and third order non-centred schemes for non-linear hyperbolic equations, *AIAA J*. 11, pp. 189-195.

WASHIZU, K. (1968). *Variational Methods in Elasticity and Plasticity*, Pergamon Press, Oxford.

WASSERSTROM, E. (1973). Numerical solutions by the continuation method, *SIAM Rev*., 15, pp. 89-119.

WATSON, G.A. (ed.) (1974). *Numerical Solution of Differential Equations*, Lecture Notes in Mathematics, 363, Springer, Berlin.

WEARE, T.J. (1976). The economic disadvantages of the finite element method for two-dimensional tidal hydraulics, Hydraulic Research Station Notes, 18, Wallingford, England.

WHITEMAN, J.R. (1967). Singularities due to re-entrant corners in harmonic boundary value problems, M.R.C. Rept. 829, University of Wisconsin, Madison.

WHITEMAN, J.R. (1968). Treatment of singularities in a harmonic mixed boundary value problem by dual series methods. *Q.J.Mech.Appl.Math.* 21, pp. 41-51.

——— (1970). Numerical solution of a harmonic mixed boundary value problem by the extension of a dual series method. *Q.J.Mech.Appl.Math.* 23, pp. 449-455.

——— (ed.) (1973). *The Mathematics of Finite Elements and Applications*, Academic Press, London.

——— (ed.) (1976a). *The Mathematics of Finite Elements and Applications II, MAFELAP 1975*, Academic Press, London.

——— (1976b). Some aspects of the mathematics of finite elements. pp. 25-42 of Whiteman, J.R. (ed.) (1976a).

——— (1978). Numerical treatment of a problem from linear fracture mathematics pp. 128-136 of Luxmore, A.R. & Owen, D.R.J.(eds.) (1978).

——— (ed.) (1979). *The Mathematics of Finite Elements and Applications III. MAFELAP 1978*. Academic Press, London.

——— & AKIN, J.E. (1979). Finite elements, singularities and fracture, to appear in Whiteman, J.R. (ed.) (1979).

——— & PAPAMICHAEL, N. (1972). Treatment of harmonic mixed boundary problems by conformal mapping techniques, *J.Appl.Math.Phys.(ZAMP)* 23, pp. 655-664.

——— ——— & MARTIN, Q. (1971). Conformal Transformation Methods for the Numerical Solution of Harmonic Mixed Boundary Value Problems, pp. 353-358 of Morris, J.Ll. (ed.) (1971).

——— & WEBB, J.C. (1970). Convergence of finite difference techniques for a harmonic mixed boundary value problem. *BIT* 10, pp. 366-374.

WIDLUND, O.B. (1966). Stability of Parabolic Difference Schemes in the Maximum Norm, *Numer.Math.*, 8, pp. 186-202.

WIDLUND, O.B. (1970). On the Rate of Convergence for Parabolic Difference Schemes II, *Comm.Pure and Appl.Math.* 23, pp. 79-86.

WIGLEY, N.W. (1964). Asymptotic expansions at a corner of solutions of mixed boundary value problems. *J.Math.Mech.* 13, pp. 549-576.

WOODS, L.C. (1953). The relaxation treatment of singular points in Poisson equation, *Q.J.Mech.Appl.Math.* 6, pp. 163-185.

YATES, D.F. (1975). A Rayleigh-Ritz-Galerkin method for the solution of one and two dimensional boundary value problems. Ph.D. thesis. University of Liverpool.

YATES, D.F., HENDRY, J.A. & DELVES, L.M. (1978). The Case for Global Elements, preprint, University of Liverpool, to be submitted for publication.

YASINSKY, J.B. & KAPLAN, S. (1967). Synthesis of three-dimensional Flux Shapes using Discontinuous Sets of Trial Functions. *Nuc.Sc.and Engng.*, 28, pp. 426-437.

YOUNG, D.M. (1971). *Iterative solution of large linear systems*. Academic Press, New York.

ZABREYKO, P.P., KOSHELEV, A.I., KRASNOSEL'SKII, M.A., MIKHLIN, S.G., RAKOVSHCHIK, L.S. & STETSENKO, V.Ya. (1975). *Integral equations - a reference text*. Noordhoff International, Leyden.

ZIENKIEWICZ, O.C. (1971). *The Finite Element Method in Engineering Science* (2nd Ed.), McGraw-Hill, London.

ZIENKIEWICZ, O.C. (1977). *The Finite Element Method* (3rd Ed.) McGraw-Hill, London.

——— , LEWIS, R.W. & STAGG, K.G. (eds.) (1978). *Numerical Methods in Offshore Engineering*. Wiley, Chichester.

ZLAMAL, M. (1968). On the finite element method, *Numer. Math.* 12, pp. 394-409.

——— (1973). Some Recent Advances in the Mathematics of Finite Elements, pp. 59-81 of Whiteman, J.R. (ed.) (1973).

INDEX

amplification matrix 176
basis (trial) functions 15
bicharacteristic 371
bilinear blending 24,35
Bramble-Hilbert lemma 39
capacitance matrix 273
Cauchy - Kowaleski theorem 2
Cauchy problem 168
collocation 19
consistency 95
constant strain condition 96
Crank-Nicolson method 172
Davidenko's method 307
deferred-approach-to-the-limit 242
deferred correction 243
dissipative schemes 375
Galerkin method 16
Gauss condition 149
$\mathcal{H}$- elliptic 28
hodograph plane 327
hypermatrix method 261
invariant imbedding 324
isoparametric transformation 30
kernel 152
Lax-Milgram lemma 28
Least squares methods 19
local similarity 220
lumping 197
method of weighted residuals 16
monotonicity 321
non-unique solutions 154
one-phase problem 351
Peaceman - Rachford method 282
Peclet number 195
problem, Cauchy 168
, pure initial value 168
quarter-point condition 33
rank-one-update 306
reduced integration 209
region of dependence 373
regularizer 148
Riemann Mapping Theorem 72
root condition 178
shape functions 14
static condensation 23
stiffness matrix 18,25
superconvergence 187
tensor product form 24
test function 17
test space 26
trial space 27
truncation error, global 171
local 171
upwinding 206
variational crimes 83
Woodbury formula 273